Introduction to Exchange Systems

Introduction to Exchange Systems

T. H. Flowers
*Formerly Head of Switching Research,
British Post Office*

A Wiley–Interscience Publication

JOHN WILEY & SONS
London · New York · Sydney · Toronto

Copyright © 1976, by John Wiley & Sons Ltd.

All rights reserved.

No part of this book may be reproduced by any means, nor transmitted, nor translated into a machine language without the written permission of the publisher.

Library of Congress Cataloging in Publication Data:

Flowers, Thomas Harold.
    Introduction to exchange systems.

   'A Wiley—Interscience publication.'
   1. Telephone stations. I. Title.
TK6211.F54   621.385'4   76—13447

ISBN 0 471 01865 1

Typeset by Preface Ltd, Salisbury, Wiltshire and printed by Pitman Press, Bath

FOREWORD

A telecommunications engineer is often forced to work within a unique set of constraints imposed by the existing network. The designer of a computer or an aircraft is restricted only by the laws of physics, the available materials and his own ingenuity. The railway engineer has to deal with additional restrictions imposed by factors such as rail gauge, track curvature and tunnel height. But in a public telecommunications network, virtually every piece of equipment has to be capable of interworking, directly or indirectly with every other piece of equipment in the network and this has, in the past, restricted the introduction of new technology, particularly in the switching field.

In the transmission field, the problems of technical advancement have not been so severe. This has come about partly because the fundamental technology (for example, coaxial cables) was often available in advance of the system requirement but mainly because the interfaces between different parts of the system were relatively simple and easily defined. Examples of such interface definitions are contained in the well-established series of transmission system recommendations produced by the International Telecommunications Union.

Switching, on the other hand, is essentially a branch of control automation, a comparatively new art. The early automatic switching systems, as partial replacements for the telephone operators, were in advance of their time and ingenious though these systems were, the implications of their adoption were not always fully understood. Thus, much of the system flexibility possible with a human operator was lost with the introduction of automation. Switching, because of its fundamentally interconnected nature, involves many interfaces and this has severely restricted the development of the network. Often it has been found that a new piece of switching equipment, based on technology different from that already existing in the network, would have to adapt to so many interfaces that the cost of its adoption became prohibitive.

The slow progress in the exploitation of new technology in switching equipment has been matched by a lack of suitable textbooks on switching. This book not only fulfils a need in providing a general treatise on switching but, more importantly, it explains in some detail how and why the switched telephone network has evolved from the operator controlled systems. Such information is invaluable for today's telecommunication engineers. The speed of advance of modern technology is such that any new system is almost certain to be obsolescent by the time it is connected into the network in significant quantities. The engineer is therefore forced to consider in greater depth the basic system philosophy before starting on the design of equipment. The analytical approach to operator-controlled and electromechanical switching systems adopted in the early chapters of the book will assist him in this approach. Moreover, for those engaged in the design of new equipment, a knowledge of the basic principles of existing electromechanical systems will enable a better appreciation of the interworking constraints to be gained.

The progress of technology in the late 1950s led to the realization that it was possible to design an electronic or semi-electronic switching system that was economically competitive with existing systems and at the same time permitted changes in operation to be made easily during the lifetime of the equipment. The particular developments were closely allied to the development of digital computers and the technique became known as 'stored program control'. The operation of the exchange could be changed or extended merely by loading a new program into the control processor and such could be made in exchanges all over the network within the space of a few minutes. The stored program control technique was a very powerful one but, as with any pioneering venture, the full implications of its use were not always realized by those who first designed exchanges based on this principle.

It is perhaps significant that during World War II the author of this book was engaged in the development of equipment that incorporated many of the concepts later used in the design of digital computers and stored program control systems. Much of his work such as the 'Colossus' project has only recently been cleared of security restrictions and it is now possible to appreciate the highly advanced thinking of people such as M.H. A. Newman, A. Turing and T.H. Flowers. From his early acquaintance with the fundamental concepts of stored program control, the author of this book has been able to recognize both the advantages and the limitations of the technique.

In the future, the prospect of using solid state devices for the switches as well as for the control has much to recommend it, in view of the high reliability of such components and the possibility of making substantial economies by using the switches in time-division. This approach has not so far gained widespread acceptance for public exchanges, not least because of the necessity for conforming to an overall transmission plan, which permits only a very small amount of attenuation within the exchange. The use of pulse code modulation to permit 'integrated' switching and transmission largely overcomes the transmission problem and appears to offer many advantages.

At the time of writing, the precise future of telecommunications switching is largely undetermined. The trend towards thinking of the overall telecommunications network rather than particular items of equipment will undoubtedly continue. Some measure of stored program control will probably continue to be applied in the control area but the extent to which stored program rather than wired logic control is used will depend mainly on economic factors. There is also likely to be a move away from metallic-contact switches and towards time-division multiplexing and solid-state components in the switching area. The opinions on the future development of switching systems expressed in this book may not receive general acceptance but at least the clear exposition of the comparative approaches will provide much food for thought.

Finally, it is worth noting that many Third World countries are in a unique situation with regard to their public telecommunications networks. Since they do not have the severe restrictions imposed by the necessity for interworking with a highly developed but obsolescent network, they can configure their own network almost without restriction to suit their national requirements both now and in the future. This they can do in the full knowledge both of modern technology and of the earlier, and often restricting, course of development of telecommunications networks in highly industrialized countries. This book will help engineers in Third World countries to gain such knowledge and so help them towards the realization of a highly effective telecommunications network.

C. J. HUGHES
DEPUTY DIRECTOR,
POST OFFICE RESEARCH CENTRE,
MARTLESHAM HEATH,
IPSWICH,
SUFFOLK.

PREFACE

To be introduced to exchange systems through principles and generalizations, as in this work, instead of through particular systems which is the way that many engineers come to learn the subject, leads, I believe, to a better understanding not only of the subject in general but also of particular systems when they come to be studied. Every system in practice inevitably includes some feature based on arbitrary decision by an administration or designer or is forced by local conditions, a fact which generally needs some knowledge broader than just that of the system itself to recognize. I hope and believe that many already in telecommunications will find interest in the book but being an introduction, I have had to keep students and newcomers very much in mind. For that reason I have endeavoured to make it complete in itself and to be comprehensible without any prior knowledge of exchange systems or having to refer to other works: this explains why no references are given and terms and conventions particular to exchange switching are defined and explained as they occur in the text.

I gratefully acknowledge much helpful criticism and advice which I have received during the preparation of the manuscript, from numerous colleagues and friends in the industry.

London, January, 1976.   T. H. FLOWERS

ABBREVIATIONS

a.d.c.	advise duration and charge
conc.	concentrator
c.o.s.	class of service
d	data
d.d.i.	direct dialling-in
d.l.	data link
d.l.j.n.	data link junction number
DM	driving magnet
dm	contacts operated by DM
d.n.	directory number
e.l.	exchange line
e.n.	equipment number
e.n.c.	equipment number, calling
e.n.t.	equipment number, terminating
e.n.f.	equipment number, file
e.n.s.	equipment number, switch
f.d.m.	frequency division multiplex
i.b.	international barred
i.c.	incoming
i.c.j.	incoming junction
i.c.r.	incoming junction register
i.d.f.	intermediate distribution frame
i.f.s.	interface switch
i.j.	incoming junctor
i.s.	incoming service
j.n.	junction number
l.d.b.	long distance barred
l.j.	local junctor

m	message
m-d	data message
M	memory, data store
md	memory, data store
m.d.f.	main distribution frame
MK	mark
Mk	stored mark
mk	mark
n.u.	number unobtainable
o	operational data
o.b.	outgoing barred
o.g.	outgoing
o.g.j.	outgoing junction
o.j.	outgoing junctor
o.o.s.	one-only selector
o.r.	originating register
or.j.	originating junctor
P	logic part of processor
p.a.b.x.	private automatic branch exchange
p.a.m.	pulse amplitude modulated
p.a.x.	private automatic exchange
p.b.x.	private branch exchange
p.c.m.	pulse code modulated
pk	park
p.m.b.x.	private manual branch exchange
p.m.x.	private manual exchange
pr	subscriber as processor
prX	processor X
r.t.	ring tone
s	signal
s-d	data signal
s.j.	service junctor
sm	signal-message
s.o.	service observation
s.p.	semi-permanent (store)
SPC	stored program control
s.s.o.	special service observation
s.v.i.	service interception
sw	switch

temp.	temporary (store)
t.d.m.	time division multiplex
t.j.	terminating junctor
t.o.s.	temporarily out of service
t.s.	time switch

CONTENTS

1	**Introduction**	1
	1.1 Object and scope	1
	1.2 Messages and transmission	2
	1.3 Development of telegraph and telephone services	6
	1.4 Other services and future development	11
	1.5 Local and higher level exchanges	12
	1.6 Exposition	12
2	**Telephone Exchanges in General**	14
	2.1 Introduction	14
	2.2 Station requirements and operation	14
	2.3 Simplest manual exchange	19
	2.4 Multiples and multipling	23
	2.5 Directory numbers and equipment numbers	26
	2.6 Traffic probabilities and equipment economies	29
	2.7 Minimal structure exchanges	32
	2.8 Processing and programs	35
	2.9 Manual exchanges in practice	42
	2.10 Exchange systems generalizations	46
	2.11 Data storage	49
	2.12 Wired and stored program logic	51
	2.13 Time sharing by space and time division	52
	2.14 Tariffs and accounting	59
	2.15 Systems and chronology of development	62
3	**Automatic Exchanges**	65
	3.1 Progression from manual to automatic	65
	3.2 Automatic, semi-electronic and electronic exchanges	65
	3.3 Station requirements and apparatus	66
	3.4 Class of service	76
	3.5 Exchange switches	79

4	Step-by-step Exchanges.	84
	4.1 General	84
	4.2 Dial controlled selection	84
	4.3 Delayed dial controlled selection.	114
	4.4 Reverted dial controlled selection	116
	4.5 Register-translator controlled selection.	118
	4.6 Conclusion.	128
5	Cross-bar Exchanges.	129
	5.1 Switches.	129
	5.2 Trunking	132
	5.3 Construction and operation	135
	5.4 Path search and connection	138
	5.5 Translation and translators.	155
	5.6 Data storage and processing	163
	5.7 Conclusion.	172
6	Electronically Controlled Exchanges	174
	6.1 Electronic devices and techniques	174
	6.2 Objectives for electronically controlled exchanges	177
	6.3 Structure and call connection.	181
	6.4 Analogue cross-bar semi-electronic systems	191
7	Semi-electronic Exchanges.	192
	7.1 Introduction	192
	7.2 Time division time sharing – scanning	192
	7.3 Time division time sharing – processing	202
	7.4 Time sharing – general characteristics	214
	7.5 Time division time shared wired logic	223
	7.6 Generalized system	230
	7.7 Security, costs and practical application	234
8	Wired and Stored Program Logic.	236
	8.1 General	236
	8.2 Practical examples	236
	8.3 Characteristics and cost.	250
9	Semi-electronic Exchanges with Stored Program Logic	255
	9.1 General.	255
	9.2 Number translation	259
	9.3 Map in memory path search	265
	9.4 Signal and data transmission	270
	9.5 Operation of ultimate SPC exchange	274
	9.6 Programs and security	286

	9.7	Characteristics and costs	290
	9.8	Practical application	294
10	**Electronic Exchanges**		300
	10.1	Introduction	300
	10.2	Exchange switches	300
	10.3	Peripheral equipments	307
	10.4	Structures	309
	10.5	Control	314
11	**Conclusion**		319
Index			321

Chapter One
INTRODUCTION

1.1 Object and Scope

Humans communicate with one another using their five senses. The conditions and maximum distances over which communication is possible are limited by natural laws and the construction of the human body. Communication surmounting unnatural conditions or distances is telecommunicated, which means that the communication must be transmitted for at least part of the distance between the communicating parties in a form different from its natural form. Electric transmission is the most flexible and widely used form and the only one here to be considered. Electric transmission first used for the sending of written messages by telegraph, was later applied to messages spoken on the telephone. Public services require the destinations of the messages to be variable from the sending points, which implies some form of switching. Public telegraph services developed a kind of switching now called message switching and telephone services have a different kind called circuit switching. Now and more particularly in the future, message and circuit switching of information telecommunicated in a wide variety of forms is or will be required, and some of the forms and the quantities of information to be transmitted cannot be clearly foreseen at present. It is important that the whole problem be appreciated by system and exchange designers but the whole problem being outside the scope of one publication, this work is limited to circuit switching and almost entirely to telephone circuit switching.

There are three aspects of exchange design which are interrelated but still identifiably separate. System design is the first. Telecommunication being required over an area too large to be served by one exchange, many exchanges are used with telecommunication between them in a network called a system. The area covered is commonly an entire country by a national system. System design comprises exchanges in networks, the exchanges and their interconnection, and both telecommunication and switching. Exchange system design comprises exchanges in networks, the principles on which they operate and the services which the exchanges provide. Basic change to a national or an exchange system once it is established is usually difficult and expensive but it occurs infrequently. The second aspect, that of apparatus design, is concerned with components and sometimes their mechanical but more often their electrical interconnection, the electrical interconnection of components also being known as circuit design. Apparatus design to satisfy the requirements of a system can change as new components and techniques become available, usually without difficulty if the system does not change and mostly as new equipment required for growth and replacement of existing plant. Traffic design, the third aspect, is concerned with quantities of apparatus needed in exchanges to carry the traffic expected to occur. Traffic design is thus individual to exchanges and changes as the traffic changes usually because of natural growth. The principal concern of this work is exchange systems: the salient points of telecommunication and of apparatus design are outlined but traffic design is mentioned only as it affects system and apparatus design.

Because choices exist for the apparatus to be used for its mode of operation even within the limits of one system, apparatus and traffic designs are possible in great variety and many exist and are or have in the past been in commercial production. To acquaint the reader with the details of present or past designs of exchanges is not an objective which is pursued nor is it essential to the main objective, that of promoting an understanding of why and how exchanges come to be designed in the ways now known and may be designed in the future to satisfy the needs of public communication services however they may develop. Existing practices are nevertheless useful to illustrate facts, problems and theories and are used for those purposes. Thus the subject of exchange systems is no more than introduced but presented in broad perspective as a sound basis for the detailed study of known systems and the design of future systems.

1.2 Messages and Transmission

A message is an intelligent communication between men, between machines or between a man and a machine. It may not be intelligible to the men or machines involved, they may be passing the message for those to whom it is intelligible, but that it is sense and not nonsense is obvious to them. The essential parts of the communication are the originator of the message, the message transmission and the message recipient.

The electric telegraph made possible the rapid transmission over great distances of written messages comprising the letters of the alphabet and some other characters which in the written order constituted a message meaningful to the originator and to the recipient and will be termed the message proper. To a telegraphist concerned in sending a message over distance, the message was a series of symbols drawn from a set of N symbols, namely letters and figures, which he had to send in the order written, to another telegraphist to write down so as to re-create the message proper. Between the two telegraphists was an electrical circuit for message transmission and what could be sent, transmitted and recognized at the receiving end were electrical states, originally current and no current, and continuous current in one or other of the two possible directions of transmission. The sending telegraphist, using his hand to operate a mechanical key with contacts to control the electrical states to be transmitted, coded the N-symbols of the message proper into the dots, dashes and spaces of the Morse alphabet to transmit a message which the receiving telegraphist read from the movements of a galvano-

meter needle or from the sound emitted by a sounder. In more general terms, the
N-symbol message was transduced by a transmitter, the telegraphist and his key, to
a transmitted message comprising two electrical states and time, the transmitted
message being transduced by a receiver, the galvanometer and telegraphist, back to
the message proper. The transmission was not limited to two states, in fact
Alexander Graham Bell was experimenting with a method of telegraphy using more
than two states when he accidentally discovered the telephone. He was plucking
tuned reeds to make them vibrate at their own natural frequencies and thus to
generate currents which at the receiving end caused similar reeds to vibrate and be
heard. Hence in the general case the transmitted message comprises n electrical
states and is known for that reason as n-state digital or merely digital transmission.
In Table 1.1 the components of the first electrical telecommunication system, the
telegraph service, are shown.

Table 1.1   Telecommunication system components

Message proper	Transmitter-transducer	Transmitted message	Receiver-transducer	Message proper
Telegraph—written N-symbols digital	telegraphist—hand—key	n-state digital	galvanometer—eye—telegraphist	written N-symbols digital
Telephone—sound analogue	microphone	analogue	ear-phone	sound analogue

Telegraphed messages designated telegrams were the first but are now only one
class of digital message which telecommunication systems are or will be required to
transmit. In all cases there is a message proper comprising N-symbols to be
transduced, including being coded, by a man or machine transmitter to a
transmitted message of electrical n-states which are transmitted over an electrical
circuit commonly referred to as a line independently of its construction, to a
receiver which transduces and decodes the electrical states back to the N-symbols
in a form which is comprehensible to the recipient, man or machine. The coding
between the N and the n depends on their relative quantities and existences. If
$N = n$ a simple one-for-one code may be used and the states may be transmitted
independently of time. For example, ten press-buttons may be used to send the ten
decimal digits by pressing one button at a time and with each button arranged to
transmit a different frequency. A receiver will recognize the digit being sent by the
received frequency, and the length of time which it is sent is not important
provided that it is long enough for the receiver to recognize it. If N is not equal to
n, each member of the larger group has to be represented by a combination of the
members of, or by a timed sequence of the members of, the smaller group. If N

exceeds n which is more frequent in practice than the inverse, and if the n-states can be transmitted simultaneously and independently of one another in combinations at least equal in quantity to N, each N can be coded into a combination of n-states unique to itself and which are transmitted together and independently of time. For example, with $N = 10$ for the decimal digits, if $n = 5$ for five different frequencies and not ten as in the original example, the press-buttons can each send two frequencies when pressed, a different combination of two frequencies being available for each press-button and the durations of depression not important. In all other circumstances the n-states have to be transmitted as pulses in time sequence, as in the Morse code. Each pulse comprises one state if the states are for example no current and current which cannot be sent in simultaneous combination, or it comprises simultaneous combinations of different frequencies. Relative time becomes a parameter of the coding, and the N-states are said to be pulse coded. Real time is always a parameter of the transmission to define the maximum and minimun durations and rates of change of the N-states and therefore also of the n-states. The number N of N-states and the rate at which states have to be transmitted are characteristic of the message proper, and the number n of n-states, the maximum duration of a state and the maximum time permitted for a change from one state to another are characteristic of the transmitted message.

The telephone made possible the transmission of sound over great distances. It was also the first means of communication which required transmission over distance of a message which is infinitely and continuously variable, namely sound waves, and designated analogue to distinguish it from the digital kind which changes abruptly from one discrete state to another. Sound is only one class of analogue message now required to be telecommunicated. The principle of telecommunication, shown in Table 1.1, is the same as for digital messages. The originator provides the message proper, air waves in the case of speech, which is transduced to the transmitted message which at the receiving point is transduced back to the message proper. The transmitted message is an electric quantity which can be transmitted and is at all times proportional to the magnitude or to the rate of change of the magnitude of the message proper. The transmitted message is thus continuously variable and analogous to the message proper and is also designated analogue. Fourier analysis of the proper and the transmitted messages shows them to be composed of sine waves of frequencies ranging over a band of frequencies which is characteristic of the messages.

Electric transmission over distance is a natural phenomenon which is continuous and linear, which means that for an input at any one frequency, the output is continuously proportional to the input and is thus the analogue of the input and the transmission is designated analogue. The proportionality varies with frequency, the output decreasing for constant input as the frequency increases. Telecommunication transmission lines include added equipments such as amplifiers and passive networks which modify the transmission over distance so that overall the transmission is commonly a maximum and uniform over a band of frequencies which is the chief characteristic of the transmission medium. As demonstrated by the telephone, an analogue message proper is easily transduced to an analogue transmitted message of the same frequency content, transmitted over an analogue transmission line the bandwidth characteristic of which does not have to be the same but must at least include that of the transmitted message, with finally the transmitted message transduced back to the message proper. The transmission of the electric states of a digital message also covers a band of frequencies to which the analogue transmission over distance must be suited.

Analogue transmission lines may contain n-state digital relays which respond to only and transmit only digital n-states. The transmission over distance is analogue but the transmission overall is n-state digital and unable to transmit analogue messages. If the transmitted message can change state at any time, as with hand-keyed telegraph messages, the transmission is asynchronous digital: similarly, if the relays of digital transmission lines change state at any time that the message changes state, the transmission is asynchronous digital. Some transmissions can change state only at regularly recurring instants determined by some kind of clock: they are synchronous digital and transmission lines with the same characteristics are synchronous digital transmission lines.

A transmission within a band of frequencies may be translated to a similar band called a channel in a different part of the frequency spectrum and transmitted along with other transmissions in other channels, by frequency division multiplex f.d.m. equipment, but without affecting the problems of circuit switching. Analogue transmissions may also be transduced to pulse coded digital transmitted messages by pulse code modulation p.c.m.. A channel of transmission being the pulses of a message, and the pulses being synchronous, many similar synchronous digital transmissions are multiplexed for transmission over one line, by interleaving in time the pulses of the different channels, to produce p.c.m. time division multiplex p.c.m. − t.d.m. synchronous digital transmission and some new problems for circuit switching. Multiplex channels both f.d.m. and p.c.m. transmit in one direction only and separate channels transmitting in opposite directions are necessary to bothway transmission.

The details of message transmission over distance are the responsibility of transmission engineers. For decades the provinces of transmission and switching were thought to overlap so little as to be separate subjects each with its own experts. It was possible to maintain this situation so long as the switching of transmission circuit to transmission circuit was always accomplished by metal contact switches. The development of electronic and particularly of semi-conductor devices has provided not only an alternative to metal contacts for switching but also a means of integrating transmission and switching to the benefit of both. The integration includes the switching of p.c.m. − t.d.m. channels as multiplexed channels, briefly mentioned in the chapter on electronic exchanges but a requirement which metal contact switches are unable to fulfil. No longer is it possible to separate transmission and switching but herein transmission can be considered only so far as it directly affects exchange systems and apparatus. For that purpose, it is sufficient to understand that transmissions are of two kinds, analogue and digital, and transmission lines are of two kinds also analogue and digital, and that there are differences within each kind which are most simply described as differences of class. The class of an analogue message is largely defined by its frequency content and that of an analogue line by the frequencies which it transmits sufficiently well to be useful to message transmission. The class of a digital message is defined first by whether it is asynchronous or synchronous and then largely by the number of states n which it needs together with the maximum duration of a state and the minimum time between changes of state. The class of a digital transmission line is similarly defined as asynchronous or synchronous and then by the number of states n, the maximum duration of a state and the minimum time between changes of state that it can transmit.

A transmitter and a receiver permanently connected together over a transmission line provide point-to-point unidirectional telecommunication of messages and is satisfactory for some practical purposes. Bothway telecommunication comprises transmission paths in opposite directions between two points for the transmission in the two directions of related messages exclusive of other communications. A two-wire line is an electrical circuit which provides unidirectional or bothway transmission. A four-wire line comprises two two-wire lines, one for each direction of transmission of a bothway communication. Point-to-point operation between points transmitting one kind of message of one class is the simplest of telecommunication systems and finds application for some private services, telegraphy between railway signal boxes for example. In public services, messages proper originate from many points and have to be transmitted each as directed to one of many points, which is the role of the switching.

1.3 Development of Telegraph and Telephone Services

The electric telegraph was used during the nineteenth century to start the first public electric telecommunication service. The technique of the time being point-to-point bothway circuits transmitting one kind of message of one class, namely messages in Morse code, and the transmission requiring skill beyond the powers of general members of the public, the public service problem including that of switching was solved by three operational devices, namely

(a) the establishment of strategically located telegraph offices staffed by skilled telegraphists able to transmit and receive messages over point-to-point circuits between the offices;
(b) manual transport of each message written on paper, namely the telegram, from the originator of the message to the nearest telegraph office, and manual delivery of the telegram to the recipient from the office nearest to him;
(c) transmission of each message from the office nearest to the originator to the office nearest to the recipient directly between the offices if a point-to-point circuit existed between them and if not, by message switching as it is now called to distinguish it from circuit switching which was a later development.

Message switching comprises storing a message received over one point-to-point circuit and re-transmitting it over another, by which means the final destination of the message is variable depending on the circuits over which it is transmitted and re-transmitted. Message switching operation requires the message to be preceded by a control message, namely the address of the destination, which in the case of telegrams is part of the message provided by the originator. At each receiving point on the way, the address is examined to see if the destination office has been reached or not, and if not, a selection is made of a point-to-point circuit in the line of advance of the call toward the destination. The analogy with letter post is obvious. If a message is re-transmitted it is in effect switched from one circuit to another, hence the term message switching even though with manual operation no physical switches are needed. In manual telegraph offices the messages are stored as telegrams, that is as writing on paper, by the telegraphists who receive them. They or other telegraphists read the addresses and route each telegram for manual delivery, or for re-transmission to another office also by a telegraphist.

Telegraph offices staffed with telegraphists were interconnected by transmission lines terminated in the office on hand-operated transmitters and on receivers in the form of sounders which responded to the received messages by producing sounds which the telegraphists could read as coded messages and decode to plain texts. A telegraphist given a telegram to send, read the address and sent the message on an appropriate line. On lines which were continuously staffed it was sufficient to send just the message. Lines not continuously staffed had to be made operational before sending a message, by sending characters which by operating the sounder would attract the attention of a telegraphist who would send some characters in reply to indicate that sending of the message could commence. The message to alert its receiving point is a calling message and that to indicate that sending may commence is a proceed-to-send message.

Developments from the first simple manual operation included machine sending to increase the speed of transmission, and keyboard operation by the telegraphists to increase their productivity and reduce the skills needed. To increase the speed of transmission the telegraphists, instead of sending messages directly, punched them as holes in paper tape which, subsequently run through a machine transmitter, caused a machine receiver to print the dots and dashes on paper tape for a telegraphist to transcribe back to the message in plain text. Keyboards relieved the telegraphists of having to code the messages for transmission and of having to transduce the coded messages to transmitted messages. The output from the keyboard machine was either punched paper tape or transmitted message applied directly to the line. In one method of direct sending, called Baudot after the inventor, twelve or more keyboard transmitters shared one line which being switched to each transmitter in regular order by a continuously rotating distributor, transmitted the characters of the various messages interleaved in time, a similar distributor at the receiving end separating the characters to their own messages. This was the first time division multiplex transmission. These developments and practices, which continued to require skilled telegraphists, overlapped the invention and development of the telephone and telephone service by many years with hardly any contact between the two services until teleprinters reduced the skills needed to send and to receive messages to levels which are within the powers of members of the public. The analogue transmission lines used had to transmit frequencies from

0, for steady d.c., to 10 Hz or so for hand sending and up to 200 Hz or so for machine sending. Digital transmission speeds ranged from about 5 to 200 bits, that is 2-state or binary digits, per second. The achievements were modest by modern standards but they established punched paper tape message storage and transmission and time division multiplex transmission, both of which have become important again in modern times.

The development of the telephone service clearly had to be different from that of the telegraph. Analogue transmission of speech requires a frequency band now accepted as 300 to 3,400 Hz. Digital p.c.m. transmission, not invented at the time, requires at least 56,000 bits per second per channel to achieve an equivalent standard of transmission. Originators and recipients of messages must themselves operate the necessary transmitters and receivers without intermediate storage. Private messages are always bothway conversations between parties who are alternately originator and recipient and need both transmitter and receiver and must themselves guard the confidentiality of the messages if that is necessary, by low-level speaking into a transmitter close to the lips and listening to a receiver held close to the ear. Telephone equipments which incorporate a microphone, that is transmitter, and an earphone receiver for bothway private conversation are called instruments. Open or conference messages use room microphones and loudspeakers, and are not necessarily bothway transmissions. Members of the public using a public service are commonly called subscribers: they may be said to subscribe to the service by using it, and they may also subscribe in the sense of payment, which is probably the origin of the term but payment is not universal for all services. Other parties to conversations may be telephone operators providing a service such as enquiries. Occasionally synonyms such as customer are substituted for the term subscriber and user for any person using the service. Installations which include instruments for the use of subscribers are called stations. The equivalent of telegrams transported to and from a public office being impossible, persons requiring public telephone service have to establish their own individual subscriber stations from which they make calls as they require them and receive calls made to them by other users of the service, or they may go to a station specially provided for casual use with payment through a coin-box or to a station attendant, such being termed coin-box and pay stations, or very occasionally to a station provided for some special purpose such as calling the police or a hotel reservation desk. The stations have to be connected to the equivalent of telegraph offices by transmission lines, see Figure 1.1, the offices being exchanges and the transmission lines exchange lines. Based on United States usage, the term office is internationally accepted as synonymous with the term exchange. Herein the term office is used only for a manual telegraph office, arbitrarily but justified perhaps by 'exchange' conveying more precisely the main function than does 'office'. Transmission lines between exchanges and performing the same function as lines between telegraph offices, are designated junctions.

Ordinary subscriber stations, Figure 1.1, comprise one instrument terminating one exchange line termed an ordinary exchange line. P.B.X. subscriber stations have more than one exchange line and more instruments than exchange lines, and a

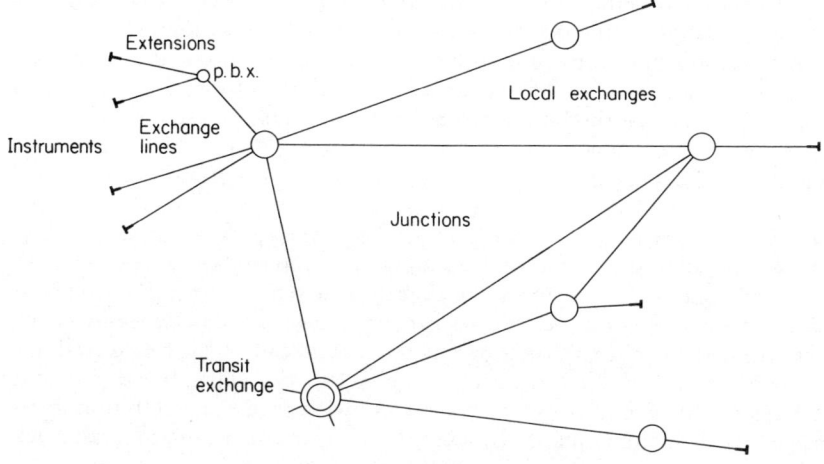

Figure 1.1   Public telephone service switched network

private branch exchange or p.b.x. to which the instruments are connected by transmission lines called extensions, the p.b.x. making connections between the extensions and between the extensions and the exchange lines for calls both to and from the public exchange.

Exchange lines are normally two wires laid up together with a twist, each wire providing a continuous metallic path from the exchange to the station. The cables and other equipments providing the exchange lines are designated local line plant, and exchanges on which local line plant terminates are local exchanges. Originally all junctions were a pair of wires as for exchange lines. Short junctions are still the same but other forms of construction are more satisfactory and economical for the longer junctions. Local exchanges have junctions terminated on them as well as exchange lines. Exchanges having only junctions terminated on them are transit exchanges. Many transit exchanges are co-located with a local exchange to form a combined local and transit exchange which is usually a higher level exchange as will be described in section 1.5.

In addition to subscriber stations there are stations concerned with services continuously available to the subscribers, stations provided for administrative purposes and stations used by the maintenance staff. Some of these other stations, which in practice are rarely called stations but designated by their functions, may be located in the exchange itself while others may be at great distances from the exchange relative to the distances of subscriber stations. At some service stations the service is rendered by a human operator, and at others by a machine, a clock speaking the time for example. Stations which are not subscriber stations have one or more lines to an exchange as for subscriber stations except that the characteristics and operation of the lines may resemble subscriber exchange lines or junctions depending on the functions and locations of the stations.

Continuous transmission from an instrument at one station, the calling station, using its exchange line or one of its lines, to the exchange line or one of the exchange lines of another station, the called station, is established by circuit switching at the exchanges on the route between them, the exchanges being equipped for the purpose with exchange switches. The transmission circuit so connected and established is designated a connection or a call. The calling party defines the connection to be made by sending over an exchange line to his local exchange the address of the station to which connection is required. At the exchange the line is connected by the exchange switches to an exchange line of the called station if it is connected to the same exchange, a call between two exchange lines on the same exchange being designated a local call. If the called station is not on the same exchange, the local exchange connects the calling line to a junction to an exchange in the line of advance of the connection required, and transmits to that exchange over the junction itself, except in the most recent of exchanges, the address of the wanted station. That exchange connects to an exchange line to the wanted station or to another junction and so on until a continuous connection exists as required. That message and circuit switching use basically the same method of routing messages is obvious: it is the method of transmission of the messages over the circuits of the routes which is different. With circuit switching the exchange switches are operated manually by manual operators in manual exchanges and in automatic exchanges by remote control by the calling parties. The addresses which are sent on individual circuits depend on the characteristics of the exchanges and of the calls, and vary from partial to complete addresses of the called stations. As none of the exchange lines is continuously operational, a calling message must be sent and the return of a proceed-to-send-the-address message awaited before the address of a called line is sent over the calling line, and the same is true of junctions on modern exchanges. When all the circuits required for a connection are connected and the called station has answered, that is, a party at the called station is ready to converse, the calling party must be informed of the fact, for which purpose the called station is required to send a proceed-to-send-the-message-proper message to the caller. Preferably this last message is the address of the station which has been called and has answered, as this serves the double purpose of message and check that the connection is the wanted one. This is a practice established for circuit switched teleprinter connections by the answer back feature which causes a teleprinter automatically to transmit the address of the station at which it is located when it is first connected for an incoming call, and by telephone subscribers who answer incoming calls with the announcement of the address of the station.

When teleprinters became available, many of the practices by then established for the telephone were adapted to the telegraph service. Telegraph offices now have exchange switches for the circuit switching of connections between teleprinters in the offices whereby telegrams are transmitted only once, from origination to destination offices, without the complication and delay of storage inherent in message switching. A subscriber-operated telegraph service called telex and analogous to the telephone service is well established, using circuit switching exchanges to serve teleprinter subscriber stations as for the telephone. A problem for the future is to integrate as much as possible the telegraph, telephone and all other services to share administration, building and plant costs in the interests of economy and good service.

1.4 Other Services and Future Developments

Circuit switched networks for public telephone and teleprinter services being well established in most countries, the telecommunication of information in other forms if provided at all is adapted to use the plant of one or other of the two established services, with not very satisfactory results in some cases. The effect has been to suppress to some extent the demand for other services but to what extent cannot be measured very accurately. For example, picture telegraphy for the telecommunication of pictures of all kinds and line drawings, is possible over the telephone service lines but difficulties discourage small users and encourage large users of such services to rent private circuits not subject to switching. Two developing services which are unable to use the already existing public service plants are videophone, which is public service television, and high speed digital data transmission mostly between computers or apparatus associated with computers. The difficulty is that even though some plant able to transmit such information may exist in the networks, it is not always accessible through the existing exchanges either or both because it is not suitably connected to the exchange or the exchange is not capable of switching the transmission. Some new provision of line and exchange plant is thus inevitable and provides the opportunity to introduce new techniques and new designs of plant.

Two aspects of public services have profound influences on the design of and installation of the plant. One is that the services are continuously growing, which means that new plant has to be added to take the growth as it occurs and in fact before it occurs to avoid delays in the provision of service when it becomes required. The second aspect is that service must be continuous. Even interruptions due to plant faults must be prevented from causing complete break-down of service at exchanges more frequently than the order of an hour per century per exchange. When advances in technology such as electronic for electro-mechanical equipment make radically new designs of exchanges possible, the new system equipment has to extend an already well-established network without interruption of service to the users. A new design does not merely have to satisfy the service requirements and be competitive with an existing system on equal terms. To supersede an existing system for installation in a network, in the absence of some special circumstance a new system has to be substantially cheaper for the same or better service, or give substantially better service at no greater cost. That the existing service is no longer adequate may justify higher cost for better service but a deterioration in service would be very unwillingly accepted even at substantially lower cost. The administrative and plant upheavals and changes inherent in a change of exchange system have economic and network consequences which must be taken into account in the design and introduction of a new system and they prohibit change to gain a trivial advantage.

1.5 Local and Higher Level Exchanges

The many exchanges which go to make up a large telecommunication system are not all the same with regard to their transmission, switching and control functions. Some provide only local switching between exchange lines and between exchange lines and junctions but not transit switching between junctions. Some have both local and transit switching and a small proportion have transit switching only. Local exchanges perform such functions as are practical and economic for them to carry out, and more complicated and difficult functions are delegated to fewer but better placed exchanges which in turn use still fewer other exchanges to perform functions which they cannot conveniently execute themselves and so on. Hence there is a hierarchy of exchanges at different levels of importance to transmission and switching, and the requirements and economics of exchanges at different levels vary considerably. Nevertheless for manufacturing and other practical reasons, the design of the exchanges must be basically the same for all, except possibly those at the highest level, such exchanges being so few in number and small in cost relative to the whole network that they are sometimes treated as special cases not part of the general problem. Exchange design as it is considered and developed in the following chapters applies to the lowest levels of exchanges unless higher levels are particularly specified.

1.6 Exposition

The transmission, switching and operational practices and the terminology of the previous sections are the base from which the exposition in the following chapters starts. The exposition is concerned with the design of modern and future exchanges which are the present end of a long chain of developments from simple manual exchanges through electro-mechanical to electronic automatic exchanges of great complexity. To be ignorant of the past leads to narrow-mindedness which is a danger to a good design, yet too detailed a study is neither useful nor interesting to any but technical historians. What is important in the present context is that as much should be extracted from the past as is useful to the present.

The first essential in the design of telecommunication exchanges, as in all problems, is to understand the problem and this can look deceptively easy but prove surprisingly difficult particularly at a time of radical change such as has been experienced in recent years in the telecommunications industry. The first part of the expostion is directed to establishing the problem and with this is combined a study of historical systems not in detail but sufficient to establish and develop the problem and the techniques which have emerged in its solution. This is possible because although the problem has developed in the direction of greater complication, it has not changed radically with changes of technique available for its solution. Thus the problem unifies all systems and its study through past systems is not only useful to present design but, it is hoped the reader will come to agree, a worthwhile education in itself.

Chapter Two
TELEPHONE EXCHANGES IN GENERAL

2.1 Introduction

The practices used and to be used in public circuit switched telecommunication services have mostly originated from the telephone service which is the most complicated in operation and likely to remain so and to be the major service in terms of traffic and commercial business for a long time to come. For these reasons the study and analysis of telephone service and its stages of development are useful not only to the understanding of the telephone problem but also to the general problem of public circuit switched telecommunication services and the ways in which new services may be developed and put into commercial operation. Telephone development has taken place in three stages overlapping in time, namely manual, automatic and electronic. Despite the fact that the latest systems look very different from the earliest, nearly if not all the basic problems occurred and their solutions can be recognized in manual systems the analysis and description of which also provides the opportunity for the establishment of many of the terms and ideas in common use and peculiar to the industry.

2.2 Station Requirements and Operation

Telephone service started with and is still very much influenced by stations and their requirements and equipments.

For a subscriber station equipped only with telephone instruments to be operational to incoming calls, the subscriber would have to have the receiver pressed to his ear continuously to hear speech when it occurred, which is impractical. The subscriber requires to be free to pursue his normal activities and to listen and talk on the telephone only when communication with another subscriber is actually required. He can go to the telephone to originate a call himself, but must be called to the telephone possibly from a considerable distance for an incoming call from another subscriber. There is no way by which the receiver itself can be employed to call the subscriber by speech or like currents transmitted to it. The normal speech level is governed by the requirements of telephone conversation with the result that the range of hearing of sounds from the receiver is only a few inches. There is no technically satisfactory way of increasing the sound level out of the receiver, and if there were, the danger that the subscriber would have the receiver close to his ear and suffer pain and even ear damage when the high sound level occurred, would prevent its use. Clearly, what is required to attract the attention of the subscriber to an incoming call is a calling device separate from the telephone instrument, to produce a considerable volume of sound from an electrical input transmitted from the exchange over the exchange line to the calling device. The problem divides into three parts, namely the transmitted power, the transmission attenuation and the efficiency of the calling device in transducing from electric to acoustic power.

The maximum power which may be applied to telephone exchange lines is limited by cross-talk which results from unavoidable imperfections in the line and exchange plants and causes the transfer by electro-magnetic induction of some of the alternating current power in one transmission circuit to other circuits in close physical association. The quantity of power transferred is very small compared with that transmitted but it is nevertheless highly significant. The maximum power which may be transmitted varies in a somewhat complicated way but so far as the present question is concerned, the permitted powers are high at very low frequencies and decrease as the frequency increases. Transmission over exchange lines consisting of just a pair of wires suffers attenuation which rises quickly as the frequency increases, with the overall effect that the power which can be delivered to a calling device at the end of a long exchange line can be very high for very low frequencies but decreases very rapidly as the frequency rises. On the other hand, a transducer to be an efficient converter of electrical to sound power must be increased in size as the frequency is decreased, which means that for a ringing transducer to be of practical size, its sound output must be mainly at a high audio frequency or frequencies, which are frequencies at which the maximum electrical power which can be received over an exchange line is of necessity small. Also the calling device must have no output for small inputs which it will experience from cross-talk from other circuits, but when it is operated its acoustic output must be very nearly independent of the electrical input or the intensity of the sound produced will vary greatly with the distance of the station from the exchange. Thus the performance specification for the calling device is not simple, but it is satisfied by a simple device, the magneto bell, invented by Watson very soon after the invention of the telephone itself by Bell, Watson being his assistant at the time. The bell requires a high power input but at a low frequency, about 20 Hz, obtained from about 70 volts applied to the line, originally from a small hand-turned magneto generator which was a valuable feature in the days when public power supply practically did not exist and the telephone had to rely on primary batteries which by modern standards were inefficient and very expensive. The bell has an armature polarized by a permanent magnet so that it is moved in opposite directions by the alternate half waves of the input current, the efficiency of transducing from electric current to mechanical movement being increased by the polarization, a fact already known to Watson from the existing telegraph practice. The armature is not moved except by inputs exceeding some value and when moved it causes a bell gong to be struck by a hammer with a blow which is almost independent of the level of the input current. The gongs vibrating at their own natural high audio frequencies are small but efficient producers of acoustic power and the bell is cheap to construct. Thus all the requirements of a satisfactory device are achieved, so much so that to this day nearly one hundred years after its first use, the magneto bell is still standard practice for alerting subscribers' stations that an incoming call has been connected and so well established that it is difficult to change. Nevertheless a change is overdue to suit electronic exchanges and is a topic of section 10. 3.

Although the use of a calling device separate from the telephone receiver avoids the possibility of the device producing an excessively loud sound close to the subscriber's ear, if the high voltage low frequency current intended to ring a bell becomes applied instead to the telephone receiver, because of its low efficiency at the low ringing frequency, the output from the receiver is of much lower level than from a bell and not dangerous to hearing, but is is nevertheless painfully loud if the receiver is held close to the ear. The possibility of that occurrence has to be minimized by the exchange operation and control.

Watson's magneto generator and bell established a new principle in telecommunication systems. The telegraph operator used one key and transmission circuit to send both messages proper and control messages, the two types of message being thus of the same kind and class. The calling message which operates the magneto bell is a two-state digital transmission of frequency below that of the analogue speech messages. It is therefore different in kind and in class of transmission, which creates two distinct but complementary transmission and switching systems, one for messages proper and another for control messages which are called signals to distinguish them. Ringing the bell is a calling signal but it is usually referred to as ringing or as a ringing signal to distinguish it from the calling signal given by the station itself to alert the exchange to the start of a new call from the station and called originating, calls received by the station being designated terminating.

Although it is impractical for subscribers to be listening all the time for incoming speech, it is possible for operators to do so and a system used for junction operation and requiring an operator to be listening all the time is described in conjunction with Figure 2.10. It was in fact attempted in very early exchanges for subscriber lines equipped only with telephone instruments. Operators had a number of lines to serve and all the lines which were not connected for calls through the switchboard were paralleled in the exchange to the operator's telephone instrument. If free and wishing to be connected to another subscriber, a calling subscriber spoke the identity of his line and that of the line to which he required connection, and the operator hearing made the connection which automatically disconnected the lines involved from the common connection to her own telephone. The operator had to listen continuously and this not proving satisfactory, it was soon replaced by continuous visual observation of electro-mechanical indicators, an indicator being connected in parallel with each exchange line at the exchange end and mounted on the switchboard in full view of the operator. A subscriber generated calling signals and also clearing signals to tell the operator that calls were finished, by turning the handle of a magnetor generator, of the type already available for bell ringing. Turning the handle also operated a contact which substituted the generator for the speech transmitter, the high voltage low frequency output of the generator thus being applied to the line. The current received at the exchange produced movement of the indicator which the operator was able to observe at all times even though she had many hundreds of indicators to watch, and of course without the mutual interference between signals which made the original speech method so difficult. In effect the indicator was a substitute for the ear of the operator, and the generator for the peripheral station transmitter.

With modern electronics, indicators sensitive enough to respond to speech currents received over telephone lines are possible and have been produced and employed for some special purposes. With still greater sophistication it is or may be possible to construct a machine to recognize decimal numbers spoken over telephone lines, a possibility which is seriously proposed and pursued periodically, but so far remains in the realms of phantasy and seems to have little practical advantage even if it were to be achieved.

Coming back to the manual system with indicators, the turning of a handle was not a very convenient or successful operation. Both the calling and the called parties were required to turn handles at the termination of a conversation for the information of the operator or operators who had to disconnect the call, and this the subscribers frequently failed to do. Also they sometimes turned the handle when the operator was listening, the uncomfortable consequences of which made the system unpopular with the operators. The lesson which had to be learned was that the only control operations which subscribers can be relied upon to perform are those which they cannot fail to execute to get the services which they are requiring. Actions which subscribers cannot fail to perform when starting and terminating a call are the picking up and the laying down of the instrument. Providing a cradle for the instrument, the cradle being constructed so that the actions of removal from and return to it of the instrument respectively close and open a pair of metallic contacts, produces from the contacts reliable signals for every call which is made. For subscriber originated calls the signals are designated call and clear and for calls incoming to the subscriber the same signals are designated answer and clear. The terms off-hook and on hook are also in common use for the instrument off and on the cradle respectively and for the corresponding signals, and the two actions are often described as taking off and hanging up the instrument or receiver. The contacts being connected in series with the exchange line, the lifting of the receiver completes a metallic loop from the exchange to the station: current flows through the loop from a battery at the exchange. The current is detected at the exchange and also its interruption when the instrument is replaced. A complete installation comprising instrument, cradle switch, bell and any other associated equipment is called a telephone set. The signalling described is designated loop signalling and proves doubly useful in that the supply of current from the exchange during periods of telephone use also provides the power needed by the transmitter containing carbon granules. In fact the value of the current which is in common use was chosen and arranged to suit not the transmitter but the signal detector in the exchange, an electro-mechanical relay controlling a lamp to substitute the indicator. The transmitter was designed to suit the current thus determined thereby to establish a practice for exchange line transmission and signalling which persists to this day and is difficult to change. Change is required for electronic exchanges which do not use electromechanical relays and are handicapped by the line currents of present practice. Whatever value it may be, current and no current in the loop constitute a binary digital signal used for calling and clearing signals for originated calls and for answering and clearing of incoming calls. Although the speech message analogue currents and the control signal digital currents traverse the same physical path, namely the exchange line pair of wires to the exchange, they make use of different parts of the transmitted frequency spectrum which amounts to transmission over separate paths as for the ringing signal. Effectively separate and individual paths for message and signal transmissions occur not only with exchange lines but also through the exchange switches and over junctions between exchanges. If the two paths run inseparably parallel by being co-existent or otherwise permanently associated, the control signal path is said to be speech path associated, or message path associated in the general case, otherwise it is disassociated or separate.

Administrations are not concerned with the meanings of messages proper, only with their adequate transmission. The meanings are the responsibilities of the parties using the system. On the other hand, the administration has to be aware of the meanings as well as to provide for the transmission of signals. In practice the term 'signal' is frequently applied to both the meaning and the transmission of messages proper and of signals, and in other contexts it may intend only a control signal. To be specific and brief, the term 'message' will denote a message proper or more usually its transmission since its meaning is not of interest to the exchange system. The term 'signal' will apply to a control message, its meaning or its transmission, over a path different from that of the message and either message path associated or disassociated. If the meaning is important to the context it is added, for example a calling signal, and similarly the transmission may be specified, as for example a loop signal. Signals which are the same kind and class as transmitted messages and transmitted over the same circuits, which is the case for all telegraph signals and for telephone system signals which are spoken or heard as speech or tones, will be termed signal-messages to distinguish them from messages proper and from signals on paths other than message paths.

So long as the transmission lines terminated on exchanges are interconnected by metal contact switches, the adequacy of message transmission is controlled mainly by the performances of the lines. The exchanges are concerned only to minimize the transmission losses and distortions of messages transmitted through them, and to interconnect some four-wire lines by four-wire switching which allows the overall losses of the connections made to be reduced below that which can be achieved with two-wire switching alone. With the interconnection of the transmission lines by electronic switches, the line and exchange transmissions can be integrated to the benefit of overall transmission, which is referred to again in chapter 10. In general signal transmission makes use of transmission lines designed and provided for message transmission and the adequacy of performance is the responsibility of the exchange equipment designer.

The loop signals which have been described as generated by the cradle switch contacts of subscribers' telephone sets are basic to all telephone exchange lines although not always generated by telephone cradle switches. They may also be generated by key and relay contacts on manual and automatic switchboards and other equipments on which exchange lines may be terminated in subscribers' premises instead of telephone sets directly. Such equipments usually do not have magneto bells but other devices to respond to the low frequency ringing current which is universal for all lines as a calling signal to the stations. Additional non-basic or secondary signals are also required by some exchange lines, being those with special characteristics or offering special facilities such as party lines serving more than one station and lines for which the cost of calls made is collected by a coin-box. The result is the need to extend the binary loop signals transmitted in one direction over the pair of conductors of the exchange line and the binary low frequency a.c. signal in the other direciton to at least three-state in both directions, a problem which is discussed in chapter 3.

2.3 Simplest Manual Exchange

Figure 2.1 shows a telephone system with the exchange lines terminated in telephone sets at the stations, and on one exchange with one operator, which is the simplest system having any practical use. Some stations may have more than one

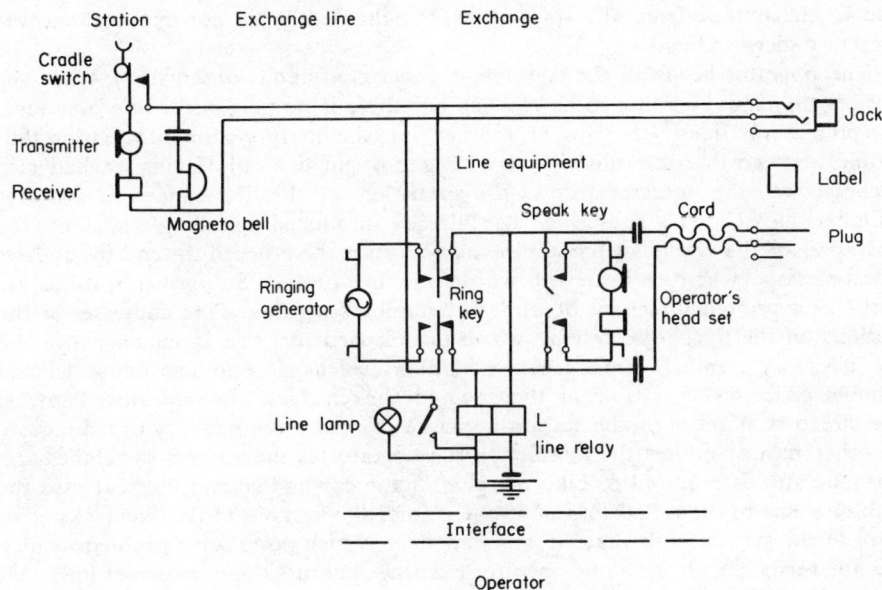

Figure 2.1   Simplest manual exchange

telephone set and exchange line. The sets provide loop calling, answering and clearing signals generated by the cradle switches and they are called from the exchange by ringing signals which operate magneto bells. Each telephone set is connected by its exchange line to equipment individual to the line and comprising a jack and line relay L to both of which the line is connected directly, and a flexible cord terminated in a plug to which the line is connected via capacitors which allow a.c. speech currents to pass but not d.c. signals. The construction and assembly of the jacks, cords and plugs which together constitute an exchange switch, is such that any plug may be inserted into any jack not already occupied by a plug, for the two wires of the plug to make metallic contact with the two wires of the jack. The two contacts of the plug and jack are known as the tip and ring and a third contact, shown unconnected in Figure 2.1, as the sleeve. A common battery with its positive pole earthed supplies current to the lines via their line relays L which operate to loop current to light their respective lamps, the states of the lamps indicating visually the states of the loops. Each line equipment also includes a speak key which when operated or 'thrown' connects the operator's instrument, which she wears as a head-set, to the line, and a ring key which when thrown connects a ringing current generator to the line to ring the station bell. The line equipment components which also include a label, are placed in close juxtaposition on the operator's switchboard so as to be easily identified and reached by the operator.

In Figure 2.1 and in later diagrams the transmitter and receiver are shown in simple series. In practice the circuitry is more complicated to reduce the side tone, which is the level of the input to the receiver relative to that of the line and both due to the output from the transmitter, but the details are not important to the present understanding.

The operator keeps all the equipment under continuous observation. When she notices the lamp of a line to be glowing but there is no plug in the line jack nor is the plug of the line in the jack of another line, she interprets the information thus visible to mean that the subscriber is calling to originate a call. Having reached such a conclusion, the operator throws the speak key of the line concerned and says 'number please'. The subscriber has already produced a calling signal by the operation of the cradle switch and he now receives the proceed-to-send-the-address signal-message when he hears the words 'number please'. Subscriber stations are listed in a printed directory of directory numbers which are the addresses of the stations on the telephone system, which in this particular case is one exchange and the directory numbers are exchange numbers, exchange numbers being decimal number addresses of stations on their own local exchanges. The subscriber knowing the directory number of the station he wishes to call, from memory or a directory or other means, quotes the number to the operator as the address signal-message. The operator is required to take the plug of the calling line and insert it into the jack of a line to the called station, if one is free. She first has to find the jacks of all lines to the station with that directory number, which poses some problems which are not very difficult for a one-operator exchange because the quantity of lines that one operator can look after is small and the line locations and other details are within her memory capacity aided by the labels. So given a number the operator will know which are the corresponding jacks, or that no connection can be made because no station with that number is connected to the exchange. If no connection can be made the operator informs the caller that the number is unobtainable and expects him to clear by hanging up. Although the line lamp may continue to glow for some time until the subscriber hangs up, the operator remembers the previous events and will not again ask for a number unless the lamp extinguishes and re-lights. If the lamp continues to glow for a very long time the operator may throw the speak key and ask the subscriber why he is still waiting and in the absence of a reply, conclude that he has forgotten to clear. If the number given by the subscriber is a valid one, the operator will examine the jack or jacks of the line or lines to the station with that number to see if one is free; that is, it has neither its plug nor its jack in use. If none is free, the operator says 'number engaged' and expects the caller to clear and call again later. If a free line exists, the operator inserts the plug of the calling line into the jack of the free line which she rings by operating the ring key of the called line, for short periods at ten second intervals or so. The operating required from the time of completion of a connection until it is broken down is called supervision. During supervision the operator infers the state of the call from the two line lamps concerned, and if in doubt at any time, she can listen on the connection and speak, if she wishes, to the participants, to ascertain the situation. If the called line does not answer within a reasonable time the operator tells the caller 'no answer' and disconnects the call. If and when the called line answers, the operator takes no further action until the termination of the call when normally both parties clear and the operator takes the call down as soon as both lamps are dim.

A description such as has just been given of all the possible events for one connection is valid for the system as a subscriber who can be concerned with only one call at a time sees it, but not for the operator. The operator cannot give all her attention to each call from the time that she answers the calling signal until the completion of the call, to the exclusion of other calls. Demands for service occur at random times in consequence of which the quantity of subscribers expecting service simultaneously varies according to the laws of chance. The operator must adjust her actions and divide her time among the callers and the established calls to give the best possible service to all equally. At any one time there will be a quantity, variable with time, of calls in progress and in states almost independent of one another. For individual calls the operator acts according to the description given for one call but must interleave in time the actions for different calls so that they may all proceed simultaneously and independently. Mistakes and irrational behaviour not covered by the description will also occur, by the operator and by the subscribers, and with which the equipment and the operator must cope. The equipment having been installed, it is the operator who has to solve the problems of varying traffic, overlapping calls and other circumstances, which she learns to do during her training and thereafter relies more on her memory than on logical reasoning from instant to instant.

In the simple system of Figure 2.1 many of the elements of a public telecommunication system of any size and complexity can be identified. Missing, but to be introduced later, are the transmission lines which are junctions between exchanges. With some slight exceptions in respect of stations, the equipments of the system, comprising the stations and transmission lines and the exchanges, are provided by the administration. The control which causes the system to function originates from the subscribers, who control their stations directly and the exchange remotely by signals and signal-messages transmitted over the exchange lines. The exchange is very obviously in two parts: one part is an engineering construction which is theoretically a machine but is more conveniently designated a structure and the other part is the operator. Less obviously perhaps is a third part, the space which is an interface between the structure and the operator. The structure comprises switches which when suitably controlled, connect together for through transmission from one to the other via the switches, exchange lines in Figure 2.1 but message transmission lines of all kinds in the general case. The switches will be termed exchange switches in the general case, by which term those of Figure 2.1 have already been identified. The control of the exchange switches and of the connections which are made by them is exercised by the operator who is continuously collecting data concerning all the message transmission lines for which she is responsible, and applying logical processes to the data to arrive at conclusions on which to act. The data include signal-messages spoken by, and signals generated by, the subscribers, and the states of plugs and jacks more generally referred to as the state of the structure. The data have to be transmitted from their points of origin to the operator, that is to the control data processor, which means transducing every time the transmission medium changes. In Figure 2.1, the cradle switch transduces the subscriber actions of picking up and laying down the telephone instrument to current for transmission over the exchange line, to the exchange where a relay transduces the current to contact state which controls a lamp to produce light for transmission over the interface to the operator. The operator is not aware of all the data all the time. She selects circuits, which may be individual lines or parts of connections, for processing one at a time according to the traffic and states of the calls as previously described, then collects data for the selected circuit and processes the data: if the conclusion which results from the processing is that action is required, she takes the action across the interface to the selected circuit. Effectively therefore, the interface is a switch by which the processor is selectively connected to circuits to be processed. Obviously the switch has much in common with exchange switches: the basic difference is that the interface switch transmits only data whereas the exchange switches are designed primarily for message transmission with which data transmission may or may not be associated. The difference results in very different designs of switch being used in some systems for the two purposes, for which reason and to separate the different functions, the terms exchange switch and interface switch will be used for the two types. The data transmitted by the interface switch comprise signals and signal-messages from and to the peripheral stations, the states of the structure and of the calls with, in addition, the actions which the operator takes across the interface, namely the insertion and withdrawal of plugs in and out of jacks, which are effectively data transmitted to the structure to change its state, as will more clearly be seen to be the case when the operator is replaced by a machine processor. For convenience and precision in later discussion, the terms signal and signal-message will be limited to data originating at or transmitted to a peripheral station, and data which are instructions to the structure to change its state will be intended when the term operational data is used. With this convention, one item of information of any kind is a datum, signals and signal-messages and operational data are particular kinds of data, and any other datum may be qualified according to its meaning, loop datum for example.

The concept derived from the simple system of Figure 2.1 of an exchange as composed of a structure and a data processor associated through an interface can be expanded to explain and to provide an unlying unity for exchange systems of all types. The primary function of the structure is the circuit switching of message transmission circuits, and that of the operator is signal and data reception, processing and transmission. The interface is primarily a switch which selectively connects the operator to parts of the structure not only for the control of the structure but also for signal and data reception and transmission which is message-path associated.

Systems of the simple kind represented by Figure 2.1 find little use in practice for economic and operational reasons. One severe limitation imposed by having only one operator is that the traffic which she can handle is finite and the quantity of lines which she can serve is far less than that which most exchanges have connected to them. Two or three operators might be employed on one switchboard without getting into one another's way too much, but with more than that number the physical size of the equipment would cause severe difficulties. The cord lengths would become excessive and the task of an operator in observing the states of connected lines the equipments of which were widely separated on the switchboard would be a practical impossibility.

2.4 Multiples and Multipling

The problem of operators, or processors or equipments of any kind, having access to many circuits without getting into one another's way too much is solved by multipling. The system of Figure 2.2 illustrates multipling in a manual exchange.

The exchange line jacks in a manual exchange are located in the upper part of the vertical face of the switchboard, as in Figure 2.3, in rectangular blocks which may contain up to ten thousand jacks. Similar blocks end to end along the switchboard have their corresponding jacks connected in parallel, as shown in Figure 2.2 and which is called multipling, the jacks being called collectively a multiple and each block of jacks an appearance of the multiple. Operators spread along the length of the switchboard may connect a plug-ended cord to any line through any appearance of the multiple and sufficient appearances are provided for the necessary number of operators. Each operator is allocated a length of the switchboard which, together with some associated equipment. is known as a position. Positions are designed to give sufficient working space for each operator and the multiple is dimensioned so that each operator has at least one jack of every line within her reach without moving from her chair other than having to stand up and stretch for some: but for this to be possible, space does not admit of individual labels for the jacks. Equipment individual to each position is mounted on a horizontal shelf below and forward of the vertical face of the switchboard. In the example of Figure 2.2, the equipment mounted on each position comprises the line equipments of a quantity of exchange lines, the originated traffic of which is answered and dealt with by the operator at the position. Each of the operators being thus responsible for only a limited quantity of exchange lines and traffic, their work loads can be regulated to practical and acceptable levels. A line originates a call by being looped and thus operating the line relay L and lighting a calling lamp as in Figure 2.1. The operator at the position answers by throwing the speak key of the line and, using the plug and cord of that line, she can connect the line via the multiple to any other line on the exchange, but she can no longer see at a glance and with certainty that a line is free or not, nor can she read a label with data individual to that line. Nor can an operator see at a glance that one of her own exchange lines has been connected for an incoming call, so that if the line lamp were to be controlled only by the line loop, she could not quickly and easily ascertain that the lamp glowing meant a new originated call or the answering of an incoming call. Thus some of the logical processing which the operator of the exchange of Figure 2.1 performs, if not impossible for the exchange of Figure 2.2, is so difficult and time consuming that some other solution is a necessity. One possible alternative is a telecommunication system between the operators whereby each could inform others affected by her own operations what those operations were and could ask other operators questions concerning the states of the lines under their control. Then, using their memories to retain information communicated and answers received, no operator need move or observe over more than the section of switchboard immediately in view and all the mechanical and operational problems are solved. A solution of this kind is possible for some machine processors but for human operators it is impractical. The solution adopted for human operators uses telecommunication but achieved in another way. The operators communicate with one another via the structure, using a third wire sleeve connection through the plugs and jacks as a communication path. To the equipment of Figure 2.1 is added the multipled jacks and the relays K,J and LB of Figure 2.2. To the sleeve contact of each plug is connected a relay J to battery, and the plug of a calling line inserted into the jack of the called line operates the J relay of the plug and K relay connected to the sleeve of the jack via a contact of its J relay which at that time is not operated. A lamp lit by the K relay when operated could tell the operator to whom the line is assigned for originating calls, that when the calling lamp lit it was the subscriber answering an incoming call from another operator, and not an originating call. The operator is saved the effort of processing these two data, the result of which is to tell her to do nothing, by using the K relay not to light a lamp but to cut off the L relay and thus to prevent the line lamp from lighting. By this simple example is illustrated a very important principle, namely that processing performed by processors (operators) common to many lines and calls, and designated common processors, can be transferred to processors (J, K and

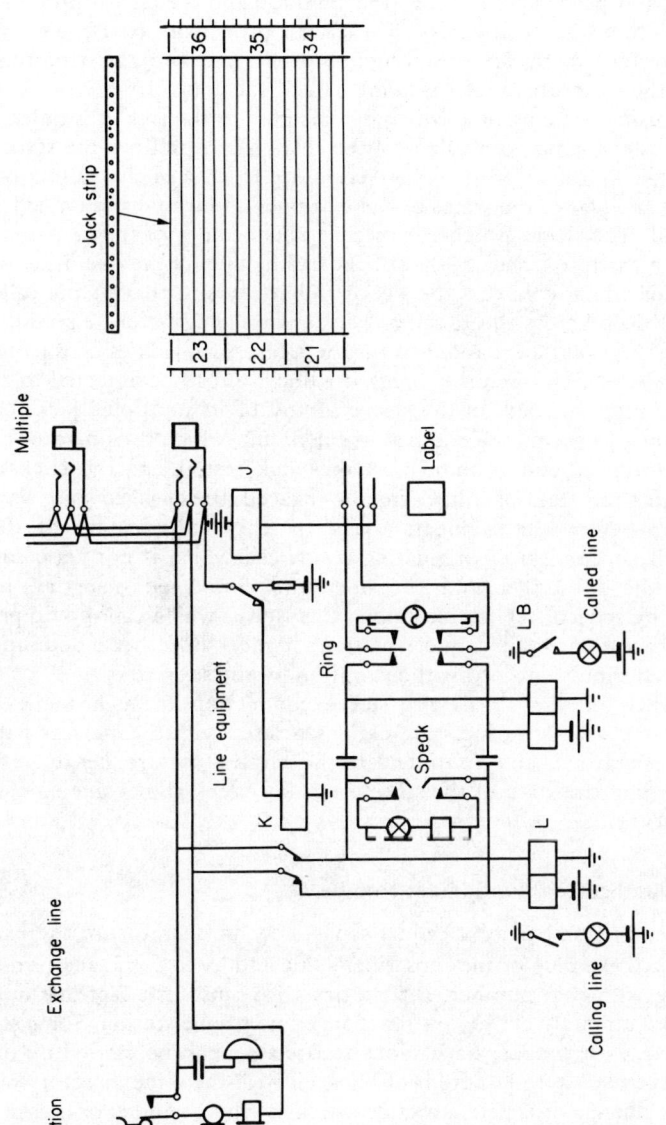

Figure 2.2 Manual exchange with multipled line jacks

L relays) individual to lines or to calls, and termed circuit processors, and that the division of the total processing between the common and the circuit processors has at least an element which is arbitrary. The circuit processors are on the structure side of the interface between the common processors and the structure, and become part of the structure as well as being part of the control.

An operator using the plug of a calling line inserted in the jack of a called line to connect a call has the lamp controlled by the relay LB to tell her the state of the loop of the called line. The lamp being very close to that of the calling line, the operator can see at a glance the states of the two lines connected and is easily able to supervise the call. The sleeve potentials of the jacks of both of the lines connected are changed from earth to negative, that of the calling line by the operation of the J relay a contact of which connects the sleeve to battery and that of the called line by the potential drop across the K relay. The tip of a plug before insertion into a jack being at earth potential, if touched on the sleeve of a jack of an unconnected line produces no effect. On the other hand, if a line is already connected to another line, the tip of a plug touched on the sleeve of any of its multipled jacks causes a current disturbance in the message transmission circuit, which the operator hears as a click in her receiver, and is known as the click test. Using the click test an operator ascertains the state of a line, free or engaged, the engaged state also being known as busy, before she connects to it via the multiple, except that the subscriber may be in process of originating a new call which is not yet connected. In that event, the subscriber and the operators concerned being all in telecommunication by speech, can decide among themselves which call should proceed. For machine processors in the same situation, it must have been determined in advance which call should succeed without option by the subscriber.

The significance of the third wire sleeve connection through the exchange switches is that it is a data transmission path associated with the message path, and by its use each operator is able to extend to the whole structure, her range of data gathering concerning the state of the structure and thereby to become independent of every other operator.

2.5 Directory Numbers and Equipment Numbers

Given a directory number by a calling subscriber, an operator on the exchange of Figure 2.2 takes the plug of the subscriber's line and looks for a jack terminating an exchange line with that number. Difficulties arise out of the fact that not every directory number uniquely defines one exchange line to one station. Some stations, particularly those with p.b.x.s, have more traffic than can be carried by one line and therefore have two up to hundreds of lines, all with the same directory number. Moreover with a change of traffic, up or down, lines may be added or ceased at any time. Party lines serve two or more stations, designated X, Y and so on, and each with its own directory number, and some lines have more than one directory number by which they may be called, for example a regular number and a night service number. Some numbers available for growth are not yet allocated to lines, and if one of these numbers is requested, or a number not within the range of directory numbers available for allocation to stations, it must be a mistake and the caller told 'number unobtainable'.

Stations which have only one line and are never likely to require more than one present no difficulties, nor in general do lines with more than one directory number. If a line has two or more numbers, more than two being rare, it is connected to the multipled jacks corresponding to all the numbers. These connections may be made and changed at any time without previous planning. The greater problem is planning for the lines of stations which have or may have in the future more than one line, all the lines of one station having to have the same directory number and the quantities of lines being different for different stations and subject to changes which can not be exactly forecast in advance.

If every jack could be labelled with the directory number of the line connected to it, theoretically an operator could identify all the lines with a given number by searching through all the labels. This, physically and operationally impossible for an operator, is worth noting because an analogous method is possible with electronic processors. The pre-electronic method first used in manual systems is also worth describing as it still has some value in electronic control systems. Clearly operators must have some quick and ready means of identifying the directory numbers of lines connected to jacks, by some systematic association between jacks and numbers. As indicated in Figure 2.2, the jacks are manufactured in strips of twenty jacks per strip and mounted horizontally in the multiple in vertical columns of strips. Blocks of five strips in the columns are identified each by a two-digit decimal number written on one of the supports for the strips. The jacks within the blocks are each identified by a two-digit decimal number which preceded by the block number produces a four-digit decimal number called an equipment number. The equipment numbers of multipled jacks are the same in all the multiple appearances. The problem of identifying a jack by the directory number of the line connected to it is solved by equating the equipment and directory numbers, which presents no difficulties except for directory numbers which apply to more than one line. An operator given an equipment number of four digits identifies a block of jacks from the first two digits; using the last two digits she selects one of the five strips in the block and then, aided by dots at intervals along the lengths of the strips, she identifies the jack with the given equipment number, all of which a skilled operator can accomplish in a fraction of a second. Lines all with one number are connected to consecutive equipment number jacks, starting with the equipment number corresponding to the directory number, a coloured line being drawn under all the jacks to indicate to the operators that the lines are all in one group. The directory numbers corresponding to the jacks other than the first in a group are thereby lost as station numbers.

Data other than directory numbers must also be visible to the operators. Stations on party lines have to be called individually, for example by ringing the bells according to a code comprising various durations or numbers of rings. An operator given the directory number of a station on a party line has to know it is a party line and which party, so as to apply the appropriate ringing. Each of the parties being called through multiple jacks corresponding to their individual directory numbers, it is possible to allocate blocks of directory numbers and their identical equipment numbers all to X parties or all to Y or other parties of party lines and to indicate the parties to the operators by group marks against the blocks. An operator given one of the numbers observes as she selects the block prior to inserting a plug into a jack that it is a party line and also the kind of ringing which she has to use to call the required party. Special service numbers such as night service numbers are similarly treated and identified. Multipled jacks with no line connected or with a line connected but not allowed to be called for some reason, are indicated by pegs inserted in all the multipled jacks thus physically preventing the calling of the lines the reasons for which are conveyed by the colours of the pegs. Lines which are expected to be only temporarily out of use are 'plugged up' using a special cord, the plug of which when inserted in the jack of a line busies the line against use. A line which is continuously looped, due to a fault or the subscriber having forgotten to hang up, will produce a permanent glow of the line lamp which is commonly known as a PG, until a special 'test and plug-up' plug and cord circuit is connected to one of the line jacks. The line is then out of service until the loop is broken when service is restored even before the test and plug up circuit is disconnected. Equivalents of all the features described are to be found in all automatic systems.

Equating the directory numbers and the equipment numbers of the jacks is the only practical solution to the problem of the quick and minimum effort identification of exchange line jacks from given directory numbers, but it is not without practical and economic disadvantages. The chief practical disadvantage derives from the uncertainty of the quantity of lines to stations which require or may require more than one line. The administration has to estimate for each such case how many lines are ever likely to be needed, and equipment numbers, that is jacks, have to be reserved for that quantity. If the estimate is too great, some jacks will be wasted, and if too small, some other lines may have to have their positions on the switchboard and therefore their directory numbers changed to make jacks available for the excess lines of the group, and subscribers do not like having their numbers changed; or the lines of the group have to be moved to another part of the switchboard where there is room for them, also with a directory number change. An economic disadvantage of equating equipment and directory numbers results from lines with more than one directory number having to use an equipment number and therefore a set of multipled jacks for each number, the line being connected to all the sets.

Because of the relationship between directory and equipment numbers, the maximum capacity of an exchange measured as the quantity of connected exchange lines is less and can be much less than the quantity of directory numbers.

A random association of equipment and directory numbers whereby any equipment number could be used for any directory number and for more than one directory number, would achieve the maximum utilization of the jacks without ever compelling directory numbers to be changed. To operate such a system, each operator would be given an indexed file of directory numbers in numerical order and showing for each number the equipment number or numbers of a jack or jacks in each appearance of the multiple to which exchange lines with that directory number are connected. This is called directory to equipment number translation.

Any other information relevant to the lines, such as party or night service, would also be shown. An operator given a directory number could find from the file all the information needed to complete a connection to a line with that number with the maximum economy of equipment costs, but the method would not even be contemplated for manual systems because there would be an increase in the quantity of operators required which would increase the running costs far more than was saved in equipment cost. The situation is different and the method is used in some modern systems using computer-like processors capable of storing and retrieving at high speed vast quantities of data, but even these systems tend to include some systematic relationship between directory and equipment numbers to reduce the quantity of information storage needed.

2.6 Traffic Probabilities and Equipment Economies

With the equipment of Figure 2.2, when a subscriber initiates a new call the operator may be engaged with other calls and unable to answer at once. Thus the subscriber experiences delay in having his call answered but if he waits long enough he will surely receive attention. The probability that any subscriber will have to wait for at least x seconds before any call which he makes is answered depends on and is calculable from the traffic which is offered to and is expected to be carried by the operator. Once answered any connection demanded can be completed subject only to the called line not being already connected or in process of making another connection. If a call cannot be completed as demanded, it is said to be lost: such calls must, however, be completed as calls by informing the originators of the situations and expecting them to clear. If a subscriber wishes to persist with a call lost on one attempt, he should make another attempt by calling again a little later. In the system of Figure 2.2 the probability of lost calls depends on the occupancy of the exchange lines. In all systems there is a probability of delay in originating calls being answered and of terminating calls being lost, and in the simple system of Figure 2.2, these are the only sources of delay and loss

Considering the time that an exchange line is in use is small compared with the time that it is idle, possibly as low as in the proportion of one to ten even during the busiest time of the day, it is clear that all of the equipment of Figure 2.2 is very inefficiently utilized when measured as the proportion of time that it is performing useful work. Practical systems include means and features having the objective of increasing the utilization of equipment provided and thus to decrease the quantities required and the overall costs of exchanges. They also have the effect of introducing the possibility of calls being delayed or lost at many points between their starting and terminating points. The probability that a call operation may be delayed by at least some given time or a call lost altogether is designated a grade of service, a not completely satisfactory term as the service becomes worse as the quantitative value of the grade of service increases. The delays and losses generally have small probabilities which are independent of one another and hence may be added arithmetically to define the overall grades of service, the delay and loss grades being of necessity treated separately. The grades of service result from and are controlled by the quantities of plant installed and are arbitrary. In practice the values chosen are the highest that administrations who choose them believe the customers will hardly notice and will in consequence tolerate. The measurement of traffic and grades of service, and the determination of the quantities of plant and its arrangement to satisfy specified grades of service at different points in a switched telecommunication system are the main tasks of what is termed traffic engineering, and its economic value is in ensuring that just enough and no more plant is bought than is needed.

Figure 2.3 shows the system of Figure 2.2 modified to reduce the quantity of equipment needed and therefore the total cost, with as penalty an additional cause of delay in answering new call demands by the subscribers. The station equipments, exchange lines and their line relays L and lamps, cut-off relays K and multiple jacks are unchanged, but instead of a plug and cord per line, there is an answering jack and label per line, the jack being connected in parallel with but in no particular location relative to the multiple jacks. The answering jack, label and calling lamp of each line are mounted close together not in association with a particular position but in the vertical face of the switchboard and below the multiple as shown in the perspective drawing of Figure 2.3, the equipments for all the lines being spread out along the length of the switchboard. Any call may be answered by any operator with a cord long enough to reach the answering jack, although operators do not normally have to exceed their comfortable reach which is ensured by limiting the span of each multiple appearance to not more than three positions, the multiple being extended over one dummy position beyond the ends of the suite of working positions for the benefit of the operators on the end positions.

The equipment which is individual to each operator's position comprises a quantity of cord circuits each consisting of a double-ended cord with a plug at each end, a speak and a ring key and two lamps. The lamps provide information for the operator to supervise the calls which she deals with, one lamp indicating the state of the loop of a calling line to which the answer plug is connected and the other the state of the loop of the called line to which the call plug is connected. The plugs and cords, keys and lamps are mounted on the horizontal shelf of the position so as to be readily visible and accessible to the operator.

An operator answers a new call demand indicated by the lighting of a line calling lamp, by taking the answering plug of a free cord circuit if she has a free one, and inserting it in the answer jack of the line concerned. Both the answer and the call plugs of the cord circuit operate the K cut-off relay of any line to which they are connected. The answer plug of a cord circuit in the answer jack of an exchange line transfers the line loop supervision to a relay LA controlling the calling line supervisory lamp in the cord circuit, which means that after accepting a new call by the insertion of the answer plug in the answer jack of the line concerned, the operator may proceed as for the system of Figure 2.2. In addition to the delay grade of service due to no operator being free to take a call when it originates, there is a delay grade of service due to the possibility that although an operator may be free, she may have no free cord circuit with which to answer a subscriber who calls. The first depends on the traffic generated per position and is adjusted by the

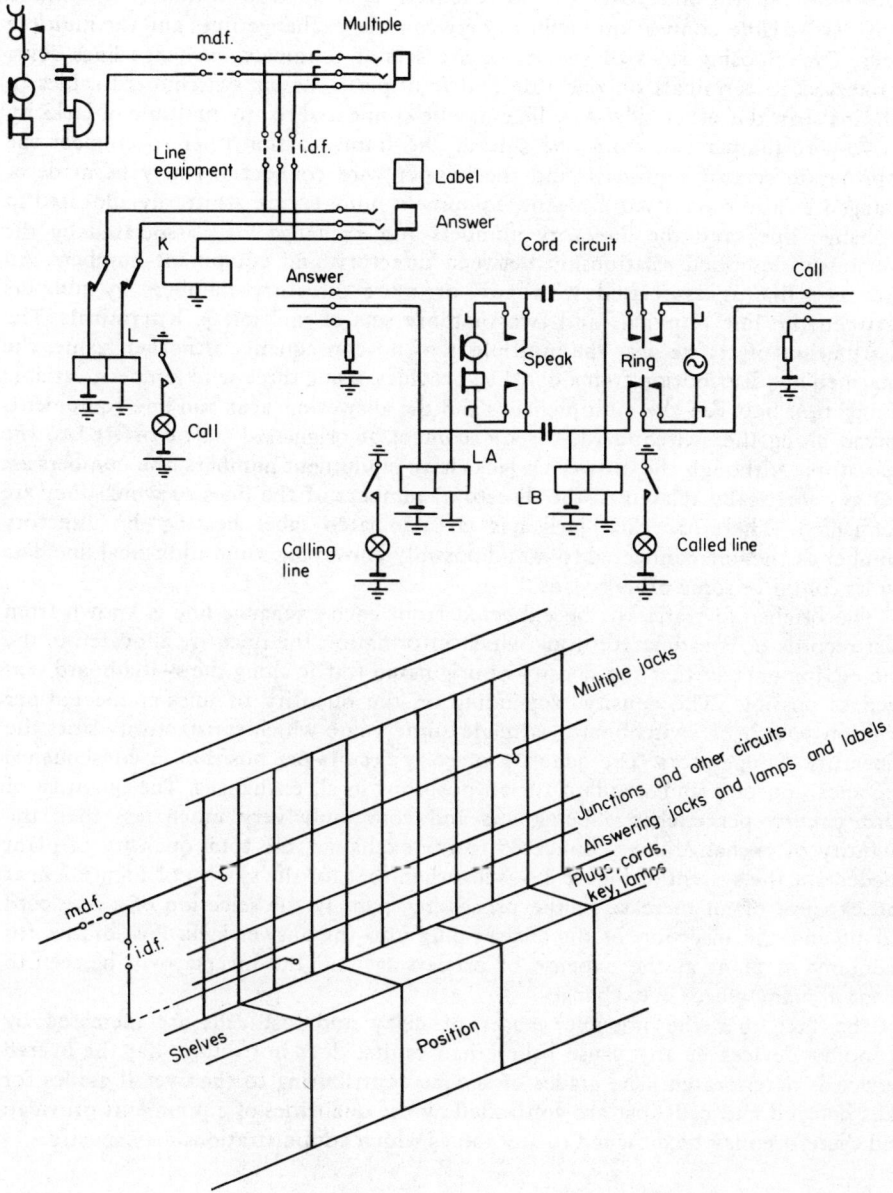

Figure 2.3 Manual exchange with double-ended cord circuits

distribution of the lines over the switchboard. A main distribution frame m.d.f. provides variable connection facilities between the exchange lines and the multiple jacks. On opposing sides of the frame are sets of terminals, exchange lines being connected to terminals on one side and multiple jacks via switchboard cables to terminals on the other side. Any line may be connected to any multiple of jacks by a two-wire jumper run from one side of the frame to the other to connect the appropriate sets of terminals, and the jumper wire connections may be made or changed at any time. By this means, equipment numbers are arbitrarily allocated to exchange lines and the directory numbers and exchange lines associated by the previously described relationship between directory and equipment numbers. An exchange line is associated with two or more directory numbers by jumpers between the line terminals and two or more sets of multiple jack terminals. The distribution of traffic over the multiple is of no consequence. Another frame, the intermediate distribution frame or i.d.f., provides, using three-wire jumpers, variable connection between the multiple jacks and the answering jacks and line equipments spread along the switchboard, for the control of originated traffic offered to the operators. Although the answering jacks have equipment numbers, the numbers are not systematically related to the directory numbers of the lines to which they are connected. Therefore each jack has an associated label bearing the directory number of the line connected to it and possibly conveying some additional line data by its colour or some other means.

The originated traffic to be expected from each exchange line is known from past records or is estimated, using which information the lines are allocated to the line equipments so that the density of originated traffic along the switchboard is as even as possible. The density, depending on the quantity of lines connected per position length of switchboard, is made some value which satisfactorily suits the generality of operators. The quantity of cord circuits per position in consequence becomes constant and standard for all positions in all exchanges. The quantity of cord circuits per exchange being less and commonly very much less than the quantity of exchange lines connected to the exchange, the total quantity of plant needed for the system of Figure 2.3 is less than that for the system of Figure 2.2, at the expense of an increase in the processing, namely the selection of a free cord circuit and the insertion of the answer plug into the answer jack. Possibilities for economy of plant at the expense of processing, and the inverse, will be seen to occur in many places in exchanges.

The fact that the possible causes of delay and lost calls are increased by economy devices or any cause other than faults, does not imply that the overall service is deteriorated. The grades of service contributing to the overall grades for calls delayed and calls lost are controlled by the quantities of equipments provided and therefore may be designed to any values which administrations may specify.

2.7 Minimal Structure Exchanges

The systems of Figures 2.1, 2.2 and 2.3 have been chosen to illustrate in simple ways some basic concepts and features of circuit switching exchanges. The next step is toward an understanding of the division of the processing between circuit processors integral with the structure and common processors associated with the structure through the interface switch, to which purpose some hypothetical but not impossible designs of manual exchanges are examined.

For a manual telephone system the exchange lines are decided as metal conductors with station calling from the exchange by low frequency ringing of magneto bells, the common processors are to be operators and the exchange switches jacks and plugs suited to manipulation by operators. An exchange having only those elements, as in Figure 2.4, is capable of rendering basic service. In Figure 2.4 each operator has a quantity of double-ended cord circuits with which to make connections between exchange lines, and she has one plug and cord for ringing and one for speaking on exchange lines. There are no signals other than to ring bells and no data transmission other than message transmission over which signal-messages may also be conveyed. An operator detects new calls by plugging into each of a quantity of answer jacks allocated to her, cyclically one after another, using her one speak cord and plug, listening to hear if speaking is taking place and if not, saying 'number please'. If there is no reply she goes on to the next line. If there is a reply and she is given the number of a line to be connected, she uses her speak cord to listen on the line and thus to test if the line is already engaged. If it is not, the operator uses her ring and speak cords alternately to ring the called line and listen for an answer. An answer being received, a connection is made between the calling and called lines, using a double-ended cord, and supervised by occasional listening with the speak cord plugged into a multiple jack. The system is theoretically possible but impractical for telephone service even without the complication of junctions in the connections. Signal and data transmissions additional to the message and signal-message and ringing transmissions, are necessary to realize a telephone exchange structure which is workable

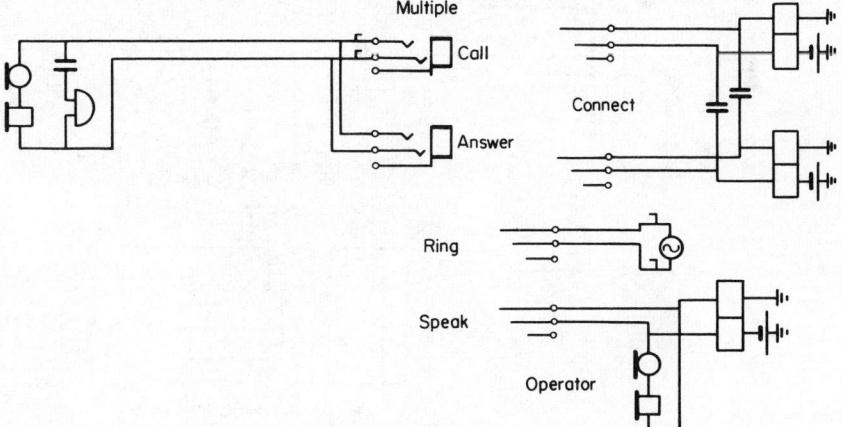

Figure 2.4   Minimum structure manual exchange

in practice. The term minimal structure will be used to define a minimum structure which is usefully operable in practice, most if not all of the control being vested in common processors, and correspondingly maximal structure implies a practical minimum of control by common processors, most if not all of the control being transferred to circuit processors in the structure.

Figure 2.5 represents a minimal structure manual exchange. All of the control is located in common processors, namely the operators, aided by signals and signal-messages over the message transmission paths and data over the third wire paths associated with the message paths through the exchange. The exchange uses loop signals to control line relays L and lamps, and cord circuits with relays LB and lamps to signal the states of the called line loops. Each operator has a quantity of double-ended connect cord circuits and one speak and ring single-ended cord circuit with a ringing key as well as an LB relay. The sleeves of the multiple and answer jacks are connected through resistors to earth and the sleeves of the plugs through resistors to battery, to provide the operators with click test means of determining if lines are free or busy. An operator observing that the line lamp of an answer jack within her reach is alight but that there is no plug in the jack, she uses the tip of the plug of her speak and ring cord to make a click test of the line. If she hears a click, she knows and remembers that the line is answering an incoming call and requires no originating call processing. Otherwise she will insert the plug of the speak and

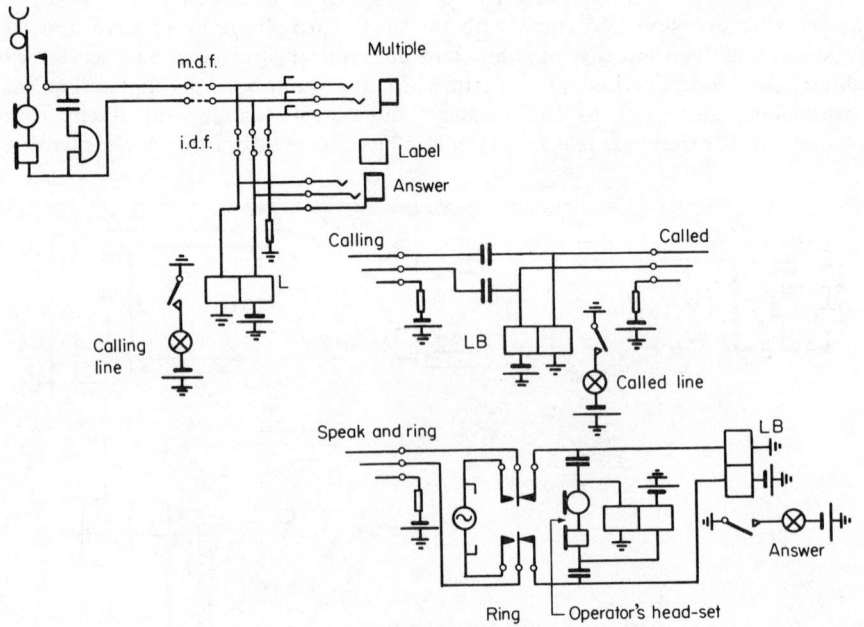

Figure 2.5   Minimal structure manual exchange

ring cord into the answer jack of the calling line and ask for the number required. To call another line, the operator uses the speak and ring cord first to test if the line is free, then to ring and listen alternately until the subscriber answers or she assumes no reply. On receiving a reply, she connects the calling and called lines with the appropriate plugs of a free cord circuit and supervises the call using the calling line lamp and the called line lamp of the cord circuit to tell her the states of the lines. The memory required of the operators of the states of lines and of calls, and the manipulation by them of plugs and cords, are so considerable as to make the system unacceptable for operators although not impossible. Theoretically equivalent systems are practical and are coming into use with electronic machine processors, thus illustrating the important fact that the design of systems is very dependent on the processors available.

Progression from minimal to maximal structure by the transfer of processing from common to circuit processors is a subject of the next section.

2.8 Processing and Programs

The behaviour of an operator comprises assembling in her mind all the data relevant to a particular situation, processing the data and arriving at a conclusion which may include actions to satisfy the subscribers concerned. Machine processors to achieve the same result comprise data memory, logic switching and means to implement decisions arrived at. The data comprise

(a) the services, commonly called the facilities, which the subscribers have learned to expect;
(b) the exchange structure, its construction and properties;
(c) the state of the structure at the moment of decision;
(d) the instructions of the subscribers;
(e) exchange line characteristics;
(f) states of lines;
(g) states of calls.

The operator learns during her training the services and the structure data important to operating, which means that the data are stored in her memory. The state of the structure, the cord circuits in use for example, the operator senses with her eyes and ears, selecting the required data by moving her eyes to ascertain the existence of plugs in jacks and so forth and operating switches with her hands so that she may hear. The instructions of the subscribers are conveyed to the operator by signals which she sees as lamps glowing or not, or hears as signal-messages which she must remember or write down. Exchange line data the operator selects and reads from labels, coloured lines and indexed files on the switchboard and possibly some from her own memory. States of lines she observes, for example lines out of order by pegs in jacks. States of calls are indicated by or deducible from the states of lamps, for example both cord circuit supervisory lamps glowing in the system of Figure 2.3 indicates that the call is still in progress and no action is needed: some states of call require memory by the operator of previous events, as illustrated by the called line supervisory lamp not being alight may mean that the line has been called and has not answered or has answered and is now cleared. Having collected all the data relevant to one situation, an operator makes a logical decision as to what action to take and acts accordingly and a machine processor behaves in the same way. Machine processors use the assembled data to derive binary decisions, that is to give yes or no answers to questions representing the data and posed in serial order and it is convenient to assume that operators' minds work in the same way. The order of posing the questions and the output decisions constitute a program. The art of the designer is required to pose the questions in such order of his own devising as to include all the possible conclusions in practical programs. Figures 2.6, 2.7 and 2.8 are flow diagrams for the programs used by operators to control the system of Figure 2.3 which is chosen in preference to that of Figure 2.5 which would give the more continuous argument from minimal to maximal structure, because the programs are shorter and more useful to later descriptions. Even so, the programs do not include all the possible operations and are to some extent simpler than would occur in practice but without detracting from their value for present purposes.

A complete program is made up of operational programs one of which is in progress at any given time. To every program there is an entry and an exit. The operator enters the operational program for detecting new calls, program 1 of Figure 2.6, by looking at the call lamp, Figure 2.3, of the first line in a group of lines available to her for originated calls. If the first line lamp is not lit and the line is not the last in the group, the operator transfers her attention to the next line in the group and observes whether the call lamp is glowing. If it is not and that line is not the last in the group, she goes to the next line and continuing in this way a call lamp which is lit may be found or the end of the group reached. Finding a call lamp lit and the line not pegged out out of service, the operator chooses a free cord circuit (if she has none the operator will not scan for new calls), then she inserts the answer plug into the answer jack of the line, throws the speak key, confirms that the line is still calling by observing the calling line lamp of the cord circuit, and says 'number please': if the line is not still calling the operator restores the cord circuit and proceeds to the next line if there is one. Having said 'number please', the subscriber may not be listening, or the line call lamp lighting may be the result of a line fault. So if the operator receives no response within some time assumed to be 3 seconds, she may say 'number please' again after checking that the line is still calling, or she may restore the speak key and attend to another line. In the second case she will return later to the first call, and after checking that the line is still calling, will throw the speak key and again say 'number please'. If having thus challenged a number of times assumed to be three, she still receives no reply, the operator abandons the call by removing the plug from the answer jack and restoring the speak key of the cord circuit to normal. Clearly the call lamp of the line will light again and need operations dependent on the operator remembering the state of the line but which are not included in the program shown in the diagram. The remainder of the program is not difficult to follow from the flow diagram.

At suitable intervals the operator returns to the calls in the ringing state to see if

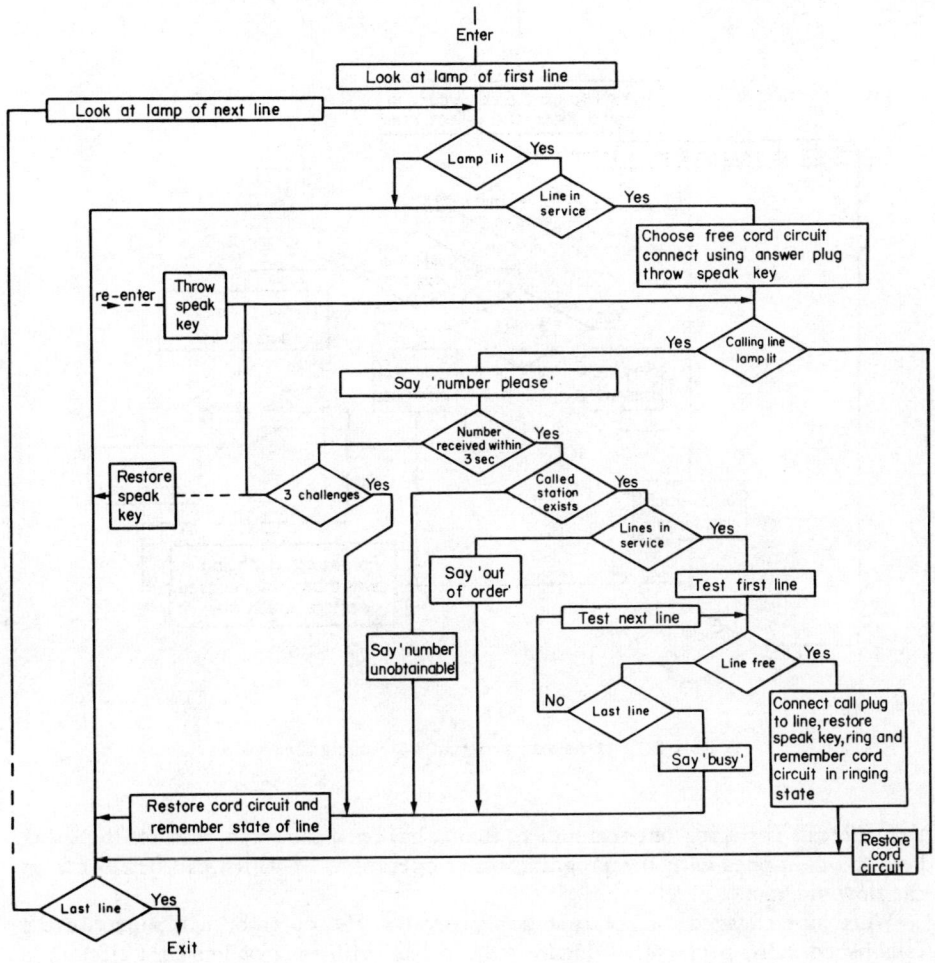

Figure 2.6 Operator program 1 — new call detection

they have been answered, progam 2 of Figure 2.7. Recalling to mind the cord circuits in the ringing condition and selecting the first, she observes the calling line supervisory lamp of the cord circuit to confirm that the calling subscriber is still on the line. If not, the operator restores the cord circuit to normal and erases the memory of the call from her mind. If the caller is still holding and the called line has answered, the operator will throw the speak key and listen for conversation. If there is none, the subscribers may not realise that call is complete: so the operator says 'go ahead'. Having done so or if she hears conversation, the operator forgets

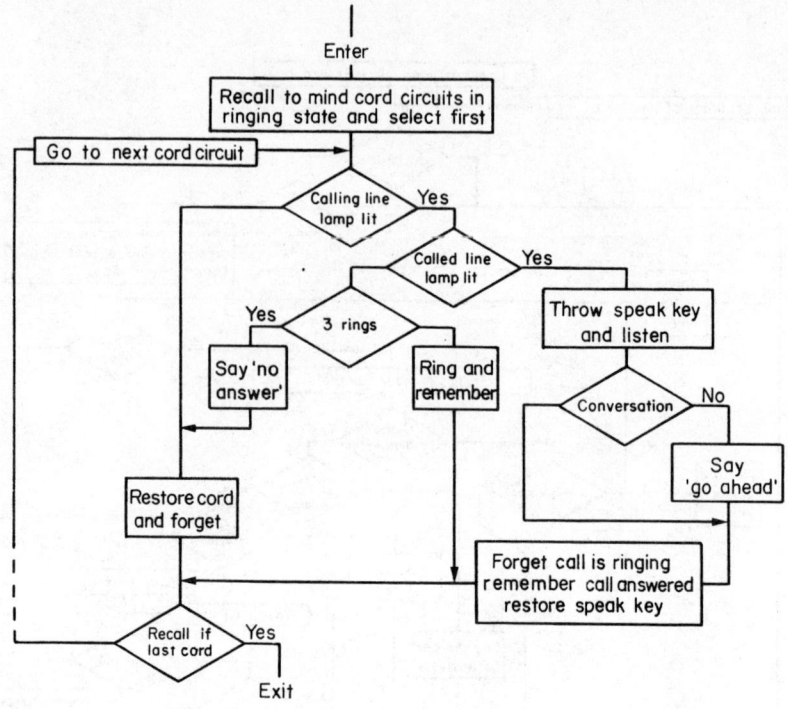

Figure 2.7   Operator program 2 – ringing supervision

that the call is ringing but remembers that it has been answered, restores the speak key and continues with the program, other operations of which can be read from the flow diagram.

When not engaged on the first two programs, the operator will supervise the established calls, program of Figure 2.8, dealing with each of her cord circuits in turn. If an answer plug is not connected but the call plug of the cord is connected, some operating error has been committed which the operator corrects by removing the call plug. If on finding an answering cord connected, the operator observes that the calling plug is not connected, she will know that she is still dealing with that cord circuit under program 1, and so on for the remainder of the program.

What action to take if either but only one of the supervisory lamps is glowing presents some difficulties, for which reason the decision box in the flow diagram is left blank. The operations implied by the box provide an example of an operator getting out of difficulties by conversation and the problem of a machine processor having to cope with the same situation. If both or neither of the cord circuit supervisory lamps is glowing, there is no dubiety regarding the state of the call. If only the calling line supervisory lamp is alight and the call has been answered, there

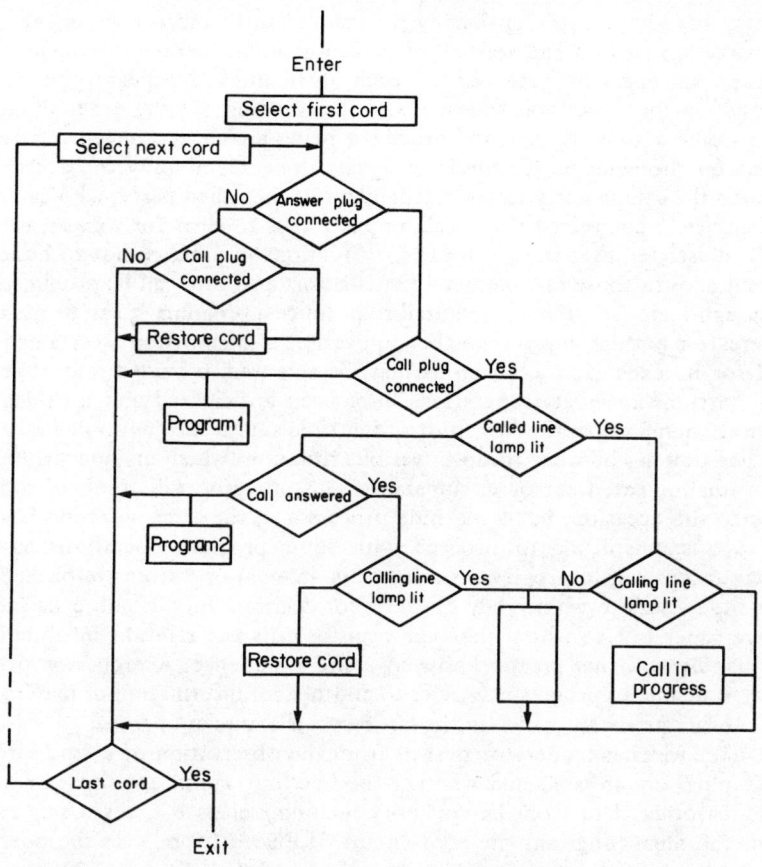

Figure 2.8 Operator program 3 — cord circuit supervision

are circumstances in which a called line can be cleared for a time of the order of minutes before being re-answered, for example while waiting for someone not the one who first answered to pick up a parallel connected instrument. On the other hand the same lamp supervision state may be due to the call having finished but the calling party not clearing, clearing not infrequently being being delayed for a minute or more and sometimes forgotten altogether. It is also possible that the call has been properly cleared but the calling line has followed on with a new demand for service so quickly that the operator has not noticed the end of the previous call. Thus it is not safe for the operator to clear the cord if only the calling lamp is alight, because the call may not have finished, nor is it safe not to clear because the calling party may have forgotten to hang up and is thus holding the other party, or is following on with another call. If only the called line lamp is alight, the calling party may have hung up intentionally, or inadvertantly and is coming back, or the called party has cleared and started a new call without the operator having noticed the clear. An operator gets out of such difficulties by speaking on the line concerned, saying 'have you finished' and thus finding out the exact situation on which to take action. A machine processor being unable to converse, must make decisions on the evidence of the loop signals alone. Commonly the decision is to clear when the calling party clears independent of the called party, who can then be unwillingly held connected if the calling party fails to clear for some reason. This example illustrates that it is possible for situations in practice not to be logically consistent and for those no completely satisfactory program can be produced.

Although logical thinking is required to produce a program, it can be memorized and thereafter nothing more than the application of memory to ascertained data is needed for its execution as often as may be required, provided only that all the possible circumstances and operations have been anticipated and included in the program. A human operator for most of her time acts according to memory gained during her training of what to do in various situations which may occur. Presented with an unanticipated set of circumstances, an operator will think of something suitable to the occasion, but a machine processor in the same situation is liable to come to a standstill or to proceed with some program operations having no relevance to the call in progress. Nevertheless, manual operators are backed up by section supervisors to whom they can refer for decisions on difficult calls and by an exchange supervisor to whom they can transfer calls and relevant information, for her to handle with her greater knowledge and experience. A processor given data which it is unable to process may refer to another for information or may route the call to another processor which in the last resort is a human operator.

The data which an operator derives from the observation of a cord circuit are whether plugs are in jacks and whether the line loop lamps are alight or not. She contributes other data from her memory including elapsed times. Using relays to perform the same functions the cord circuit of Figure 2.9 relieves the operator of nearly all the logical processing of programs 2 and 3 of Figures 2.7 and 2.8. The machine processor effectively asks the questions in the same order as the operator is assumed to ask them, executes the logic and indicates the operational output decisions for the operator to carry out. Referring to Figure 2.9, if neither plug is in use, the cord is free, which is indicated to the operator by a lighted lamp. The call plug in use and the answer plug not is a mistake requiring the cord to be restored to normal. The answer plug in use and the call plug not indicates that the call is in the new call program 1 state. Both plugs being in use, when the called line answers the relay LB operates and in turn operates relay CA which locks to one of its own contacts until the cord is taken down, thus memorizing that the called line has answered. With both parties on the line, the call is in normal progress and requires no operator action. Neither party on the line indicates that the call is finished and the cord needs to be restored. For the called line to be rung, both ends of the cord must be in use, the calling line looped and the called line not yet answered, which conditions cause the relay RG to be operated and apply ringing to the called line. To imitate the periodic ringing by an operator, the ringing current applied to the called line is derived from a ringing machine which controls switches to interrupt the machine output so that it is periodic much like operator ringing. The ringing is tripped, that is cut-off, by the answering of the called subscriber and the operation of the relays LB and CA, the locking of the CA relay to one of its own contacts ensuring that ringing is not re-applied if the LB relay is released. Ringing is also cut-off if the cord circuit is returned to normal. A timing device is shown as relay DT with three inputs, an earth connection to any input causing the relay to operate after a suitable time interval of possibly 30 seconds. Inputs occur while the called line is being rung and when either of the parties clears while the other still holds on. If the delay time is reached, some action is required of the operator as indicated by one of three lamps, one lamp telling her to say 'no answer' and then clear the cord, and two being the equivalent of the blank box of program 3 Figure 2.8, and requiring her to say 'have you finished' and to take some appropriate action dependent on the answer, or no answer, which she receives. Thus an operator having responded to a new call demand to the point of

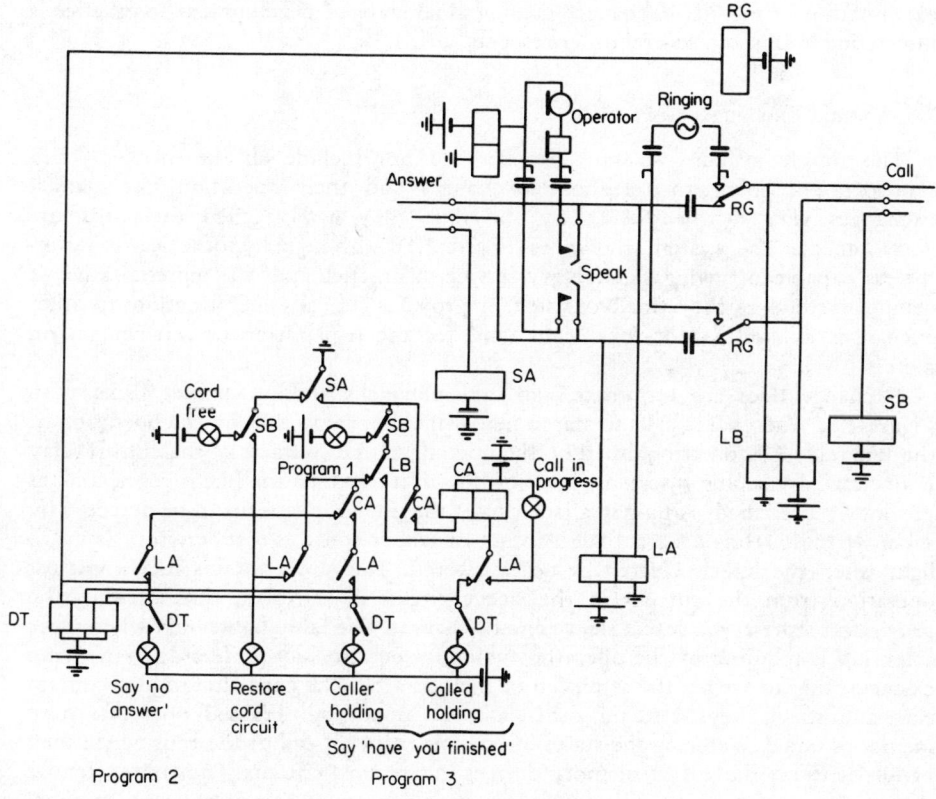

Figure 2.9   Cord circuit for maximal structure manual exchange

insertion of call plug into a jack, has nothing more to do for that call until one of the four lamps light to give her an operational instruction.

Given cord circuits as in Figure 2.9 for the system of Figure 2.3, the operating programs would be reduced to two, one for new call detection and one for connected calls, with the processing load on the operator minimized by the transfer of processing and memory to the cord circuits and therefore to the structure which becomes maximal. The greater cost of the structure would be off-set to some extent at least by fewer operators being needed.

The examples which have been given do not include all the operations which an operator may have to perform in a manual exchange, nor the complete development of the transfer of processing from the operator to the structure but the omissions are unimportant. The important point is that starting from the minimal structure and therefore maximal common control, processing can be progressively transferred from the common control, to the structure to the point of maximal structure and minimal common control. The system designer's problem is to determine the point of maximum advantage between the two extremes. An examination of manual exchanges in their final state of development in practice is interesting in this and several other respects.

2.9 Manual Exchanges in Practice

The simple systems so far described do not include all the problems and solutions pertaining to telephone exchanges and their operation but manual exchanges very nearly did so by the time they had reached their ultimate development. The system shown in Figure 2.10 although hypothetical is nevertheless capable of giving satisfactory service in practice and it is representative of manual systems as they finally existed. It provides for calls over junctions to other exchanges as well as for local calls, and for the registration of call charges on meters.

Exchange lines are terminated on multiple jacks and on answer jacks as in Figure 2.3. When a line calls to start a new call, a line lamp is lit by the operation of the line relay L and extinguished by the opening of the contacts of the cut-off relay K operated by a plug inserted in any of the jacks to which the line is connected, as previously described. A plug in a jack causes the line loop current to be detected by a cord circuit relay LA or LB a contact of which controls a supervisory lamp to light when the line is cleared or not answered. The lamp obtains its current for operation from the current in the sleeve circuit of the plug, thus avoiding the provision of a relay to detect sleeve circuit current. The lamp lights only when some attention is required of the operator, which is operationally preferable to the lamp extinguishing to attract the attention of the operator. The cord circuits also contain ring and speak keys but no other aids to operating. Trained operators have memories equal to storing the states of line and states of call of the calls which they handle without the aid of memory devices in the cord circuits. Thus the system as shown for operators handling subscriber originating traffic is near to minimal structure.

To make calls to lines on other exchanges the operator must connect a calling line to a junction. Junctions are shown in the figure as terminated on multiple jacks, with only resistors in the sleeve circuits through which to operate the cord circuit supervisory lamps. To obtain connection to a line on another exchange it is necessary to precede the directory number of the line on its exchange with a directory number which specifies the exchange and is called the exchange code. The exchange codes were originally not numbers but names usually associated with the locations of the exchanges. A name could be communicated to an operator as a signal-message and was, at the time, the most reliable and convenient means of identifying exchanges. A caller on one exchange wishing for connection to a station on another exchange quoted to his operator the exchange name and number on that exchange of the required station. On the switchboard all the junctions to one exchange were connected to a group of jacks in the multiple, and the group marked by a group label bearing the name of the exchange. An operator given an exchange name and number would use the call plug of the cord already selected and connected to the calling line, to connect to a free junction in the required group, the groups being arranged in alphabetical order to facilitate identification. An operator at the distant exchange would connect her end of the junction to the called line to complete the connection. If there were no direct junctions to the wanted exchange, the operator would know or could find out from an indexed file the route to take through one or more intermediary exchanges at each of which an operator would be given the name of the wanted exchange and would extend the connection to that exchange or yet another on the route. Any operator finding all the junctions on a route to be engaged would say 'all lines busy' to distinguish that condition from the subscriber busy condition.

The precise way in which connections over junctions were made as described between operators in different exchanges varied according to the sizes of the groups of junctions and to some other circumstances but are not now of interest other than that shown in Figure 2.10 which is worth attention because not only was it the most advanced method of working between manual exchanges but also after a long period of disuse, it is now revived in an analogous form for exchanges with electronic control. No signals were passed over the junctions themselves for the establishment of connections to them: connections were made as the result of signal-messages passed over a disassociated message transmission circuit called an order wire. For every group of junctions to which an operator had access and used the junction operation shown, there was a press-button which when pressed connected the operator's headset telephone to an order wire to that exchange. The many operators in one exchange all had access in this way to the order wires to other exchanges. The far end of each order wire was terminated in the headset telephone of an operator called the B operator in the distant exchange, the others being A operators, and a B operator did nothing else but process calls incoming over one group of junctions from another exchange. No other operator in her exchange handled calls over those junctions, which were terminated each on a single-ended cord circuit on the B operator's position. As a result the B operator could see with a glance at the cords, which junctions of a group for which she was responsible were free and which were not. An A operator making a call through an exchange to which there was an order wire, pressed the order wire button and listened to hear if

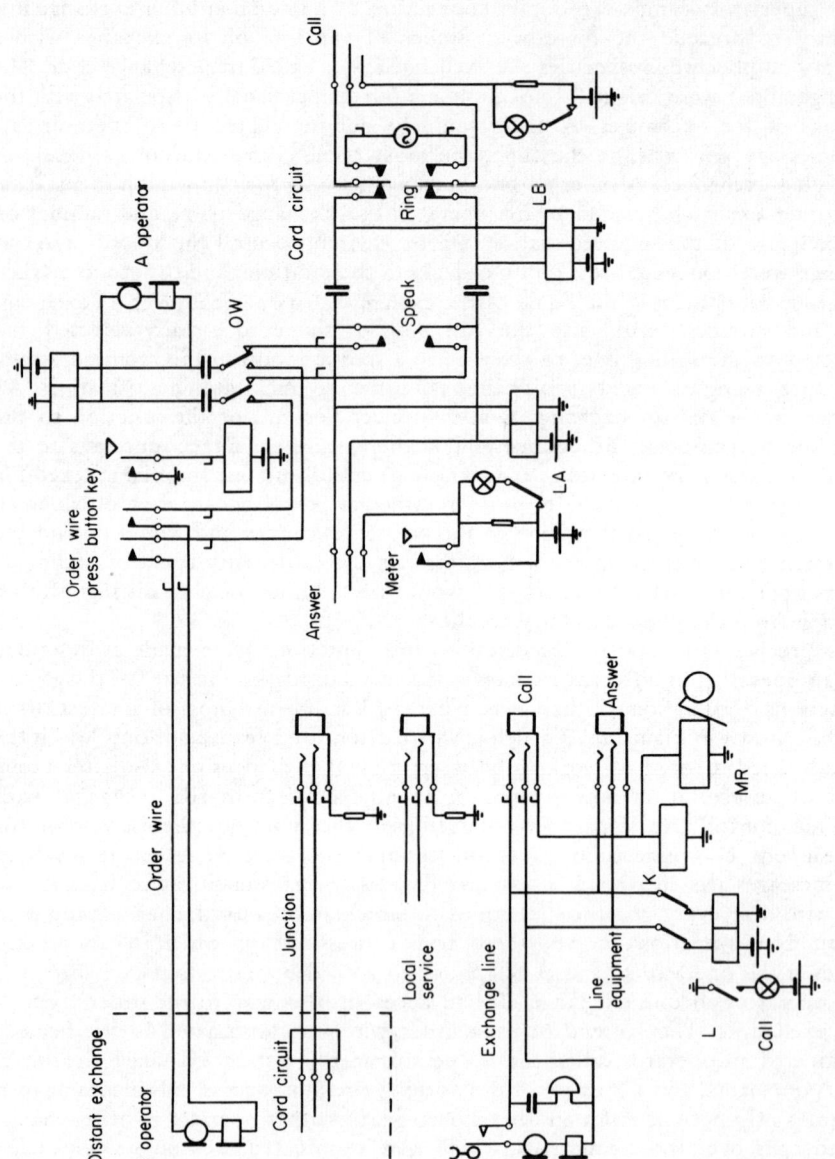

Figure 2.10 Manual exchange in practice

any other operator was speaking on the order wire. If or when not, the operator spoke the number of the station on the B operator's exchange to which she required connection. The B operator looked to see if there was a junction free, and if so quoted a number identifying the junction to the A operator, the junction being identically numbered at the two ends. The A operator inserted the call plug of the cord circuit carrying the call, into the multiple jack of the indicated junction. At the same time the B operator used the cord of the junction to test and find, if free, a line to the wanted station, the line being rung and the ringing tripped automatically as in the cord circuit of Figure 2.9. The junctions provided associated message and signal transmission for supervision after connection, both operators receiving cord lamp signals indicating the states of the connections, the B operator being concerned only to take connections which she had made down when finished as indicated by clear signals from both sides. Thus in the case of order wire junction operators, as much as possible of the processing was transferred from them to the cord circuits which they operated.

Connections have to be made not only between subscribers on the same and on different exchanges but also between subscribers and services on the same and on different exchanges. Services which are called by subscribers include directory enquiries, time announcements and many others. Figure 2.10 shows such services to be available via the multiple, for connection in the same way as subscribers' lines are connected. Calls over junctions to services provided elsewhere mostly use normal subscriber traffic junctions but some services need junctions special to their purposes. Some service calls are made to subscribers' lines usually for administrative or maintenance purposes and these may be connected and operated as local, incoming junction or special junction calls.

The cord circuit of Figure 2.10 also provides for the operation of meters for the recording of units of charge for completed calls. A meter, or message register, is an electro-mechanical cyclometer type decimal digit counter. Each exchange line circuit has the operating coil of a meter MR connected in parallel with the K cut-off relay with which it shares the current in the sleeve circuit. The normal supervisory current is sufficient to operate the K relay but not the meter. The current is increased and the meter operated when a press-button contact in the answering side of the cord circuit is closed to short-circuit the resistor and the lamp if alight, in the sleeve circuit of the plug. Before removing the plugs the operator presses the meter button once for every unit of charge incurred for the call, the meter count being advanced one unit for each depression of the button. The meters are read periodically and the results used in the rendering of accounts to the subscribers. Some calls can not be treated in this way and other bases of charging and recording are used. These and the general principles of call charging are described in section 2.14.

A practice much used by subscribers was 'flashing' which comprised operating the cradle switch by hand so that the calling or supervisory lamp, depending on the state of the call, flashed on and off. Much of the use was mere impatience but it was also useful to recall the operator to a call to correct errors such as the wrong line being connected and to re-connect calls encountering faults such as message transmission being too poor for satisfactory conversation. Flashing recall was lost as a facility when automatic exchanges eliminated operators but interestingly enough, is being introduced again for facilities which new automatic systems make possible.

2.10 Exchange Systems Generalizations

It was made clear in section 1.1 that it was not intended to describe any particular exchange system in detail, but some details of exchange practices will be described to explain and to establish types of exchanges. Types of exchanges in production have been subject to long term changes as will appear in section 2.15, such changes having their origins in equipment and control technique developments. Differences in exchange systems of one type have been due to differences of objective, for example exchanges most suited to small or to large quantities of traffic, or to commercial interests mostly of competition. Exchanges now being designed for the future have no superficial resemblance to the first exchanges but with a proper understanding can be seen to be no different in principle as a solution to a problem which has not changed basically with time. For the same reasons that a people can be properly understood only in relation to their history, exchange design is best understood from the way that it has developed from the first beginnings. Therefore the stage by stage development is traced during which the practical requirements and problems to be solved will be established.

The stage by stage development of exchange design will be illustrated by generalizations of two kinds, the first of exchanges from the outside, that is without specification of their internal constructions, which generalizations include all types of exchanges, and the second of exchanges according to their internal constructions, each generalization including exchanges all of one type. Figures 2.11(a), (b) and (c) are of the first kind and Figure 2.12 is of the second kind for manual type exchanges. Other generalizations of the second kind are given in later chapters describing exchanges of different types. The basic function of an exchange is the controlled interconnection of the message transmission lines terminated on the exchange, which requires a structure comprising exchange switches. The basic function depends for its implementation on a control system comprising data generation and data storage and retrieval, the transmission of data between processors, the processing of data by processors, and transmission of operational data to the structure to control its operations by changing its state, all of which should be comprehensible from the examples and descriptions of the foregoing sections. Exchanges can therefore be generalized by symbolic representations of the message transmission lines and paths and the exchange switches, together with the data transmission paths and equipments and the processors.

Figure 2.11 shows exchange line and junction message transmission lines but it has to be understood that service and other circuits exist and are included. With only message transmission through the exchange switches, the structure of the exchange of Figure 2.11 (a) is minimal and possibly minimum, with all the control concentrated into common processors with access to the structure members via an interface switch i.f.s.. Data transmission for control comprises signals s over exchange line signal paths, signal-messages sm over message paths, data d between

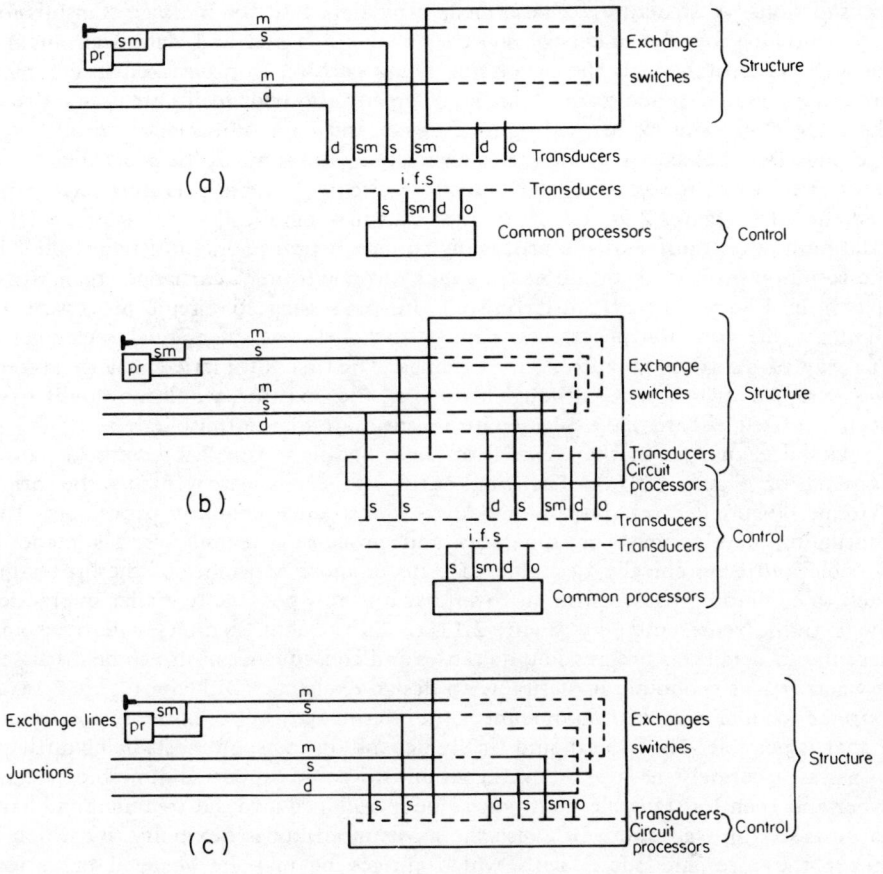

Figure 2.11 Exchange generalizations
(a) minimal structure
(b) general
(c) maximal structure

processors, state of structure data d to the processor and operational data o from the processor to change the state of the structure, with transmission transducers possibly on one and possibly on both sides of the interface switch. Signal-messages originate and terminate at common processors in the exchanges and at exchange line peripheral processors pr which are the subscribers concerned with the connections. There being no signal or data paths associated with the message paths through the exchange switches, signals between the peripheral stations and a common processor are transmitted via an interface switch and no other signal transmission is possible. Data exchange between common processors in different exchanges also has to go via the interface switches independently of the message transmissions. A structure, as described, providing only for message transmission and switching by the exchange switches and not signal and data transmission through the switches, all the processing being perforce concentrated in common processors, is an extreme form of exchange organization not realizable in practice in the early days. The exchanges required signal and data transmission through the exchange switches as well as message transmission, and some of the processing to be transferred from the common to individual circuit processors as illustrated with reference to Figures 2.9 and 2.10 and shown symbolically in Figure 2.11(b). Continuing the transfer of the processing to circuit processors until none is left in the common processors, produces the other extreme form of exchange organisation shown in Figure 2.11c. Transferring all the processing to circuit processors to eliminate the common processors, namely the operators, of manual exchanges is one way of designing an automatic exchange. The first automatic exchange system was designed in this way, although Strowger, the inventor, would no doubt have been surprised had anyone explained his invention to him in those terms.

Manual exchanges in their final form described in section 2.9 were near to the extreme of Figure 2.11(a). The first automatic exchanges went to the other extreme, Figure 2.11(c), but soon had to re-introduce common processors. The continuing development of apparatus and processing techniques has made it possible and economically desirable to transfer more and more of the processing back to common processors, so much so that it is now possible to realize in practice the extreme represented by Figure 2.11(a). Subsequent chapters will trace and describe in detail this progression, its causes and consequences. Also to be discussed is whether it is economic or desirable to design exchanges with the Figure 2.11(a) extreme form of organization or some form intermediate between the two extremes if that is possible. The issue should finally depend on costs, but costs being difficult to assess accurately or looking unfavourable to some otherwise desirable system, other and even less tangible factors become introduced into the argument and have to be taken into account. Of these the most important is flexibility, by which is meant the ease and speed with which processing may be changed or varied. Processes performed by circuit processors are difficult and expensive to change or to provide in variety where different options have to be available. Common processors being less numerous have in these respects an advantage which gives a bias toward common processors. What should be clear from the foregoing examples and discussions is that there is no best solution to the exchange design problem in the sense of a unique design which meets all the requirements at the lowest cost, and which anyone possessed of the necessary knowledge and skill could arrive at by abstract thought: but some designs are more successful in practice than some others, and how to design an exchange which will not be less successful than any other is another statement of the exchange design problem.

Referring to the internal organizations of exchanges, Figure 2.12 is for the manual exchanges of Figure 2.10, a symbolic representation which can be extended to all kinds of exchanges for the circuit switching of all kinds of message circuits. Figure 2.12 and subsequent generalizations elaborate those of Figure 2.11 except for the transducers in the signal and data paths which it is not possible to show but the possible existence of which has to be understood. In Figure 2.12, stations have a message transducer, the telephone instrument, transmitting and receiving signal-messages over the m path and signals s over a signal transmission path s. An exchange line provides the m and s paths to the exchange at which the line is terminated on line equipment which is a circuit processor for the line and on multiple and answering jacks which are the exchange switches. Signal and data paths from the line equipment via the interface to common processors, the operators, transmit a lamp calling signal and data which can be read from labels: and from the multiple area of the exchange switches data d such as directory and equipment numbers are available. Operators, by making connections through the switchboard, effectively transmit operational data o to the exchange switches to change their states. Junctions provide m and s transmission paths between exchanges, and local services terminating in the exchange comprise circuit processors connected to the exchange switches over associated m, s and d transmission paths. Data concerning junctions and services are also available to the operators as common processors from the exchange switch labels, pegs and so forth. The jacks, plugs and cords which are the exchange switches provide variable connection of the associated m, s and d paths of the exchange lines, junctions and service circuits via cord circuits which are loop trunks or the incoming ends of junctions. Temporary connections which are made as part of the control of the message circuit connections are made through operating trunks which are the answering halves of loop cord circuits or single-ended cord circuits like the speak and ring cord of Figure 2.5. Circuit processors in the lines and trunks control the interruption or the through transmission of the m, s and d transmission paths and provide for sm, s and d transmissions through the interface switches to and from the common processors. The m, s and d paths of the diagram generally indicate bothway transmission although transmission may in operation be only one way.

Circuit processors are part of the exchange structure, their signal-message, signal and data transmission and reception includes transducers if required, and they comprise logic switching equipment P for the processing of data in conjunction with a data store or memory M. Common processors similarly comprise signal-message, signal and data transmission equipment including transducers where necessary, memory M and logic P equipment, and they may have access to common data, in the case of operators in indexed files, in common data stores as indicated in Figure 2.12. If there is more than one common processor, mutual interference which is otherwise possible must be avoided by ensuring that either no two can be connected at the same time or no two can be connected simultaneously to the same paths through the interface switches. The circuit and common processors share the total exchange control between them to achieve the best result judged by cost or other relevant factor.

2.11 Data Storage

The division of data into the classes s, sm, d and o is useful mostly for data in transmission. Other divisions describe the characteristics of the data and of the stores used to hold the data in memory.

Data with indefinitely long lives are semi-permanent, directory number to

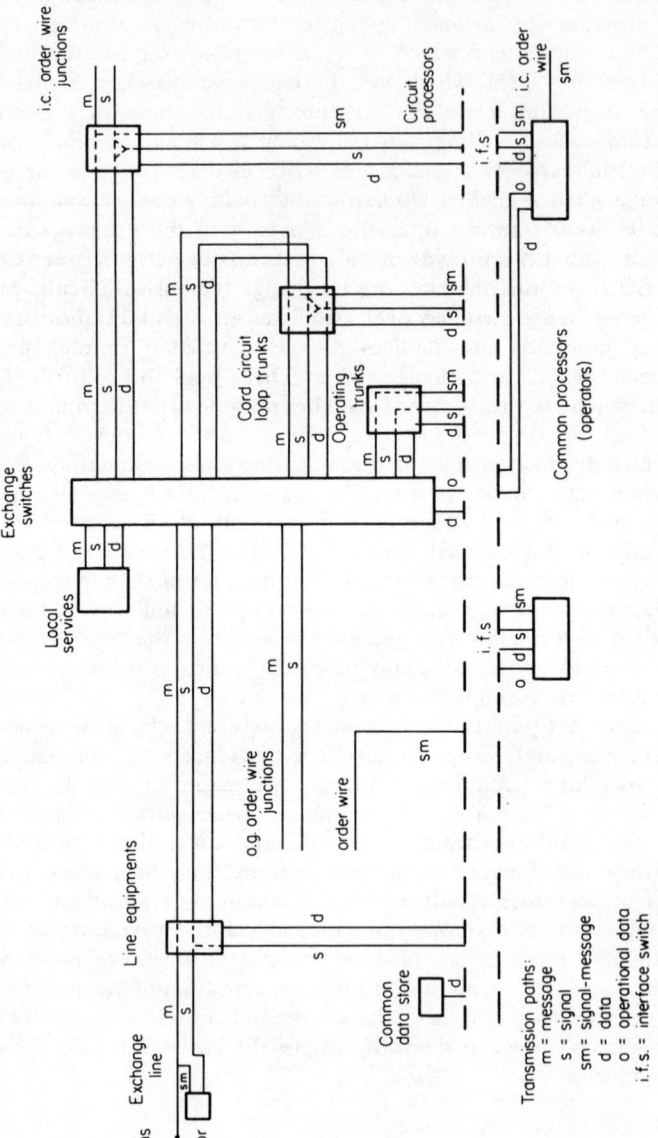

Figure 2.12 Generalized manual exchange

equipment number translations being an example. The data have to be very secure against accidental change but intentional change when required need not be made very quickly.

Temporary data have brief but definite lives, the data usually pertaining to particular calls in progess. It is essential that data may be erased or new data written very quickly, but very high security over long periods is not essential.

Data with lives which are definite but range in duration from fractions of seconds to months, and which must be instantly changeable but nevertheless be very secure against accidental change, will be called period. They are mostly concerned with exchange line state of line and call accounting.

Data storage in systems without operators has presented many problems and difficulties in practice. A store in which data could be written and changed very quickly and be very secure over indefinitely long periods, and in addition be cheap for any quantity of data to be stored, could be used as a universal store for data of all kinds. For lack of such a store, different types have been used, the stores having the characteristics described for semi-permanent, temporary and period data and being identified by the same designations. Data and data store may be expected to be both of the same type but it does not follow of necessity that they are in practice. Temporary stores are frequently adapted to the storage of period and semi-permanent data, by making frequent checks of the stored information and correcting any in error in order to satisfy the security requirements. Equipment and practices with regard to data stores and their uses in exchange systems have not yet reached stability and may be expected to change with the development of new and more reliable stores.

The designations of the categories of data and stores have arisen out of computer and machine processor design and operation but that they were equally evident in manual exchange practices is not difficult to see as coloured line and equipment markings on the multiples, the memories of the operators and the exchange line meters to mention only one example of each of the three kinds of data and stores.

2.12 Wired and Stored Program Logic

Logical processing by an operator can be represented as a program of binary decisions and actions to be executed, as described in section 2.8. A program of operations which an operator has to perform being stored in her memory together with semi-permanent data acquired during training, she collects data selectively by eye and ear and out of her recent events memory to move step by step through the program to an operational decision. This is logical processing by memory stored program. A machine taking over processing otherwise performed by an operator as in Figure 2.9 may have binary relays operated or released to transmit binary data, by means of contacts, to wires carrying currents, the output currents providing the operational data outputs. This is also logical processing by memory stored program, the relays with their contacts and wiring storing the same program and making the same deductions and decisions as previously made by an operator but constituting a processor very different physically and operationally from an operator as processor.

More exactly, all processing uses memory stored program logic, but there is more than one way in which data and the remembered program may be related. The human operator provides one way, and a machine comprising wired binary devices another which is called wired or hard-wired logic to distinguish it from a third which is stored program logic using wired logic in conjunction with memory and decision taking methods which are in many respects analogous to those of the human processor as explained in chapter 8. The three kinds of logic, human, wired and stored program, have each in turn had a profound influence on exchange systems as subsequent chapters will show.

2.13 Time Sharing by Space and Time Division

Referring to the generalized exchange of Figure 2.12, the exchange lines each have circuit processors which are the line equipments permanently connected. The loop trunks with their processors are fewer in number than the line equipments because they are included in connections as the connections are made and serve the connections only while they exist. The time of each trunk is thus shared between connections and the total traffic is carried by a group of identical trunks just numerous enough to attain some specified grade of service. The trunks are said to be space division time shared, which term is applied to any group of trunks or equipments of any kind which individually may be connected as required to the trunks or equipments of another group, for periods long enough to complete some immediately needed function and then to be released and thus available to serve other trunks or equipments.

Still referring to Figure 2.12, an operator as a common processor does not give her time exclusively to one circuit or one connection throughout its duration but divides her time between all the circuits and connections under her control by periodic attention to each. A processor or other equipment operating in this way is time division time shared among the equipment items which it serves.

Space and time division are applicable to the time sharing of all kinds of equipments, not only processors. The essential difference is that space division comprises items each continuously connected to circuits for as long as the circuits may require the services of the items, whereas time division comprises one item periodically connected to circuits for short periods during which its services may be used if required. Time division time sharing is in general impractical with electro-mechanical equipment, and consequently it was almost completely absent from the first telephone transmission and switching systems. Electronic equipments have made it feasible for both transmission and switching equipments and an important feature of systems now available or to be developed. A thorough understanding of the attributes and fields of application of space and time division time sharing is thus essential to the design of future systems, and is included in chapter 7.

A circuit or space division time shared processor acting for one call at a time, the cord circuit of Figure 2.9 for example, is for most of its time in a state of watchful inactivity. It springs into activity when some change occurs to disturb it and continues until it can again lapse into inactivity. More exactly, it is continuously receiving data which it processes in conjunction with stored data to a conclusion possibly with an operational output, activity being induced by a change of input data and continuing until processing is completed. A time division time shared processor such as an operator operates similarly but without the periods of inactivity. Even if an operator pauses for some time on one connection, waiting perhaps for a subscriber to answer, she is mentally timing how long she has been away from other lines and calls and is scanning with her eyes for lamps to change condition. A machine processor performing the same function must behave in a similar manner. Connected to a circuit, it assembles data relevant to that circuit at that instant, processes the data to a conclusion possibly with an operational output, is disconnected and connected to another circuit and so on. Having attended in turn to all the circuits under its control, the processor starts at the beginning again. The question which arises is, how frequently must the processor be connected to each of the circuits? Considering one of the circuits, the result of processing the data is the same as it was last time if, since the processor was last connected, no change to the circuit data has occurred. If there has been a change of datum, action and a new stable state of circuit are required: but if more than one change has occurred, the first not having been processed before another occurred may cause later processing to be seriously corrupted. Therefore one criterion of how often a time division time shared processor must be connected to circuits which it controls is, at least often enough to ensure that serious corruption of the operation is not possible due to more than one data change in any one circuit between consecutive connections. There is a delay after a change of data occurs until the processor becomes connected to deal with it and which gives rise to a second criterion for the frequency of connection, that the maximum possible delay shall be within some acceptable limit either arbitrarily decided by the administration or convenient or necessary to the operation of some other part of the system.

The two criteria just stated apply to the frequency of connection of time division time shared processors to the equipments which they control. Another question is the rate at which processors process the data offered to them. Considering a continuously connected circuit or space division time shared processor, it must have completed the processing caused by a change of data before another change can occur to corrupt the system operation: and the delays in responding to changes of data must be within acceptable limits. In other words, the criteria for the processing speed of continuously connected processors are similar to those of the frequency of connection of time division time shared processors. But if the processing speed of a time division time shared processor is only just sufficient to satisfy the criteria for a continuously connected processor, the time sharing is limited to one equipment and thus vanishes. Hence time division time shared processing requires a higher processing speed and the quantity of equipments among which a processor can be time division time shared is, all programs and other conditions remaining constant, proportional to the operating speed of the processor.

To illustrate the relevance of the two criteria for the frequency of connection of a time division time shared processor to a circuit or an equipment, consider the three programs of Figures 2.6, 2.7 and 2.8. One program detects new calls each originated by a subscriber looping his exchange line. The subscriber must wait for the operator to answer but if he does not wait but hangs up, no harm is done. Hence the criterion is what maximum delay should be permitted before the operator answers, and 10 seconds is assumed. Every line should therefore be examined for new calls at least once every 10 seconds. With lines thus examined at regular intervals, the average delay in answering is slightly more than half the interval between examinations. Program 2 rings called lines until they answer or no answer is assumed. Having rung a line, the operator cannot wait for the subscriber to answer but turns her attention elsewhere before returning later to see the state of the line. If the line has still not answered she may ring again or tell the caller that there is no answer. If the subscriber answers and clears before the operator again gives attention to the line, the operator may ring again and inappropriately. Hence the longest time that the operator should leave the call unattended is the shortest conversation time of about 5 seconds. The third program, that providing supervision of established connections, depends on the calling and the called lines' loop signals. When a line clears by hanging up on one call, how soon after is a new call and loop signal likely to occur? Based on practice a time of about 2.5 seconds may be assumed. Therefore all the cord circuits should be supervised at least once every 2.5 seconds. To satisfy the conditions enumerated, an operator would have to divide her time so that during every 10 seconds she would scan for new calls once, observe every cord circuit in the ringing state twice and every cord in the call supervision state four times. Figure 2.13 shows an executive program which would satisfy these conditions. The circle represents the 10 second cycle, with four equally spaced periods during each of which the supervision program 3 is executed. The new call program 1 occupies one of the intervals between two supervision periods, and the ringing program 2 is divided between two diametrically opposite periods. No human processor would or could maintain indefinitely such a repetitive cycle. Nevertheless that is what the operator should have in mind and could possibly achieve with super effort for a short time, so let the examination continue on that assumption and the consequences of the operator not keeping to that or to any preconceived program be considered later.

The quantities of exchange lines, cord circuits and connections which one operator could deal with on the basis of the system of Figure 2.13 would be very small and their determination subject to the inherent difficulty that the times of completion of programs from first entry to final exit are not constant and predictable such as would enable them to be fitted efficiently into a regularly cyclic executive program such as Figure 2.13. For example, during the new call detection program, program 1, if a line is not originating a new call the operator can see at a glance that that is so, but a new call needs many times as long as one glance for its processing which is itself dependent on if and how quickly the subscriber responds to the 'number please' challenge and so forth. The quantity of lines found, during one glance or scan of all the lines of a group, to be originating new calls is not predictable but may be 0, 1, 2 ... up to the number of lines scanned. The times of

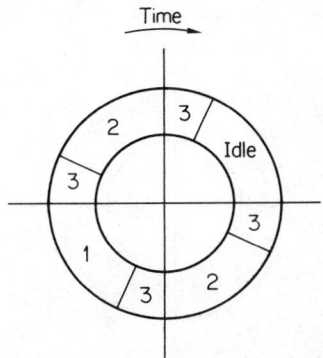

Figure 2.13 Simple executive program
1 = new call program
2 = ringing program
3 = supervision program

completion of the processing for any one line or for a group of lines processed one after another continuously without pause, are dependent on random incidences of traffic and processing. The times can be expressed as probability statistics, which means that the mean times for the processing of one line and for processing a group of lines, the times being based on many lines processed during many scans, can be stated exactly and also the probabilities of any given times being exceeded. If maximum times need to be known, that for one line is ascertainable from the program and is clearly much greater than the mean, and that for a complete scan is the product of the maximum time for one line and the number of lines and is similarly much greater than the mean. What has been described for program 1 applies in general to all programs. So, returning to the program of Figure 2.13, maximum times must be assumed for the operation programs to be sure of completing them in the allocated times, which means that on the average for most of the cycle time the operator is idle having completed one program and waiting for the time of commencement of the next program, in addition to the part of the executive program which is shown as idle.

With the program of Figure 2.13, one operator could handle the traffic from no more than about ten exchange lines. To increase the quantity and thus reduce that of the operators, that is common processors, which has to be employed, requires faster operating processors, or better use of the time of the processors or a reduction in the processes of the operation programs, or some combination of all three. In this example there is no possibility of increasing the speed of operation of the processors, they being human, and little variation in the programs to be performed is practicable for the same reason. Better use of the processor time is, however, possible by the use of a more efficient executive program. The first step is to eliminate the idle interval between two of the supervision periods, program 3,

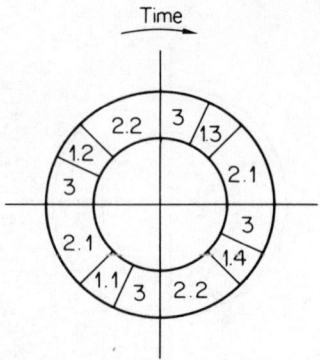

Figure 2.14   More efficient executive program

which may be accomplished by a different division of the cycle. Figure 2.14 shows the cycle time fully utilized by division of the exchange lines to be scanned into four groups scanned one group at a time during the periods 1.1, 1.2, 1.3 and 1.4, the calls in the ringing state being attended to in two groups in the periods 2.1 and 2.2 and all the cords being supervised four times per cycle as before. The proportion of the time which the processor is usefully occupied is, however, still low because of the necessity of assuming maximum times for the operation programs to accommodate the inherent difficulty that the actual times are unpredictably variable. Despite the low utilization of the processors, executive programs of the kind of Figure 2.14 are sometimes satisfactory with very fast operating machine processors. The processors are time-controlled to dwell on each circuit or equipment to be processed long enough to include the maximum processing time which being very short, the quantities of circuits or equipments in the cycle are usefully large enough. For an operator as processor, idle times occur between the completions of programs and the strictly timed commencements of the next programs in the cycle. So still using the same executive program, let the operator, when she has finished an operational program for a group of circuits or equipments, immediately go on to the next in the executive program cycle. The frequencies with which the programs 1, 2 and 3 are executed are still in the ratio of $1:2:4$, but the periods of the executive program cycles are not constant but dependent on the traffic intensity variations. The quantity of processors needed to serve a given exchange is reduced but there is no quantity which will guarantee that the maximum periods between the connections of a processor to equipments for the execution of programs 1, 2 and 3 will be 10, 5 and 2.5 seconds respectively, as postulated for satisfactory operation. A way out of this kind of difficulty sometimes possible is to make the probability of exceeding what should be the maximum so small that effectively the maximum is not exceeded. Applied in this case, the processors would all be working continuously but the average executive program cycle time would be much less than 10 seconds, and the effective use of the processors as measured by the quantity of traffic processed by each would still be low.

If to achieve adequate effective use of the processors the postulated time for the executive program cycle is bound sometimes to be exceeded, consider the effects. If the 10 second cycle time for answering new calls, program 1, is exceeded, some subscriber may get annoyed but he will get service in time. If the 5 second time for program 2 is exceeded, difficulty will occur only if a call is answered and cleared very quickly and the subscriber originating the call does not clear quickly or clears and re-calls quickly, which means that the probability of difficulty occurring is the product of small probabilities and is therefore very small. With program 3, either of the parties to the conversation clearing and immediately calling for a new connection can cause difficulty, the probability of which is the sum of the two probabilities concerned and not negligible. Thus the consequences of the maximum times being exceeded are very different in the three cases but with the executive program of Figure 2.14, the probabilities of their being exceeded are all the same. So let the programs be arranged in the order of their tolerance to delay in execution, which is the order 3, 2, 1. Let their order of execution be program 1 as base, which means that program 1 is executed when and if none other is in operation: then under the control of clock generated timing pulses, let program 2 be commenced every 5 seconds interrupting program 1 and program 3 be commenced every 2.5 seconds interrupting any other program in progress. The points at which programs are interrupted being remembered, and more than three programs generally being involved, when any is completed the most important of the interrupted programs is resumed at the point of interruption. Thus the periods between successive executions of the most important program are constant and those of other programs are constant except for the risk of interruption, which risk increases with reducing order of importance of the programs and varies from very small to possibly near certainty during periods of heavy traffic. In fact in practice some programs, test programs for example, are commonly not attempted except during slack traffic times. In this way the overall consequences of failure to meet the theoretical maximum periods of repetition of the various programs are minimized and kept to acceptable levels and the effectiveness of the processors is maximized.

A human operator behaves intuitively in much the same way as has been described. She selects for execution what she thinks to be the most urgent task, interrupts one task in favour of another which seems to be more urgent, scans for new calls when the pressure of existing calls eases and so on. If she gets into difficulties, and she does at times because she cannot work fast enough to keep up with the changes of states of the circuits and calls for which she is responsible, then she extricates herself by conversation with the subscribers concerned and placates with explanations and soothing words those who become irate. It is arguable that an operator can do two things at once, such as scan with her eyes for lamps changing state while listening for a subscriber to answer, and in this and other respects is different from a machine processor. It is not important to establish whether this is so or not. What is important to exchange design is a clear understanding of the relative importances of services and operations and their dependences if any on the equipments used. Differences and changes of practices over the years are then more easily related to the characteristics and limitations of the equipments which have been available and the best uses of new equipments are more reliably determined. When the first ideas for eliminating operators were produced, the only conceivable equipments were electro-mechanical relays and ratchet-driven switches and counters. With these, time division time sharing is impractical because they cannot work fast enough nor are they durable enough to stand up to the continuous operation necessary. So the first invention, that of A. B. Strowger of the U.S.A., transferred all the control to circuit processors which were part of the structure, to eliminate operators, that is common processors, and also the time division time sharing which they performed. No doubt Strowger would have been surprised a second time to have had his exchange explained to him like that, but that is the way of progress. Inventions are made first and fully understood much later. With the advent of better equipment and increasing complexity of basic requirements, some of the processing was transferred back from the structure to common processors, as was outlined with reference to Figure 2.11, but with space division time sharing with which the equipment could cope, as will be explained later. The electronic computor is a processor approaching the human original in its memory and logical processing abilities and capacity for continuous operation, which coupled with its much greater speed of operation, makes it possible to conceive of exchanges controlled much as they were originally by operators but with only one operator working at the speed of a Disney cartoon character. There are inevitably some differences, a machine cannot converse its way out of difficulties with the subscribers for one, but the description is nevertheless a fair one for some modern systems with electronic control.

Time sharing of processors as discussed in this section has the object of increasing the effective use of processors so that the quantities required and therefore the total equipment costs are reduced. Time sharing and its advantages are not limited to processors but apply also to transmission and other equipment, nor is it the only way to economy. Processing speed and the programs chosen to be used also have important effects which could not be illustrated by the simple manually operated systems described but are fully dealt with in later chapters, chapter 8 in particular. A machine processor working at infinite speed could operate any set of programs linked by any executive program and satisfy any size of exchange: clearly the simplest of executive programs would be chosen with no interrupts, and only one time division time shared processor would be needed. Nevertheless, the quantity of equipment is not thereby of necessity minimized. The quantity also depends on the individual programs which in turn are very dependent on the type of logic used, wired or stored program, as described in chapter 8. In fact the design of a control system for an exchange system involves so many related factors that there is no quick and easy way of deciding the best and cheapest even when the structure has been decided, and the structure and the control also being interrelated makes the problem still more difficult. In the following chapters the characteristics and influences of the various factors are described as a general introduction to exchange systems design which in any specific instance must take into account all those factors and the particular circumstances and specification of requirements of the system to be produced.

2.14 Tariffs and Accounting

Pay station charges must be collected as they are incurred, by coins inserted into coin-boxes or money handed to station attendants. Private subscribers may also pay through coin-boxes if they wish but for most of them charges are aggregated for payment at periodic intervals through accounts rendered. For each item of service the charge should cover the cost of rendering the service or under-charged items must be subsidized by others if overall there is to be no loss on running the enterprise.

The cost of providing a telephone service may be divided into two parts. The first is independent of the use made of the service and includes the cost of station, exchange line and line and other equipments individual to exchange lines in exchanges. The second part is dependent on the traffic in the network and includes the cost of those parts of exchanges used only for connections through the exchanges and of transmission plant used for connections between exchanges. Most administrations use a rental payable independently of the calls, plus a charge per call, as the means in principle of relating charges to costs but have to admit relaxations for a variety of practical reasons.

The cost to the administration of a call at peak traffic times is dependent mostly on its duration and the distance between the terminal exchanges. At other times plant which would otherwise be idle may be used at little cost and this leads to favourable tariffs being offered to encourage the use of plant outside of peak traffic hours. The per call charges and their collection commonly lead to much complication in the tariff structures.

Measured rate applies to calls the charges for which are measured in units which are aggregated for all calls over a period between the rendering of bills, with no record available to the subscribers of individual calls made. Bulk charging and bulk billing are terms also used to describe the method of accounting. Each call is charged separately a quantity of units which originally were units per effective call but are now more often charge units per unit of effective duration of a call. The definition of effective is arbitrary and decided by the administration: usually a call becomes effective when the called line answers and the effective duration is from then until the calling line clears no matter what the called line may do.

Regular payment of an agreed sum to cover the cost of all calls within a specified and usually small area no matter how many calls are made and of what durations, called flat rate, is offered by many administrations to their subscribers. Frequently the subscribers have a choice of area, each with a different rate, over which the flat rate applies.

Private subscriber calls which are not bulk billed or covered by a flat rate are 'ticketed', which means that for every call a detailed record is made of at least the directory numbers of the calling and called stations and the duration of the call, which details may be supplied to the customer with the bill. Originally the records were made by operators using a printed ticket for each call but later machines recording in various ways became available for automatic exchanges and manually assisted services. Occasionally, for lack of other means, measured rate calls are ticketed and the records used to prepare measured rate bills not including the details of the calls.

The methods of charging and the rates may not be uniform for all the subscribers on an exchange. The rates may be adjusted by the administration according to the business interests of the subscribers or the use which they are expected to make of the services. Some administrations offer a number of different tariffs from which the customers may choose. Thus exchange lines may be in different classes with regard to charging and the class must be known for every call made in order that the appropriate method and amount of charge may be used and made. For measured rate calls the actual charge has to be assessed and metered. For ticketed calls from private stations the details of the call are sufficient, the actual charge being worked out later. Similar calls from pay stations require special and often complicated arrangements to make the charges known quickly enough for immediate collection.

With few exceptions, charges are made only for effective calls and they are debited to or collected from the calling parties. In some circumstances a called party may agree to accept the charge for a call incoming to him. This is known as a reversed charge and is very convenient to callers from pay stations. Some subscribers accept the charges for all calls incoming to their stations, usually business houses anxious to encourage custom. This service is called 'freefone'. Some charges are levied not for completed calls but for service of some kind such as personal calls, also called person-to-person service. Callers using this service specify a particular person to whom they wish to speak at a given station. An operator sets up the call and asks for the person. Normal charging starts from the time that that person is ready to speak or not at all if the person is not available, but with a charge for the service which is payable whether the call is successful or not. No charge is made for some services, including emergency calls and some enquiry calls.

Tariffs and accounting systems started with manual exchanges and operators. The characteristics thus established had to be continued in the first automatic exchanges and developed thereafter as improved techniques became available. Except for some special services, the data required for call charging are the origins and destinations of calls, the classes of service of the callers and the effective durations of the calls. A manual operator dealing with originated calls at a local manual exchange is aware of the origins of the calls and of the destinations from the directory numbers given to her to set the calls up. As she accepts a call from a line, the class of service of the line is visible to her as an individual label or as a label or coloured line or some such indication applying to a group of answering jacks of lines all with the same class of service. Different classes of service also occur for service characteristics other than charging and the possible combinations of all the classes of service can become very numerous. Because manual exchanges and operators had difficulty in accommodating more than a few classes, the classes were restricted in number. As will be seen later, the first automatic exchanges had similar difficulty and only in later exchanges are numerous classes and combinations of classes of service to be found. Also it will be seen that the directory numbers of calling lines on automatic exchanges were not readily available to the control. The registration of call charges on meters as in Figure 2.10 eliminates the need for the exchange control to know the directory numbers of calling parties but it is still necessary data for other means of registration and was a source of difficulty in early automatic systems. For meter registered calls the operator in a local manual exchange had to determine how many times to press the meter button for each effective call, which meant that she had to combine the origin, destination and class of service data with tariff information which was in her memory or readily available to her in written form usually in an indexed file on her position. The operator was expected to notice and to remember that a call which she had connected had become effective, and if it continued for an unusually long time to challenge the parties for an explanation, but otherwise she could not be concerned with the effective durations of calls. The tariffs were accordingly chosen so that for short distance calls the operator pressed the meter button not at all or a small number of times dependent on the call data available to her but not on effective time. For longer distance calls the operator had to route the calls to another operator at a position equipped with call timing means and in fact local operators had to route to other operators any calls with which they were unable to deal from their own positions. The first automatic exchanges operated in the same way, namely short distance calls were not timed but charged some quantity of units which were aggregated on exchange line message registers, and longer distance calls were routed to operators for completion together with service and other calls which could not be satisfied by simple automatic operation. But whereas a local exchange manual operator could communicate to another operator the directory number, class of service and other data referring to a calling line which was being transferred to the second operator, the automatic equipment was unable to do so and the data had to be obtained by asking the calling subscriber himself at the risk of incorrect information being given to avoid payment. As automatic exchange systems have developed and changed, so the proportion of calls requiring operator assistance for any reason has steadily reduced. The automatic metering of calls the costs of which are proportional to the durations of the calls as well as being dependent on the distances between the terminal exchanges, in conjunction with substantial reductions in the charges for the longer distance calls, has extended the practical range of bulk billed subscriber calls to all national calls and to international calls between some countries. Not all administrations, however, use bulk billing to its practical limit, but resort to ticketing before the limit is reached. The use made of operator services has declined in favour of subscriber dialled calls because subscriber dialled calls are cheaper to make, which applies to person-to-person calls, or quicker and more convenient, which is the case for reversed charge calls. To reverse a charge at small cost to himself, a person dials the call he requires and when it is answered, he asks the recipient of the call to ring him back, using a directory number which he quotes for the purpose, and of course to pay for the call thus established. The immediate collection of charges for otherwise metered calls is made possible without operator assistance by extending the exchange meter pulses over the exchange lines to operate private meters in the stations for the information of pay station attendants and for the collection of coins in coin-boxes, both of which are described in section 3.3.3. Where operator assistance is unavoidable, machine aids to the collection of charges by metering or by the ticketing of call data including the charges can be used to reduce the time and cost of manual operating.

It will be understood that tariffs and accounting are difficult and complicated subjects not all aspects of which are included in the exposition which has been given. Power, for example, is becoming a factor in the running costs of electronic exchanges which absorb power continuously independent of the traffic. Although primarily matters for the administration, the problems and requirements of tariffs and accounting have to be known for the design of exchanges. In this respect flexibility to meet future conditions and requirements is important. In future and some present exchanges, almost any system of call accounting becomes possible because of the computer-like control which is used, but the system must still be simple enough for the subscriber to understand and work out the cost of individual calls for himself. Also, accounting for the subscriber is not the only form of accounting which is required: the revenue for international calls has to be divided among the administrations concerned and gives rise to another set of problems. It is sufficient at this point to know that such problems exist.

2.15 Systems and Chronology of Development

The account given of manual exchanges and operation is not and is not intended to be historically complete. Manual operating still has a place in exchange systems operation and must be understood for that reason. Manual systems which were in service in the past have some interest from this point of view but mainly their importance lies in what their history and analysis teaches and gives to the understanding and solutions of present problems. It should be understood if not already suspected that the way in which manual systems have been used in this chapter to illustrate telephone exchanges in general has been designed to show that very nearly if not all of the problems identifiable in the most modern and complicated of systems for the telecommunication of information of all kinds and forms, can be seen to have existed and been known in some form right from the beginning of the art of exchange systems design. For this reason it has been possible to establish in this chapter and in principle at least, all of the important problems, ideas and techniques needed in future chapters to study the development of systems up to the present day.

That the problems and at least some of the solutions were inherent or have existed from the beginning is not to say that the fact has been recognized or should have been recognized at all times. Some considerable knowledge of systems and systems development is needed to see why and how the statement is true and becomes an objective and justification for works of the present kind. It is also of interest to reverse the situation and ask why if the problems and solutions are known at least in principle, it takes such a long and tedious time even now for new techniques when they become available to be applied to their best advantage. It might appear that when a new technology becomes available, all that is necessary is to design various parts of a system according to known principles but using the new technology, but that is only part of the problem. Exchanges comprise structure and control which are interdependent and interchangeable. There is no one design using a given technology but a range of designs between minimal and maximal structure exchanges, and the object of design is not to use a given technology but to satisfy a performance specification at the lowest cost. It has been shown that manual exchanges as they came to be used in practice were at a point between minimal and maximal structures which practice had decided and could not have been predicted by theory. The fact that the point was generally nearer to the minimal than to the maximal structure was largely due to the interface between the structure and the common processors coming free with the processors, namely the operators, and to the operators costing the same whether they were used to perform simple or complicated operations. The economics of machine processing can be different because of the costs of data transmission across the interface and elsewhere and of processor costs being related to the quantities of processes and processing. It has previously been remarked that the first machine exchange systems were maximal structure systems and that as technology advanced, so design moved back toward minimal structure again. Such changes are not difficult to understand qualitatively but only practice makes quantitative decisions reliably.

The time scale of exchange system developments is interesting to trace with the aid of Figure 2.15. Manual exchanges natural at first from the invention of the telephone, in about 1880, to the ultimate development of manual systems took about thirty years, to about 1910, followed by nearly another twenty years of installation before becoming completely superseded by electro-mechanical automatic systems. Strowger invented the first automatic system astonishingly early, in about 1890, and established the step-by-step method of making connections through exchanges, each step being controlled by a circuit processor using one of the digits of the directory number, no common processors being required. Step-by-step systems reached their peak of development in about thirty years and became obsolescent after another twenty years, that is by about 1940. During that time, in 1905, Molina invented arbitrary number translation which, applied in common processors called registers used in the control of step-by-step structures, extended the use of step-by-step systems for another thirty years. Arbitrary number translation led also to the invention, in about 1915, of the cross-bar switch and free-selection systems. Cross-bar systems were fully developed by the late 1930s and became universally accepted as the best design of electro-mechanical exchange in the 1950s. Electronic devices were introduced into exchanges in the 1930s but the first inventions for completely electronic exchanges are dated in the 1940s and were made practical by the invention of the transistor in 1947. Since that time, much effort has been put into the design of electronic exchanges which, on the evidence of previous history, should reach their ultimate development in the 1970s and be universal by the year A. D. 2000. The Bell System propounded in the 1950s the solution to the use of electronic devices in telephone exchanges to be a

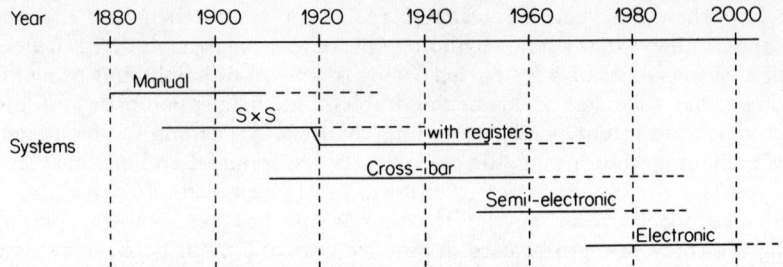

Figure 2.15   Chronology of exchange systems

near-minimal structure using metal contact exchange switches controlled by one electronic common processor working at high speed with stored program logic, the concept of a manual exchange with one operator capable of doing all the work. This solution being widely accepted has resulted in a generation of semi-electronic exchanges with electronic exchanges not using metal contact switches only now beginning to appear. A study and analysis of exchange systems should not and is not in this work limited to any particular practice if only because later system and manufacturing developments may change the situation.

Chapter Three
AUTOMATIC EXCHANGES

3.1 Progression from Manual to Automatic

Tradition has it that Strowger, a funeral undertaker in the U.S.A., was moved to invent the first automatic exchange to eliminate all operators, because he believed that his local operator was diverting his business to his rival with whom she was friendly. A more valid reason at the time was that automatic exchanges give service day and night, which is often difficult and expensive with manual exchanges, and now that they are well developed, they give cheaper and faster service particularly on long distance calls. Nevertheless the human operator has qualities such that despite the highly complicated services which modern exchanges are capable of rendering, there are still manual services to fall back on in cases of difficulty or exceptional circumstance. Astonishing as it may seen now, the battle for automatic to supersede manual exchanges was long drawn out and often bitter. When automatic operation became a real threat it was asserted that if a subscriber paid for service as he did, it was not right and was even dishonest to make him do his own operating, so semi-automatic exchanges with operators to do the dialling was the right thing: when that battle was lost for local service, it was argued that the long distance plant was so valuable that the public should not be 'let loose' on it to make their own calls. How often can lovely words be mistaken for valid argument?

A human operator scans and receives information by sight and brain, converses with subscribers and can act intelligently in unexpected circumstances, accomplishments of which no practical machine is capable: a machine because of its high speed of operation can out-perform human operators in some tasks. Because of these differences, it might be thought that manual and automatic exchanges could and would differ in some basic respect, but this is not confirmed by practice. There are some differences but the similarities between all forms of exchanges are more striking than their differences.

3.2 Automatic, Semi-electronic and Electronic Exchanges

Automatic exchanges were originally electro-mechanical in both structure and control and reached a high state of development before electronic devices suitable to replace some or all of the electro-mechanical devices became available. Exchanges in which electronics appeared only in the control were at first called semi- or quasi-electronic but electronic soon became conventional even for exchanges in which only the control was electronic and even if only partly electronic. To distinguish exchanges with both structure and control electronic from those in which only the control is electronic some term such as all-electronic may come into use, but electronic and semi-electronic respectively are more convenient and accurate descriptions, for which reason they may become conventionally accepted and they are adopted for this work.

3.3 Station Requirements and Apparatus

Magneto bells and the ringing conditions associated with them are continued with automatic exchanges but alternatives to signal-messages in the form of spoken words have to be found. In the place of speech, signal-message tones are used which the subscribers can hear and also mechanical dials which the subscribers can operate by hand to send numbers to the exchanges. These together with ringing are basic to exchange lines of all classes except the one mentioned in the next paragraph. There are in addition some non-basic signals necessary to some classes of line: the most important of these are also described in this section.

Private branch exchanges have extensions and one or more operators or attendants to provide manual operating facilities for them. Originally all the exchanges were manual and an operator made and controlled the connections between the extensions and between the extensions and the public exchange. So far as the public exchanges were aware, p.b.x. lines were indistinguishable from ordinary subscriber lines. The attendants made and received calls in the same ways as subscribers, and involved extensions as additional operations. Many p.b.x.s are manual at the present time and called p.m.b.x.s, but many are private automatic branch exchanges, p.a.b.x.s. Extensions on p.a.b.x.s connect to other extensions on the same p.a.b.x. by dialling the p.a.b.x. directory numbers of the extensions, and to an attendant by dialling code digits. Attendants connect to extensions by dialling the extension numbers. The extensions make and receive calls to and from the public exchange either through an attendant or automatically. To make a call automatically, an extension dials not the code for an attendant but a code which causes the extension to be connected directly to a public exchange line, the extension subscriber then proceeding with a call over the public network as if he were an ordinary exchange line subscriber. This called through dialling. The inverse is also possible with some difficulties with the numbering, namely for a public exchange subscriber to dial a directory number which will connect his line to a p.a.b.x., and to follow that number with digits which will connect him either to an attendant at the p.a.b.x. or to a selected one of the extensions without involving an attendant. This is called direct dialling-in, or d.d.i.: the public exchange lines to p.a.b.x.s with d.d.i. is the one class mentioned as the exception to which basic exchange line signalling does not apply. It does to calls from the p.a.b.x. to the public exchange but not to the d.d.i. calls to the p.a.b.x., for which reason it is usual to separate the exchange lines into two groups, one group for calls to the public exchange and operated as normal exchange lines and another group for calls from the public exchange and operated as junctions between exchanges.

3.3.1 Signal-message Tones

The signal-messages of manual operation listed in Table 3.1 are all standard phrases any of which may and some of which will occur on every call on any kind of exchange. For automatic exchanges the signal-messages have to be suited to machine generation as will be developed and described. Other standard phrases, used for money collection into coin-boxes and some other purposes, are not important at this stage of description.

Standard spoken phrases are readily reproduced from a sound record, and were used in that form at one time but ultimately abandoned, partly because words in a national language are not universally understood internationally. For tones to be internationally understood they must be internationally standardized, which has not yet been achieved but is the ultimate intention.

If a subscriber does not understand what an operator says to him, he will say so, and if he does understand he will usually say something or act in some way which conveys that he has understood. Alternatively, if the subscriber does not happen to be listening at the time, which is not uncommon for attendants at p.b.x.s, the operator will realize the fact and come back to the line later to ensure that words which she speaks are received, which means that the operator persists with a signal-message until sure that it has been effective and then stops giving it. The machine substitute must achieve the same results. The solution is to let the machine speak words in an international language comprising tones derived from generators of sinewave currents at various frequencies, and to go on speaking them until sure that further repetition is not needed or is useless. A problem of some difficulty is to find enough words which a machine can produce and members of the general public can learn and recognize easily. Distinctive characteristics of tones are the pitches of pure tones, combinations of pure tones harmonically or discordantly related, and pulse coding. No general public is good at identifying sounds by any of these characteristics, all of which have to be used, and exaggerated, to produce enough words which can be reliably employed. Each tone is generated by a tone generator common to the whole exchange. Three tones, namely dial, n.u. and busy, substitute directly the phrases number please, number unobtainable and busy, respectively. To imitate the persistence of the operator until she knows that the announcements have been received and understood, the machine transmits the tones continuously until a signal from the subscriber indicates that he has understood, or until the tone is no longer significant. Clearly a 'go ahead' tone may not be continued indefinitely or it will prevent conversation. Instead, a tone called ring tone is used to indicate that the bell is being rung and is ceased when the called line answers, the caller interpreting the cessation of the tone as 'go ahead' and its long persistence as 'no answer'. The current to ring the bell must be applied by the machine in bursts, to imitate operator ringing and be maximally effective. The ring tone is applied in similar bursts the cadence of which is the most distinctive feature of the tone. The

Table 3.1   Signal-messages spoken and corresponding tones

Spoken	Tone
Number please	Dial
Go ahead	Ring—cessation of tone
Busy	Busy
Number obtainable	n.u.
All lines busy	Busy for caller
	Plant busy datum for control
Out of order	n.u.
No answer	Ring—persistence of tone

dial tone is a discordant rattle applied continuously until the subscriber sends the first directory number digit as described in the next paragraph. Number unobtainable tone is low-pitched and continuous, and the same tone is pulse coded to produce a cadence which the subscribers can recognize as the busy tone. Because it is difficult to find other easily recognizable tones, number unobtainable tone is used also for 'out of order', and one busy tone is given to subscribers for both subscriber busy and the all lines busy encountered with junction operation. The difference between the two kinds of busy is not important to the subscribers but to the control which requires not tones but data which it can recognize: so that if failure to connect is due to plant shortage, the control may have the opportunity to make another attempt at connection over a different route.

3.3.2 Dials and Dialling

In manual systems there is only one signal-message transmitted from the stations to the operators, and that is the directory number in response to 'number please'. The 'numbers' being an exchange name and decimal number, were convenient for subscribers and operators, but names are unsuitable and only decimal digits may be used for automatic exchanges. Nevertheless, for the convenience of subscribers, names followed by decimal numbers were printed as station numbers in directories for many years, the names becoming decimal digits in the communicating of the directory numbers to the exchanges. The communication of directory numbers involves the subscribers in the manual operation of some form of sender which transduces the digits to n-state digital signals suitable to transmission over exchange lines. Independent n-states may comprise ten different frequencies used one at a time, or five frequencies two at a time, or four frequencies in various combinations, all of which have been used at some time or other, and so on. Using states which are independent of one another makes unimportant the durations of transmission of the digits, and hand operated press-buttons depressed one after another at any rate to suit the subscribers concerned may be used for number sending. Sending of this kind using $n = 8$ frequencies is now an international recommendation for national systems. The decimal digits and two symbolic characters are each represented by

Table 3.2 Number sending codes

Digit	Frequency code 2 times 1-out-of-4 frequencies Hz		Arythmic code 5-unit n-states S = space M = mark
1	1209	697	SMSSS
2	1336	697	SSMSS
3	1477	697	SSSMS
4	1209	770	SSSSM
5	1336	770	SMMSS
6	1477	770	SMSMS
7	1209	852	SMSSM
8	1336	852	SSMMS
9	1477	852	SSMSM
0	1336	941	SSSMM
*	1209	941	SMMMS
□	1477	941	SMMSM
–	1633	697	SMSMM
–	1633	770	SSMMM
–	1633	852	SMMMM
–	1633	941	SSSSS

two frequencies transmitted simultaneously as shown in Table 3.2. The frequencies are in two groups of four, and one in each group being used for each digit transmitted, there are sixteen available combinations of which four are not specifically allocated. The frequencies being within the band of transmitted speech frequencies, the signals are signal-messages transmitted over the speech transmission lines. Reception at the exchange requires amplifiers, frequency filters and rectifiers to transduce the signals to d.c. currents in eight wires coresponding to the eight frequencies. Using transistors, the signal-messages are generated at the subscribers' telephones without difficulty and at no great cost, and the method may become universal for number sending over exchange lines: but originally something simpler was necessary.

If n is less than 4, pulse coded signals must be used with timing at both the sending and the receiving ends. The loop signal controlled by the cradle switch is a convenient n = two-state signal to use pulse coded according to a code of at least five units illustrated in Figure 3.1(a) and in Table 3.2. The two states are designated space and mark. Space is sent continuously during idle periods and mark is sent continuously, except during number sending, while a call is in progress. The code for a character comprises five elements each sent for a unit time. The first element is a unit of spacing to start the timing at the receiving point and called for that reason a start signal. Marks or spaces in the following four unit times provide sixteen combinations as with the 2 times 1/4 frequency code, although the all-space SSSSS combination is preferably not used because of its possible confusion with the clear signal. In telephony this is called arythmic coding. It needs accurate and complicated time measurement means both to send and to receive the

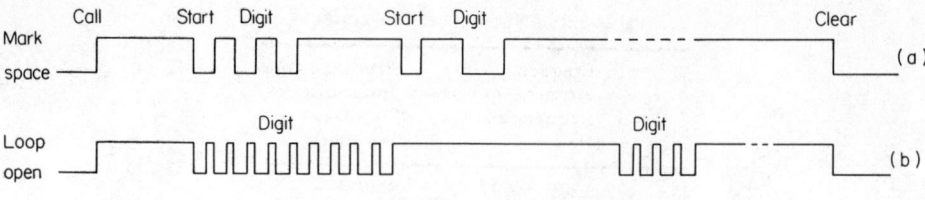

Figure 3.1 Pulse coded digit sending
(a) arythmic
(b) rotary dial

signals, for which reason it has not been used for subscriber number sending, but it has for decimal digit sending over international telephone lines, and the same kind of coding but using six instead of five elements per digit is commonly employed in teleprinter exchange systems, the teleprinters at the subscribers' stations providing the accurate timing of the elements. Because of new methods of manufacture, arythmic coded impulse sending from telephone stations, much like teleprinter sending, is possible and may be used not only instead of coded frequencies but also for alpha-numerical sending, that is letters as well as numbers, for messages as well as signals. Originally, however, and still much in use up to the present time for telephone station digit sending, the timing requirements and the code recognition are simplified to the sending and counting of pulses the durations of which and the intervals between which can vary over wide ranges and still be reliably recognized. The call and clear signals, Figure 3.1(b), are generated by the cradle switch as before and the decimal digit signals by a mechanical dial mechanism wound up an amount according to the digit to be sent, see Figure 3.2, by a finger inserted into one of ten holes in a dial plate, then rotated to a stop and let go. The dial plate restores under the control of a spring and governor to operate contacts in the loop, which is shown in Figure 3.3, and which break the loop a number of times equal to the value of the digit to be sent, ten breaks being used for the digit 0. The uniformly rotating spindle of the dial plate carries cams which operate the dial contacts in the loop. Thus the sending comprises loop breaks which are pulses of time durations and frequency dependent on the accuracy of the dial governor, the nominal frequency being commonly ten pulses per second and the durations of the breaks sixty to sixty-seven percent of the pulse period. At the receiving point the only timing requirements are means to determine when the duration of a loop break exceeds some predetermined value and when the duration of loop closure exceeds some value which does not have to be the same as that of loop break. Following the detection of a loop signal to initiate a new call, loop breaks are counted and timed and when the duration of a break exceeds a predetermined time, the call has been cleared. The make signals following breaks are also timed: one which exceeds some other predetermined time signifies the completion of a digit the value of which is that of the count up to that point. Any further breaks belong to the next digit or are the clear signal. With no close tolerances on the durations, the timing accuracy required is low, and counting is simpler than code recognition as needed for arythmic code reception.

With the change to automatic working, exchange names were dialled using alphabetical characters allocated to decimal digits, the U.K. allocation being that shown in Figure 3.2. The subscribers were instructed to dial the first two or three letters of the exchange name followed by the exchange number. Some of the names had to be changed when the exchanges became automatic because the equivalent digits were the same for more than one name. In the U.K. Hammersmith and Hampstead clearly could not both be used, nor Bank and Camden Town. With the growth of telephone networks the difficulties of choosing suitable names became acute but it was international dialling which made it impossible. The decimal equivalents of the letters were not the same in every country, even among those using the same scripts, and different scripts Latin, Greek, Arabic and so forth, make names in one country incomprehensible in some others. Hence directory numbers do not now include names of exchanges, only decimal digits. Names still appear in directories, the names being followed by numbers. The numbers are directory numbers to be dialled within areas defined by the names. To dial a number within an area from outside the area, the number must be preceded by digits which can be ascertained from a separate directory of area names and their numbers.

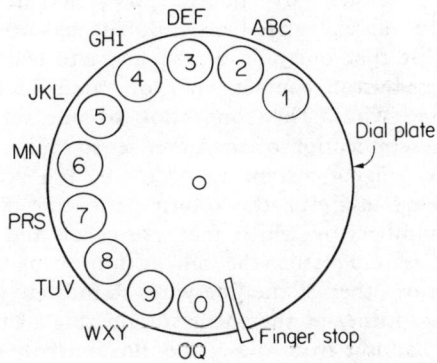

Figure 3.2   Rotary dial

All exchange systems can operate with dial pulse decimal digit sending but some only with difficulty with press-button sending. The logic of dial pulse sending being easier to explain, it is assumed in the following descriptions, except when press-button is specifically mentioned, without loss of understanding because exchange operation overall is the same with either.

The electrical circuits of the telephone sets for automatic working have to incorporate rotary dial or arythmic code impulsing contacts, which being normally closed may be included in the loop, or press-button tone generators which have to be switched into the loop whenever a button is pressed. Figure 3.3(a) shows dial operated contacts in series with the line and off-normal contacts o.n. which are open when the dial plate is in its normal position of rest but closed when it is off-normal. The closed contacts short-circuit the telephone instrument to protect the subscriber from clicks from the receiver which the dialling would otherwise produce and they also remove the resistance of the instrument from the dialling loop. Similar arrangements are needed for press-button sets.

3.3.3 Non-basic Signals

In addition to the electrical states used for basic signalling over the two wires of exchange lines, namely d.c. loop for cradle switch, rotary dial and arythmic code dial pulse signals, sub-audio frequency for ringing and audio frequencies for press-button dialling signals, others are needed for party lines, coin-boxes and some purposes which affect only a very small proportion of the total exchange lines.

Party lines give rise to many problems in manual and in automatic exchanges, for which reason they are tending to disuse, but one surviving type is of sufficient interest to be described. Figure 3.3(a) is the normal telephone set with the instrument, cradle switch and dial operated pulse contacts in series with the exchange line comprising wires A and B, in parallel with the magneto bell and capacitor. Figure 3.3(b) shows party line telephone sets at each of two stations X and Y connected to one exchange line A, B. The sets are the same as those of ordinary stations except that one side of the magneto bell is connected to earth and an additional press-button contact when operated disconnects the loop and earths one of the line wires. The connection of one set to the line wires is reversed relative to the connection of the other set so that either of the parties can be rung individually by ringing current connected by the exchange to the A or to the B wire, earth being used for the return path. The X and Y parties have individual directory numbers by which they are called and which determine the kind of ringing to be sent. Pressing the call button opens the loop and connects earth potential to one or other of the line wires to indicate to the exchange which of the parties is calling. Either of the parties originating a call starts in the normal way by removing the handset from the cradle, but must then press the call button to make the call effective. The exchange stores the identity of the calling party so that the charges incurred may be recorded against that party. The number sending can be by rotary dial or press-buttons. If press-buttons are used, the call buttons instead of earthing one wire of the line can use two of the spare codes of Table 3.2, one for each party, if the exchange system is capable of using the data in that form.

In Figure 3.3 the two wires of an exchange line are identified by the letters A and B. Alternative designations more often used are earth, positive or the + sign for one wire, and for the other wire battery, negative or the − sign, according to which pole of the exchange battery the wires are connected and as best suits the immediate context.

In many systems, the insertion of coins into coin-boxes has to signalled to the exchange, and coins of more than one denomination may be used. The exchange line loop may not be broken except to terminate the call, but the loop current may be decreased as a coin insertion signal. In Figure 3.4 the insertion of a coin opens

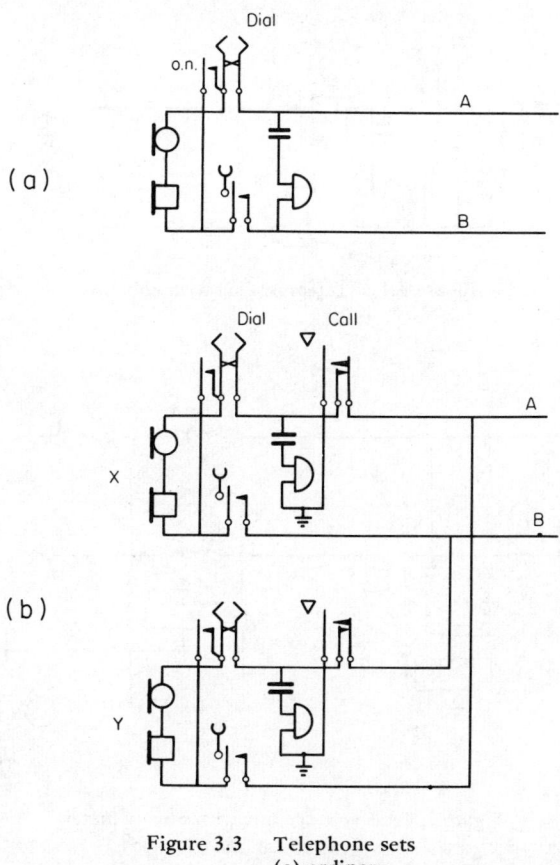

Figure 3.3 Telephone sets
(a) ordinary
(b) party line

coin contacts a number of times similar to dialling but according to the value of the coin. The resistor which shunts the contacts permits enough current to continue in the loop to hold the connection while the contacts are open. In all systems a signal is required from the exchange to the coin box to control the collection of money deposited. A polarity reversal at the exchange end of the exchange line is one possibility, a coin collection mechanism magnet CC in the coin box being short-circuited normally by the shunt diode which becomes non-conducting on polarity reversal to allow the magnet to operate. If the connection is prolonged beyond the time paid for, usually an audio tone is sent from the exchange to prevent conversation until another coin is inserted, and if no coin is inserted within a given time the connection is released by the exchange.

A small percentage of exchange lines have private meters at the subscribers' premises, which operate effectively in parallel with the meters in the exchange

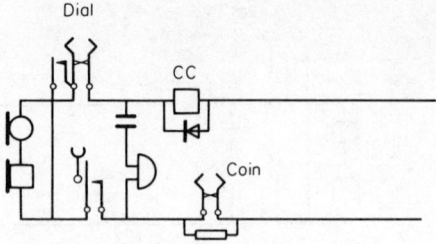

Figure 3.4 Telephone set with coin-box

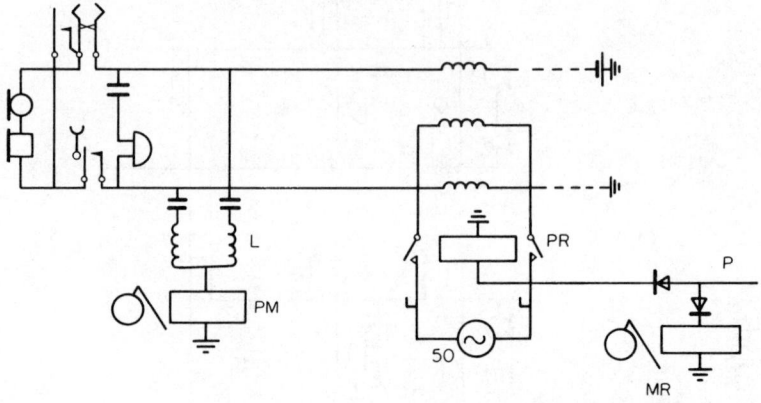

Figure 3.5 Exchange line with private meter

specifically to provide the subscribers with an immediate record of charges, mostly for collection from hotel guests or customers of various kinds. One of the requirements is that the meter must be capable of being operated during but without interference with the conversation, which is one reason why signalling over the earth phantom is used. Referring to Figure 3.5, in the exchange a relay PR individual to the line is operated in parallel with the regular meter MR, by a positive battery supply applied to the P wire which is the third wire data path of the line equipment, as described later in conjunction with Figure 4.1. Contacts of the relay PR connect 50 Hz power supply to the primary winding of a transformer, the two equal secondary windings of which are connected in series with the two wires of the line so as to be non-reactive to currents round the line loop. The transformer therefore has a negligible effect on speech transmission, and the equal longitudinal currents which it produces in the line wires to operate the private meter PM at the subscriber's premises do not produce audible effects in the speech receiver. The centre tapped inductor L prevents the private meter from causing more than a very small speech transmission loss and the capacitors prevent d.c. flowing in the inductor or in the meter.

The use of the telephone by the public gives rise to innumerable circumstances in which some particular variation to widely available services would be of value to a small and often very small proportion of the total users. An example is that of physically disabled persons able to operate the cradle switch of a telephone set possibly specially adapted for the purpose, but not a dial or press-buttons. Manual service being satisfactory to such persons and not infrequently an important part of their lives, it must be available to them through automatic exchanges without dialling any digits if they are to be able to make telephone calls unaided. The service is not dependent on a signal but on class of service to be described in section 3.4.

Incoming call transfer is a service of great importance to some subscribers, doctors and some business men particularly. Dialling the listed directory number of such subscribers normally calls a home station but when in transfer another station is called, the station having been specified in one of two ways. With pre-arranged transfer, the administration having been advised in advance by the subscriber of a number of possible stations, the one to be used whenever the line is put into transfer is specified by the subscriber when his line or lines are put into transfer. With follow-me transfer, the directory number of one station to which incoming calls are to be connected is specified at the time of first transfer and possibly changed more than once later to follow the subscriber as he moves around. In manual and early automatic exchanges, the subscriber had a key associated with his telephone set to substitute another line on the same exchange for his regular line. The key used an earth signal over the exchange line to affect the switching. More complicated transfer arrangements required the subscriber to give instructions to an operator to whom subsequent incoming calls were transferred: callers were then told by the operator what number to dial if they still wished to contact the subscriber. Although inconvenient and costly to operate, such procedures can still be found in practice. What is wanted is that subscribers have control of the transfer facilities, by dialling a code followed by one digit to define a pre-arranged transfer station to which all incoming calls should be connected, or for follow-me transfer, dialling a code followed by the directory number of the transfer station to be used. In both cases but for follow-me transfer in particular, there must be adequate safeguards against calls being transferred by unauthorized persons and there are operational and transmission difficulties if a transfer station is not on the same exchange as the home station. By using dialled codes to put lines into and out of transfer, line signals special to the service are avoided. From a home station in transfer, originated calls should be possible without cancelling the transfer of incoming calls, the line being taken out of transfer by the dialling from the home station of a code allocated for the purpose.

It is not important to design to define all the known special services, but to realize that they exist and others may be added at any time. What is important is that there should be some general way of satisfying such requirements as and when they occur. The requirement was scarcely realized for the early automatic systems which resorted to opportunism when the need for special services occurred if the need was satisfied at all. Automatic exchange services in general were modelled on manual exchange practices as far as it was possible to do so, as much for the benefit of the subscribers used to such systems as for any other reason, and not always with complete success. The equivalent of operator recall by flashing, to correct an incorrect connection or to change a connection because of unsatisfactory transmission or other reason, was, for example, not possible on automatic exchanges, and subscribers were left to release one connection and make another to achieve the same result. It is interesting to note, however, that the equivalent of flashing is now available on some p.a.b.x.s using modern systems of control, to achieve subscriber controlled so called enquiry and transfer facilities. When a p.a.b.x. subscriber receives a call on his extension, one flash of the cradle switch causes him to receive dial tone and he can then dial another extension on the p.a.b.x.. On gaining connection to that extension, the first subscriber can converse with that extension subscriber without the original caller being able to hear, this being the enquiry facility if the original connection is restored by another single flash. On the other hand, if the first extension hangs up, the incoming call is transferred to the second extension. Similar facilities are beginning to be demanded of public exchanges.

3.4 Class of Service

In section 3.3 some of the most important and generally used services available to exchange lines have been described and others will be apparent later. They are not all needed by or available to every line but are selected as required by the line or station equipment, or as instructed or paid for by the subscriber concerned or imposed by the administration or maintenance force and some other reasons. A class of service defines one service or combination of services which may be semi-permanent, period or temporary in character, and a class may be positive by defining services to which an exchange line or particular equipment is entitled, or negative in the sense of denying some service which is available to the generality of subscribers, or conditional if it applies in the absence of some other class. An exchange line may have more than one semi-permanent class of service provided that none is incompatible with another. It may have more than one period or temporary class, one or more of which may override a semi-permanent or even another period or temporary class of service.

The quantity of classes of service necessary or desirable has grown with the scale of telephone development. The labels and multiple markings of manual exchanges provided a limited quantity of exchange line data which the operators used to render some moderately varied service to subscribers. The variable connection together of transmission lines terminated on an exchange and under the control of the subscribers was a sufficient problem for the first automatic exchange designers without the complication of many classes of service. Therefore those services which could not be provided easily by automatic operation were made to bring an operator into the processing, or were not rendered at all. Subsequent developments can be seen to be the extension of the services offered to the public coupled with a steady reduction in the help needed from operators. The circuit interconnection problem has become one of the least difficult while others are more important, and class of service is one of the most important.

Classes of service, some offered to and some imposed on subscribers, differ between systems and countries but include many which are common to them all. Those described herein are to be found in the same or in similar forms in most systems but it is to be understood that it is the principles and the effects on exchange design of classes of service which are the main objects of the descrptions.

Classes of service which impose no requirements on the exchange operation other than data processing will be designated control classes of service. Others depend for their implementation on a particular physical feature in the exchange construction and are termed structure classes of service. Both kinds are explained in more detail in the next two sub-sections. The difference is important. A structure class of service involves the provision of structure members and can be changed or a new class added only by a change of structure which may be difficult and expensive. A control class of service may be changed or a new class added by a change of processing which is not difficult or expensive.

3.4.1 Structure

The loop impulsing of rotary dialling needs, in the exchange, one kind of transducer and processor to detect, count and store the impulses as decimal digits, and press-button arythmic code and v.f. pulse dialling different kinds of both transducer and processor to detect, decode and store the decimal digits. It will certainly happen as press-buttons are superseding rotary dials that, on one exchange, one kind of dialling will occur on some and the other kind on other of the lines. In such exchanges, originating call classes of service may be used to cause the appropriate kind of exchange equipment to be connected to a line when it calls. On yet other lines either kind may occur unpredictably at the exchange. To give one example, either kind can occur on some p.a.b.x. exchange lines due to through dialling from extensions some with dials and some with press-buttons, or due to the extensions having dials and the attendant's console having press-buttons. And finally, calling but no dialling occurs on the lines of disabled subscribers who can operate the cradle switch of the telephone but not a dial. Four classes thus exist, namely rotary dial, press-button dial, either kind of dial and neither, and because structure members are necessary to the provision of the first three, they are structure classes of service. They may be distinguished by four data, one for each, or a class with the most lines may be assumed in the absence of a datum and individual data assigned to each of the other three.

Party lines as described with reference to Figure 3.3 give rise to both originating and terminating classes of service, the first to direct call charges to the calling party and the second to cause the required party to be rung. All exchange lines originate calls by looping the line and expect dial tone in response, but before dial tone is given to a party line, the calling party must be signalled so that metering and

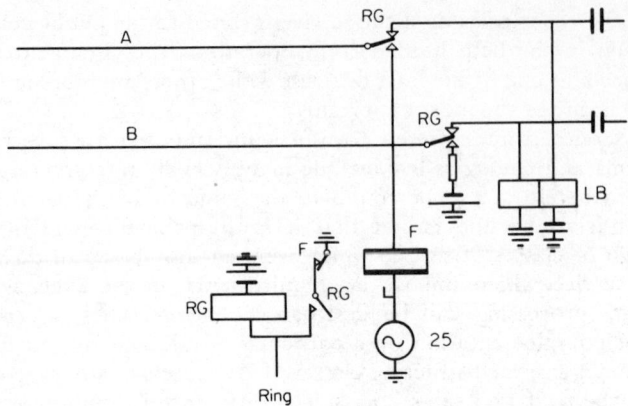

Figure 3.6 Ordinary and party line X party ringing

other processing individual to the party may be carried out later. Thus the exchange control needs a class of service indication for party lines, and it may be a structure or a control class of service as will be subsequently described.

Exchange lines other than party lines are called by ringing current balanced in both line wires. Figure 3.6 shows a common arrangement. A common ringing current supply with one side connected to earth is connected via a relay F and contacts of the RG relay to one wire of the line, the other wire being connected via RG relay contacts and a resistor to the exchange battery. The relay F is made slow to operate so that ringing current being alternating may flow through it to the line without operating it. The F relay is operated by the d.c. current from the battery when the line is looped, to trip the ringing. It will be seen from Figure 3.3 that the same ringing circuit will also ring the X parties of party lines but that Y parties require the ringing and ringing return to earth to be reversed at the exchange. Class of service indication may be used to choose between two kinds of processor, one suited to ring non-party and X parties on party lines, and another to ring Y parties on party lines; or alternatively, the class of service may select the kind of ringing to be applied by one kind of processor able on instruction to supply either kind of ringing. If the line reversal required at the exchange to ring Y parties can be made in the exchange wiring, no other class of service indication is necessary. An example of this solution is given in chapter 4, but not being general to all systems, the possibility of class of service selection has to remain. Also only two parties per party line has been considered, but some systems include lines with four or more parties for which class of service selection cannot be avoided.

An exchange line terminated with the coin-box equipment of Figure 3.4 requires structure type class of service to be indicated when starting a new call, so that the line becomes connected to a time shared processor able to receive coin deposit signals and to send coin collection signals. Such a processor is complicated and expensive but even so is not always worth time sharing, some coin-box lines being so continuously in use during the busy part of the day that time sharing would have no economic advantage.

Private meters are needed on such a small proportion of exchange lines and the equipment in the exchange to operate them, as shown in Figure 3.5, is so simple that the complication of time sharing is doubtfully useful, and an individual equipment as in the figure for each line requiring the service is usual, and class of service indication unnecessary.

3.4.2 Control

Some subscribers require the range of calls which can be made from their stations to be limited in distance, which means that longer distance calls are barred and a long distance barred (or l.d.b.) class of service is required for their exchange lines. Another class, the i.b. class, is barred international calls. Nothing special in the structure is necessary for these calls, merely data processing to connect n.u. tone to the caller if a barred connection is dialled. Hence l.d.b. and i.b. are control classes of service.

Subscriber controlled transfer obtained by ordinary dialling procedure, as previously described, would be available to all subscribers if it were not limited to those authorized to use it by a transfer class of service.

The two examples just given are of semi-permanent classes of service. Some control classes are period or temporary: for example, lines sometimes have to be put out of service temporarily for maintenance or administrative reasons. Some classes which are basically control classes of service are realized in some systems as structure classes, as will appear in later descriptions.

3.4.3 General

Some of the problems of class of service may be seen from the few examples which have been given and more will become apparent as the development of systems is described. Some classes are directly related to the public services offered, such as party lines and coin-boxes and the barring of some connections, and are available to the subscribers at their options except that some combinations of classes may be prohibited by the administration as being too costly or difficult to provide. Some classes will be seen later to be imposed by the administration or the maintenance engineering, such as lines put partially or completely out of order because of some kind of fault or irregularity.

The solutions which have been used to the class of service problem have in general been related to the systems of which they form part and will be described in the expositions of those systems.

3.5 Exchange Switches

With few exceptions the exchange switches of all systems which have been put into service have comprised metal contact switches electro-mechanically operated.

It is not difficult to see that the electro-mechanical equivalent of an operator controlling a plug and inserting it into a jack is a machine moving a metal contact relative to fixed contacts and controlled to cause the moving contact to rest in contact with a selected fixed contact. The most usual practice is for the moving contact to slide over the fixed contacts thus making contact with each in turn. The moving contact is on an arm rotated so that the contact at the free end slides over contacts fixed in an arc, as represented in Figure 3.7(a), or the contact is on a linear slide so that it moves over contacts fixed in a straight line, as in Figure 3.7(b). The diagrams show only one contact at each position of the switch and one moving contact. The moving contact on its arm is called a brush or a wiper. In practice each switch has several independent wipers and sets of fixed contacts operating in parallel. The fixed contacts considered as a unit of one switch are termed a bank of contacts, and the bank contacts of many switches are multipled and connected to message circuits to which the wipers of each switch thus have access. One problem is to make a switch of any kind large enough, up to ten thousand fixed contacts and more if connections through the exchange are made with one switch. Very early on it was attempted to reproduce the largest multiples of manual switchboards by means of switches having a hundred rows of a hundred sets of contacts, but this was soon abandoned. The solution to the problem requires switches in standard size units to be used as building bricks to attain the required size of exchange. The size of unit to emerge in practice for automatic exchanges has from one hundred to five hundred sets of fixed contacts. Rotary switches called uniselectors have typically one hundred sets of fixed contacts in one arc as in Figure 3.7(a) and are motor driven either by an individual rotary electric motor per switch or by continuously rotating shafting common to many switches and to which a wiper carriage can be clutched electro-magnetically. The symbol with the letters DM suffices for both, being the driving motor of the first and the clutch electro-magnet of the second, and which when given operational data o over the lead o, causes the wipers to move. A Strowger two-motion switch, named after the inventor, has the fixed contacts in ten arcs of ten contact sets each, as shown symbolically and in perspective in Figure 3.7, the arcs being mounted in parallel to form what are called levels. A set of wiper contacts is mounted on a shaft co-axial with the contact arcs and driven first axially to raise the wipers level by level to reach a selected one of the levels, and then rotationally contact set by contact set to rest finally on a selected one of the sets. Step-by-step axial and rotational movement is produced by individual ratchet motors pulsed with current once for each step, for which the symbol of Figure 3.7(a) is sufficient if it is understood that the DM equipment is two motors with an associated processor to cause first one and then the other motor to be operative. An equivalent two-motion switch driven through clutches and continuously rotating shafting has ten levels of bank contacts and ten sets of wipers all of which are normally mechanically latched so as not to be in contact with their bank contacts when the wiper carriage is rotated. Before the wiper carriage moves, however, an electro-mechanically operated clutch rotates a spindle through a controlled angle to select which set of wipers shall be tripped to engage its bank contacts, after which the wiper carriage is rotated by the operation of a

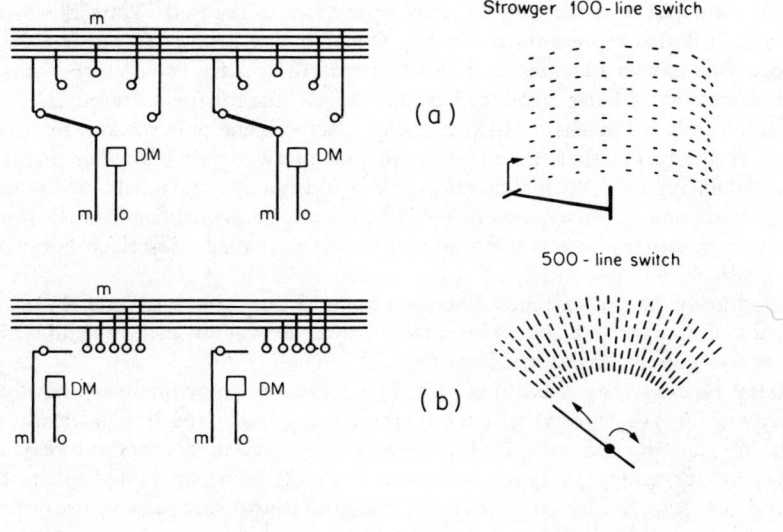

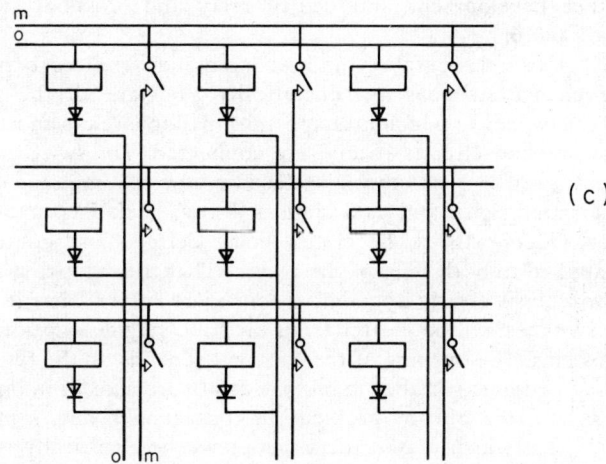

Figure 3.7 Electro-mechanical exchange switches
(a) rotary sliding contact
(b) linear sliding contact
(c) relay

second magnet and clutch. The trip spindle and wiper carriage each carry cams to operate contacts once for the equivalent of every step taken by a Strowger switch and by which the same step-by-step operation is secured. Thus the symbol of Figure 3.7(a) also represents this kind of switch if it is understood that the driving magnet DM is two magnets and the steps of the switch are counted back to the control instead of being produced by pulses into the magnets. In the L.M. Ericsson five-hundred-line switch, a stick carrying a set of wipers is rotated by a clutched motor and rotating shafting to point in one of twenty-five angular directions, as indicated in Figure 3.7(b), then the stick is slid linearly outwards for the wipers to engage with one of twenty sets of fixed contacts, the principle and operation of the switch being otherwise the same as that of the previously described common shaft driven switch.

In addition to the switches described some sliding contact systems use auxiliary switches of smaller size, generally ratchet motor driven uniselectors and with ten to twenty-five contact sets in the banks.

Relay type switches have sets of relay contacts in co-ordinate arrays of columns and rows, the two springs of each contact being multipled one vertically and the other horizontally, as indicated in Figure 3.7(c). Each contact set may have an individual operating coil as in the figure, the coils being multipled as the contacts are and one selected by current in the horizontal and vertical wires intersecting at the coil, rectifiers being required to decouple the rows and columns. Alternatively, in a cross-bar switch the contact sets are operated by common row and column magnets. The further development and use of relay and cross-bar switches is pursued in chapters 5 and 6.

There are many practical operating and economic differences between the various types of exchange switches but analytically they are all the same. An exchange switch has two sides to which message transmission circuits and associated signal and data transmission circuits if any, are connected. The switch comprises sets of contacts, and a set of contacts when closed connects a message circuit on one side and its associated signal and data circuits if any, to similar circuits on the other side, the set of contacts to be closed being defined and controlled by operational data o applied to both sides of the switch. That the definition applies to co-ordinate relay switches is clearly apparent from Figure 3.7(c). That it is true of the sliding contact switches will be clearer from the following descriptions of their uses in exchange systems. The opening of the contacts depends on the type and use of the switch. Sliding contacts whether in motion or not are, except in one type of switch specially designed to avoid the problem, in contact with one or more fixed contacts all the time, for which reason the wipers must be electrically isolated as they move over contacts to a desired contact or they might interfere with circuits connected to the intermediate contacts. An auxiliary relay is used with contacts in series with the wipers and controlled to close the contacts only when the wiper contact is required to be effective, functions which relays K and H in Figure 4.1 will be seen to perform.

For convenience of explanation and simplicity of drawing, and also because it aids comprehension, the three fixed contacts and rotating wiper symbol of

Figure 4.1 when used may represent any practical form of wiper and bank exchange switch performing the function specified, under the control of operational data on leads o, the switch having any number of fixed contacts. It is to be understood that more than one and usually many more than one switch, all similar, exists and that contacts on at least one side of every switch are multipled to similar contacts on other similar switches, as shown by the common connection symbol. Each path making up one connection through the switches is shown separately on the diagram. The switches of Figure 4.1 provide six wires through each switch from one side to the other. Wires 1 and 2 switch the message transmission paths of exchange lines on one side to trunks on the other side. That much is essential for all exchanges. The quantity and use of other wires through the Figure 4.1 switches depend on the control system employed.

Chapter Four
STEP-BY-STEP EXCHANGES

4.1 General

The first automatic exchange systems used a principle of operation which has become known as step-by-step, often written S x S. Although now obsolescent, there are many step-by-step exchanges in service and some are still being extended with new equipment of the same type. It is one of the problems of new systems that they must interwork with systems already in service and likely to continue in operation for some years, and additionally if possible to work side by side with existing equipments as extensions needed to cater for growth of established exchanges. S x S systems as described in this chapter are thus still of practical interest as well as being important in the development of exchange systems. Like the treatment of manual exchanges in chapter 2, this chapter specific to one type of exchange, will also establish some circuit and system design principles which are found in and are useful to the understanding of later systems.

The first system produced by Strowger has been subject to many variations without departing from the basic step-by-step principle. The account which will be given in this chapter will not refer to any one system in detail but will be general to all systems of the same type. Inevitably the descriptions will include circuit operations as well as general principles. Although there is a unity of underlying technique, circuitry varies greatly in detail between systems by different manufacturers. The circuits given do not necessarily apply to any one but are designed to illustrate the operations involved in them all.

Most of the problems and complications of step-by-step exchanges and to some extent of all types of exchange, are to be found where the exchange lines are connected to the exchange and this forms the first part of the following exposition. The second part is concerned with an equally important subject, that of making exchanges of any required size up to ten thousand lines and more.

4.2 Dial Controlled Selection

4.2.1 General Principles

Strowger used one two-motion switch per exchange line to produce an automatic system which was the analogue of the simplest manual system with multipled jacks, Figure 2.2, but with one operator per exchange line. Keith followed soon after with the addition of what he called line switches to produce the equivalent of a manual system with double-ended cord circuits Figure 2.3 and one operator per cord circuit, from which point systems as shown in Figure 4.1 developed. For the answering sides of the cord circuits the exchange switches can be used in either of two ways called respectively pre-selection and line finding; both are shown in Figure 4.1, with dashed line connections to indicate that one or other can be used in conjunction with one kind of switch for calling exchange lines, the switch being designated a final selector (or connector in the U.S.A.). The 'cord' is not a flexible conductor with a plug at each end. It may have short lengths of flexible conductor to the wipers of the switches but essentially it is two exchange switches with permanent wiring between them, but still known as a cord circuit.

The exchange lines each have a line relay L and cut-off relay K, as part of a circuit processor individual to the line. The circuits of Figure 4.1 are drawn mostly with the relay coils and contacts associated but some are disassociated. Used as a pre-selector, the fixed contacts of the exchange switches are connected to the answering sides of the cord trunks and the wipers to the exchange line equipments and the lines. Thus there is one switch per exchange line, usually a small one with rotating wipers and up to twenty-five contacts in the bank. One bank position is sacrificed to a home position to which the wipers are returned after use so that operation always starts from the same point. The operation comprises the rotation of the wipers when the line initiates a call, until a free cord circuit is found when the rotation stops and the line circuit is connected through the switch to the cord circuit. The operation amounts to a one-only selection of a cord circuit in the same way that an operator selects a cord circuit to connect to a calling line. If no cord circuit is free, the switch continues to rotate until either a cord circuit becomes free or the subscriber gives up and clears. At times of heavy traffic many switches may be rotating continuously unable to find free cord circuits. Rotary switches with wiper contacts permanently in contact with the bank contacts and particularly those with individual ratchet motors to drive them, suffer deterioration reflected in increased maintenance due to the continuous and prolonged rotation which occurs under these conditions, from which point of view the alternative line finder method of operation is to be preferred. Cost is another point of view which is considered later.

The pre-selector circuit of Figure 4.1 shows that a subscriber looping an exchange line to start a new call operates relay L to close a circuit through the K relay in series with the driving magnet DM of the switch. The relative characteristics are such that a relay K can operate in series with DM but DM is unable to operate in series with a relay K. Initially the K relay is short-circuited by the earth potential on the home contact of the arc 3 of the pre-selector switch: so the switch commences to rotate and continues to do so until the arc 3 wiper reaches a contact which is not at earth potential. Rotation then stops because there is insufficient energization of DM, and relay K operates. If every contact is at earth potential because there are no free cord circuits, the wiper rotation is continuous. If a free cord circuit is encountered, the exchange line is switched through to it to control

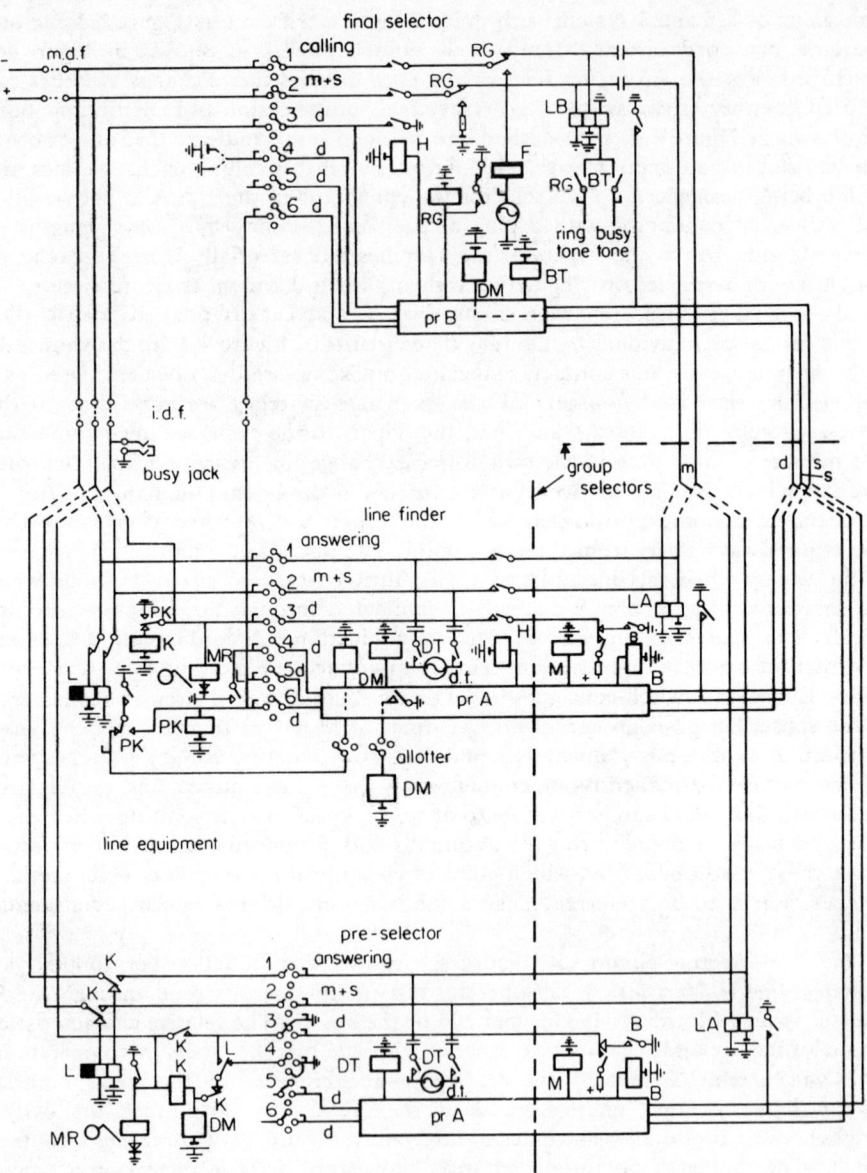

Figure 4.1 Exchange line and step-by-step selector operation

the relay LA, a contact of which relays the loop signal s to a processor prA individual to the answering side of the cord circuit. The processor responds by operating relay B until the call is cleared, and operates relay DT to apply to dial tone signal-message to the message circuit. The relay B applies earth potential to the P-wire, to hold through arc 3 the relay K in the line equipment and to prevent another exchange line from being connected to the cord circuit. The relay K must be held before the relay L releases, and this relay is made slow to release to ensure the necessary conditions. Contacts of the LA relay repeat the line loop signals not only to the processor prA but also over a connection through the trunk to a processor, the processor prB in Figure 4.1, of the final selector terminating the trunk. The subscriber on hearing the dial tone, dials the directory number of the station that he wishes to speak to. The pulses, the first of which releases relay DT and removes the dial tone, operate the magnet DM of the final selector directly, to position the wipers on a line with the directory number which is dialled. A number of two decimal digits is required by a Strowger two-motion switch, the vertical stepping DM being operated by the first and the rotary stepping DM by the second digit, as explained in section 3.5. If and when the called station answers, the exchange line loop which signals the fact is relayed by the relay LB to the processor prA which, assuming a measured rate charge has to be recorded, operates relay M for each unit charge incurred. Relay M when operated applies a positive battery potential to wire 3 to operate the meter MR in the line equipment. Using a positive battery to operate a meter connected to earth via a rectifier, makes the design and adjustment of the meter easier than in the manual exchange differential current arrangement: there were no suitable rectifers available when manual exchanges were designed. The wire 3 is analogous to the sleeve wire of the manual switchboard, serving as a path for two data, one indicating the free or engaged state of the cord terminals and the other being charge units to be aggregated on the line meter. The calling line can be cleared at any time, the processor prA then releasing the B relay to allow the K relay in the pre-selector circuit to release and complete a circuit for the driving magnet of the switch. The wipers move until wiper 4 reaches the home contact, which has no connection and is therefore not at earth potential like the others. Hence motion stops and the equipment is back to the normal state.

The equivalent of scanning is called hunting in automatic systems. If two or more pre-selectors are hunting at the same time for a free cord circuit, reliance is placed on the very low probability that two will test the same circuit within the operating time of a K relay, to ensure that no two pre-selectors select the same cord circuit.

Pre-selection of the next circuit to be used as in Figure 4.1 should be termed post selection, because selection of a free trunk takes place after a call has been initiated by a subscriber. The term pre-selection comes from the original invention by Keith which is historically interesting because it anticipated to some extent at least, and by about twenty years, the cross-bar switch invention. Keith's switches, provided one per exchange line, had the equivalent of relay contacts in sets, ten sets per switch, as in the switch of Figure 3.7(c) but arranged in the arc of a circle, Figure 4.2. The switches were mounted co-axially in columns, and their contacts

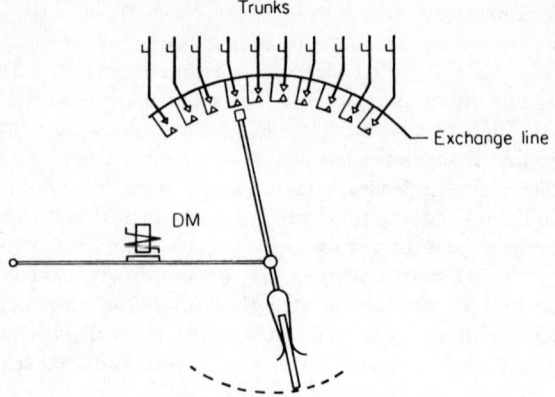

Figure 4.2   Keith line switch

multipled vertically to cord circuit trunks and horizontally each to an exchange line. Each switch had an electro-magnet DM to move a plunger radially outwards to close a set of contacts selected by the direction at which the plunger was pointed when the magnet was operated. An electro-magnetic solenoid controlled mechanism comprising a vertical blade common to all the pre-selectors in a column, pointed the plungers of all the unoperated pre-selectors in the direction of bank contacts multipled to a free trunk, which was the next to be used and therefore pre-selected. When a line originated a new call, its plunger operated to connect the line to the pre-selected trunk, the common mechanism then moving all the free plungers to another trunk outlet. The plungers were coupled to the vertical selector blade by hair-pin springs which permitted a plunger once operated not thereafter to be controlled by the selector until it was released. If there were no free outlets, the common mechanism hunted back and forth continuously until an outlet became free, but without damage because no rubbing contacts were involved. The cross-bar idea, described in chapter 5, of sets of relay contacts selected and operated by co-ordinate data and mechanisms none of which is individual to a set of contacts as with relays, is clearly seen in the Keith line switch.

In Figure 4.1, the alternative shown to pre-selection is exchange lines connected to the fixed bank contacts of exchange switches called line finders and provided one per cord trunk. When a line originates a new call, the wipers of a line finder of a free trunk move until they encounter the bank contacts of a calling line, when they stop. The switch which moves is instructed by an equipment commonly called an allotter for this particular application, but its function occurs in many places in exchange control systems and is more generally designated a one-only selector, or o.o.s., equipment. It is shown in Figure 4.1 as one exchange-type switch with two contacts per set of fixed contacts and the driving magnet DM connected to one wiper. If the contact to which the wiper is connected is at earth potential the wipers move to the next contact and so on, the switch resting only when a contact which is not earthed is encountered. The contacts are each connected to a processor prA of a line finder and earthed by the processor unless the line finder is free to accept a call. Thus the allotter switch continuously indicates one, and only one, line finder to be used for the next call, provided that at least one line finder is free. If none is free, the allotter switch hunts continuously over all its contacts, but there being only one switch, it can be specially designed and protected against excessive wear. The second wiper is earthed by a new call occurring on any exchange line and operating its L relay. The earth potential continues through the allotter to the DM coil of the line finder connected to the bank contacts on which the wipers of the allotter switch are standing. The DM coil is energized and its exchange switch wipers move until they rest on a position at which the arc 4 contact is at earth potential from an L relay contact of a calling line. Thy processor detects the condition, interrupts the DM magnet circuit to stop the switch before it can make another step, and operates the H relay. The LA, B and K relays then operating as for pre-selection, the line finding operation is complete. If more than one line is calling simultaneously, the calling lines are connected one after the other without difficulty. With two-motion switches used for line-finding, an operated L relay has to indicate both the bank level and the contact in the level to which the line concerned is connected. The allotted switch then hunts from level to level until a level with a calling line is connected, when the hunting stops and hunting within the level begins.

The choice between pre-selection and line finding is mostly a question of economics. The first needs one exchange switch per exchange line and the second one switch per cord circuit. The quantity of switches needed as line finders is thus much less than the quantity as pre-selectors, but whereas line finders must be full-sized exchange switches to achieve satisfactory traffic loading, pre-selector switches may have as few as ten contacts in the banks. The break-even point between the two methods depends on the traffic density. Other considerations are the effects of faults. In both cases a permanently looped line will seize and hold a cord circuit indefinitely unless some measure is provided to prevent it. One or two cord circuits so held may not be serious but cable breakdowns can produce many short-circuited exchange lines and seriously affect the grade of service of the other lines. In such cases the maintenance staff have to disconnect the affected lines or take some such action. With modern systems requiring less and less maintenance, the forces available to cope with emergencies by manual methods become inadequate and self-protection of the equipment more essential. Other faults can also produce undesirable effects. A short-circuited L relay contact having the effect of starting a pre-selector or line finder hunting but not completing its operation normally, can cause trouble until the fault is cleared which in unattended exchanges may be days. Automatic means of parking faulty lines are sometimes provided and are illustrated in Figure 4.1 by the additional relay PK in the line circuits served by line finders. A continuous line loop not followed within a reasonable time by dialling pulses, or the relay LA not operating when the cord circuit is seized, are detected by the prA processor which becomes involved. The processor applies an earth potential to wire 6 to operate the PK relay of the connected line, thereby inhibiting calling by or calls to the line until the loop is broken or the fault is cleared. The relay PK is thus a data store.

The trunk between the answering terminal and the calling or final selector terminal of a cord includes not only the message transmission m but also signal s and data d transmission circuits, namely the circuit which transmits the calling line loop signals relayed by an LA relay contact to the prB processor, a circuit which transmits the called line loop signals via the LB relay to the prA processor, and a circuit between the processors prA and prB for the transmission of processing data. The final selector exchange switch controlled by a processor prB is a Strowger selector with one hundred sets of fixed contacts in ten levels of ten sets, each set being allocated to an exchange line. The dial pulses repeated by the LA relay of the pre-selector or line finder are directed by the prB processor first to the vertical stepping ratchet motor DM of the final selector and then for the second dialled digit to the rotary motor DM. The wipers are then positioned on a line with the directory number which has been dialled, the equipment and directory numbers being related as for the multiple jacks of manual exchanges. The processor tests the condition of the arc 3 contact: if it is at earth potential, the line is already engaged and the processor operates the BT relay to apply busy tone to the message circuit for the benefit of the caller, and waits for him to clear. Otherwise, the line being free, the processor operates relay H to switch the transmission circuits through and pulse operates the relay RG to ring the called line as in Figure 3.6 and to send ring tone to the caller. The line relay LB is used as a transformer for the transmission of tones to the caller. The called line answering, relay F operates to the loop, relay RG releases to stop the ringing and the ring tone and allow conversation to proceed, the relay LB then being controlled by the loop and communicating the answering to the calling processor as previously described, to control the start of metering.

Both the calling and the called processors of a cord circuit are appraised of the conditions of the loops of both parties to a call using the cord circuit and must decide between them when both switches are to be released to disconnect the call and make the cord circuit available to another call. The logic of release cannot exactly reproduce that of manual service described in chapter 2, in particular of an operator who, noticing that one party is still on the line some time after the other has cleared, asks that party if something is wrong and perhaps re-rings the other party. The termination of connections may be subject to calling party, called party, first party or last party release, occurring respectively when the calling, called or either party clears or when both calling and called parties have cleared. There are no other possibilities and all have been used in automatic systems. Called party and first party release being prone to accidental clearing and last party release to accidental and some intentional holding of one party by the other, calling party release is most usual. The prA processor determines the moment of release which it communicates over the data d path to the prB processor. The calling party originates the call and usually pays for it and is to that extent in control, and the possibility that called parties may be held unable to make another call is less serious than the holding of calling parties on calls for which they are being charged.

Commonly a connection held after the called party has cleared, is released after a time interval of the order of minutes, an operation called forced release. The uncleared line then receives dial tone as a new call, or is given n.u. tone or is 'parked' out of service until cleared. Forced release was a manual maintenance operation in early systems and only in recent systems is it automatic.

The operations described reproduce the manual basic operations of connecting to a called line, saying 'number please' and connecting to a called line the directory and equipment numbers of which are the same, followed by ringing, charging and supervising the call up to its completion and release. The continuous scanning and time division time sharing of the operator for the detection of new calls are avoided by arranging that no switch moves, that is scans, unless a calling line is known to exist: then only one switch hunts with the certainty in the case of a line finder and near certainty for a pre-selector, of completing the operation in no more than one scan. Clearly an element of time division is involved but limited so that the wear and deterioration of the switches due to scanning is minimized and spread over many switches. An analogous situation in manual system operation is night service operation in which a new call rings a bell to summon an operator to the switchboard to find the calling line in one scan predetermined to be successful. Scanning and time division time sharing of the control after a call is connected is avoided by individual processors per call equivalent to one operator per call.

4.2.2 Data and Data Storage

What the operations described so far do not reproduce are the variations in the processing which a manual operator makes as the result of data which she reads with very little effort off the switchboard in the course of operating, namely the coloured lines which indicate groups of jacks connected to lines all with the same directory number, the group labels and coloured pegs and marks which indicate the services to be given, lines out of order and so forth. These are class of service data and the same principles of associating originating and terminating call data with the answering and multiple jacks respectively, and reading the data as connection is made to the lines, was carried into the first automatic exchange systems. Data are stored on bank contacts and on wipers individual to exchange lines, as illustrated in Figure 4.1 by the arcs 3 to 6 of the switches used as selectors and line finders and by the wipers of the arcs 3, 5 and 6 of the pre-selectors. As a line is connected its data are read through the connecting switch by the switch processor and reciprocally the processor may communicate data through the switch to be stored for the line connected, examples of the second being charge units to be stored in message registers and the operation of PK relays to record lines which are parked. The principle is simple and effective but limited by the quantity of data storable at an acceptable cost being small and the data not being available and processable until a connection has been made, and other means have to be used in addition to satisfy all the requirements.

The exchange line data listed in Table 4.1 are, excluding those in parentheses, typical for early automatic systems but less than are necessary for modern exchange systems. The data have to be stored in the exchange equipment using means suited to the system, at locations identified by addresses which have to be known to retrieve the data when they are required. All the data are line data but period and temporary data are specifically states of line data. Data which are relevant to originating calls are shown with a calling equipment number e.n.c. as address and those relevant to terminating calls with a directory number d.n. as address, and some relevant to both originating and terminating calls are shown against both kinds of address. Step-by-step systems use, as in manual systems, separate stores for originating and terminating call data, and use m.d.f. and i.d.f. jumpering, shown in Figure 4.1 and in Figure 4.5 with a different symbolical representation of the exchange switches, to associate exchange lines with directory numbers d.n. and calling equipment numbers e.n.c. independently of one another.

Examining the table in conjunction with Figure 4.1, a directory number being available as the address of the lines being called via a final selector, it is required to know if at least one exchange line with that number exists and if so, the equipment numbers e.n.t. for terminating service of all lines with that directory number: if no such line exists, n.u. tone has to be connected. The wipers of the selector having been positioned on bank contacts corresponding to a directory number not yet taken into use, n.u. tone is encountered, the n.u. tone supply being connected to the arc 1 and 2 contacts of all such unused contacts and heard by the calling subscriber unless the supply circuit is already engaged when the subscriber receives busy in the normal way. A difficulty is immediately apparent and one which is a penalty of using the same two wires for both message and signal transmission. The calling subscriber can hear the n.u. tone only if the F relay is operated to trip, that is to disconnect, the ringing and ring tone by the release of relay RG, and those operations are the same as in a successful call and they cause the call to be charged. This problem also occurs on other kinds of calls which are required to be answered so that message transmission may take place but without causing the calls to be charged. The solution adopted in the case of ring trip without metering includes the artifice shown in Figures 3.6 and 4.1, of making the exchange battery connections to the line during ringing the inverse of those through the LB relay when the ringing has been tripped. A normal loop trips the ringing and operates the LB relay to cause metering. An n.u. tone processor connected in the absence of a line, does not provide loop current but exchange battery potential on the A wire and a disconnection of the B wire, conditions which will be seen to operate the F relay but not the LB relay. In this way the n.u. tone circuit processor is able to trip the ringing and transmit n.u. tone to the caller without charge to him. This kind of solution is called opportunist because the designer, not having the means of solution by processing data in a straight-forward logical manner, has to look for some chance condition which allows him to solve the problem by circuitry. Opportunism characterized so much of early exchange design as to cause the switching art to be described, by some not directly concerned with it or not understanding the difficulties, as a form of low cunning.

Again referring to Table 4.1 and to the system of Figure 4.1, terminating equipment numbers e.n.t. are the same as directory numbers d.n. for ordinary class of service, that is for lines to stations with only one line, and also for the first lines of groups of lines, such groups being designated p.b.x. class of service because almost all groups are lines to p.m.b.x. or p.a.b.x. stations. Class of service data concerning groups of lines are stored on the final selector arc 4 bank contacts, by electrical states hand-wired to the contacts as required by the lines connected to arcs 1 and 2. Four states are conveniently available, namely disconnection and earth, negative and positive battery potentials. In one system, negative potential indicates the first line of a group and earth potential the last line, intermediate contacts being left disconnected. A prB processor having located the wipers of its switch on the contacts corresponding to the directory number which it receives as dial pulses, can determine from arc 4 if there is another line with that number and, if necessary, make another step to connect to it, then determine if that is the last line and if not make yet another step and so on to the end of the group. Directory numbers corresponding to the equipment numbers of the lines except the first are not usable as numbers for other stations but are used to connect to individual lines in the groups. Hunting over the lines of a group is arranged to occur only if the selector finds negative battery potential on its arc 4 contact at the end of dial pulsing, which it does if the directory number of the first line in the group is dialled. If the directory number which is dialled is the equipment number of any other line of the group, the selector finds earth or a disconnection on its arc 4 contact, and no hunting follows if that line is busy. By this means, individual lines

Table 4.1 Exchange line and state of line data

Address	Line data	State of line data

d.n.	e.n.t. or n.u.
	Class of service
	  ordinary
	  p.b.x.
	  party line X party
	           Y party
	(p.a.b.x. with d.d.i.)
	(coin-box)
		s.v.i.—intercept

d.n./e.n.c.		Busy Parked t.o.s.

e.n.c.	d.n.
	Class of service
	  ordinary and p.b.x.
	  coin-box
	  party line
	  disabled subscriber
	  l.d.b.
	  i.b.
	  (private meter)
		o.b.
		s.o.—observation
		s.s.o.—special
		      observation
		X party calling
		Y party calling
		
		Metered units of
		  charge

of groups may be called to render special service as well as p.b.x. group service. A common use of this facility is night service: normal service being available only during working hours, at other times one of the exchange lines not the first in the group is switched to a night watchman's telephone which can be called using as directory number the equipment number of the line. One-line groups, that is non-p.b.x. single lines, have their arc 4 contacts left disconnected to indicate that no hunting is required. The wiring to the arc 4 contacts is analogous to the coloured lines drawn under the multiple jacks of manual boards to indicate groups of lines all with the same directory number, that of the first line in the group. Obviously there is difficulty with groups of lines exceeding in number the contacts in one level of contacts of a two-motion switch. Solutions exist in practice but are not so important as to merit description here.

Contacts of other arcs shown in Figure 4.1 may be used to store other data, party line class of service for example. A switch being positioned on the bank contacts of a line, its processor reads the data and selects its program of operations accordingly. In the case of party line data, the data would determine the way in which ringing current was applied to the line.

A processor prB having located the wipers of its switch on the contacts corresponding to the directory number which it receives, its further action depends on the state of line as well as on the line data. The possible states of line listed in Table 4.1 are busy and parked, as already defined, temporarily out of service t.o.s. and service interception, s.v.i. As shown in Figure 4.1, arc 3 is used for the busy state of line data and arc 6 for the park data. A line is temporarily out of service when it is parked, but it can also be t.o.s. for a reason which does not automatically park the line, for example the line being disconnected or excessively noisy. Also a line may be in working order but t.o.s. for non-payment of the account or some such reason. If the processor finds a line with the t.o.s. state of line or the park, and it is the one line of a station with only one line, it is required to send n.u. tone to the caller. If a group of lines is concerned, n.u. tone should not be sent unless no line of the group is obtainable. If some lines are available for service but all are busy, busy tone is the proper conclusion.

The s.v.i. state of line is used when the subscriber has some reason to complain about incoming calls and the administration has to take some action. The most common reasons are malicious calls and frequent wrong number incoming calls. Occasionally a malicious person makes calls frequently and intentionally to worry a particular subscriber, and some exchange faults can cause frequent wrong number calls to a particular line or lines, a wrong number call being one received by a line although the number dialled was not that of the line. The administration investigates such cases by having an operator intercept all calls incoming to the line, for verification and investigation if abnormal, only the valid ones being allowed by the operator through to the line itself. This is service interception, or s.v.i. state of line.

The description so far given includes all the means of data storage available to electro-mechanical systems, namely

(a) the structure stores semi-permanent data in the construction and disposition of its components. The equivalence of directory and equipment numbers of selectors is one example and others will emerge;
(b) n-state electrical conditions semi-permanently connected to wires on terminals in or available to processors store semi-permanent data such as some classes of service;
(c) n-state electro-mechanical devices store period and temporary data by their states of stability, meters storing numbers, exchange switches the directory numbers dialled and so forth;
(d) n-state electrical conditions connected temporarily by electro-mechanical switches to wires on terminals in or available to processors, store temporary data such as busy and park.

Using the storage means enumerated, all the line and state of line data listed in Table 4.1 could be stored and the data processed, that for originated calls by prA processors, Figure 4.1, of one design containing all the relevant programs and for terminating calls by prB processors of one design, but even if it were practical to do so, it would be too costly. The practical and economic solution is to divide the data storage and processing first between the exchange equipment, the maintenance staff and the subscribers, and then to divide the exchange processing programs among groups of processors made relatively simple because of the division, as will be better appreciated from the following examples and descriptions.

Considering terminating calls for which a directory number address d.n. applies, in any one exchange the s.v.i. state of line is required for very few lines at any one time, not more than two or three and often none at all. When required a specially designed interception circuit processor is interposed between the line concerned and its exchange equipment, manually by a maintenance man. The need for t.o.s. is more frequent but more easily satisfied, by manual disconnection of the line or, more usually, by a link inserted manually into a miniature jack provided for each calling line equipment as shown in Figure 4.1. The link applies earth potential to the P-wire to busy the line and to operate the K relay to disconnect the calling relay L. Automatic parking is rare in step-by-step systems, the effects of the conditions which would lead to automatic parking being allowed to continue unless the service to other lines is much affected when the maintenance men take some suitable action, often to make the line t.o.s. The actions of the maintenance men can be seen to be data storage by the methods enumerated, namely changing the structure, or wires on terminals or the n-state electrical conditions of wires, usually by connecting earth potential to them. The data being stored, the already existing or specially connected processors produce the services intended or suppress services normally rendered.

For none of the remaining processing for terminating calls can automatic processing be avoided. The structure provides the storage for e.n.t. and n.u. data against d.n. addresses, Table 4.1, and the corresponding processing including the selection of exchange lines and the connection of n.u. tone, is incidental to the step-by-step operation. The P-wire and associated processing provide for the free-busy states of line operations for both terminating and originating calls. For the remainder, the exchange lines are organized into divisions of lines all requiring the same terminating service, as indicated in Figure 4.3. The lines of a division are connected to the banks of final selectors capable of no more data storage and processing than is needed by the lines and is designated group class of service. In Figure 4.3 there is a division for ordinary lines selected by individual directory numbers and rung with regular ringing, a division for p.b.x. lines called by one directory number per group and rung by regular ringing, and a division for party lines called by directory numbers individual to the parties and to which X or Y party ringing has to be applied in dependence on the directory numbers used. The divisions correspond to classes of service and are the same for all exchanges. The quantities of exchange lines in the divisions vary widely among different exchanges but have to be served by selectors with capacity for one hundred lines imposed by the step-by-step selection and the construction of the switches. Hence the lines within the divisions are further divided into blocks of not more than one hundred lines each for connection to the multipled banks of blocks of final selectors. The way in which calls are directed to the blocks by the directory number digits except the last two, the last two being used within the blocks for final connection to wanted lines, is described in section 4.2.3 to follow and in which the symbol used in Figure 4.3 to represent all the selectors serving a one-hundred-line block is used to represent all the selectors serving one division of lines.

The selectors for each block of party lines are divided into two groups, the processors of one group applying X party and of the other group Y party ringing. The last two digits of the directory numbers allocated to the two parties sharing one line are the same, only the initial digits being different, which means that the X party and the Y party final selector banks can be in one multiple as indicated in Figure 4.3. Connections to any line can be made by any selector but depending on the directory number dialled, the connection is made by a selector in one or other of the groups and the ringing to call the wanted party is thus selected and applied. An additional directory number is also a means of adding some special service to other more regular services. For example, the service given by a p.b.x. line in night service may be a machine which asks a caller to dictate a message which will be recorded. A person may cause the machine to play back the messages over a telephone connection by dialling a number known only to those authorized to receive the messages, the line being reached via a selector in a one-hundred-line block and through which the recording machine can be controlled. The additional numbers advantageously do not have any systematic relationship to the other numbers by which the lines can be called, and the lines of a one-hundred-line block of additional numbers are parallel connected to lines spread at random over the normal class of service divisions except that commonly additional numbers are not allowed for some lines such as party lines.

One consequence of the methods and limitations described is that only the selectors providing p.b.x. class of service need have four arcs of bank contacts. Two arcs for the message circuit and one for the P-wire free-busy state of line are otherwise the essential minimum and all that is usually provided for public exchanges. More can be found in some private and special purpose exchanges which have different needs and greater economic latitude. It is also possible to use only two designs of final selectors, one for p.b.x. lines and the other for all other lines. The two groups of selectors serving a one-hundred-line block of party lines have a common multiple: a reversal in the multiple wiring between the two groups so that the A and B line wires are connected to arcs 1 and 2 of X party but to arcs 2 and 1 respectively of Y party selectors, enables the ordinary kind of selector applying ringing as in Figure 4.1 to ring either of the parties on party lines. Thus the equipments are simplified and their variety reduced but the total quantity of equipment and its cost are not necessarily reduced. Lines in divisions with further division into one-hundred-line blocks inevitably involves wastage due to partially filled blocks, the wastage depending on the size of exchange and obviously being serious in exchanges of only a few hundred lines. The wastage is mitigated to some extent by filling up otherwise partially filled p.b.x. and party line blocks with ordinary lines connected as one-line p.b.x. groups and as party lines with X parties but without Y parties. On the other hand, the division has been reduced by limiting the classes of service offered, those shown in parentheses in Table 4.1 not having been included. P.B.X. lines with d.d.i. cannot be operated by processors for p.b.x. lines without d.d.i. and ultimately require special line signalling and processing selected by class of service, although an opportunist method described in section 4.2.3 is possible for step-by-step systems. Calls cannot be made to coin-box lines and their charges collected in the coin-boxes because such service is difficult to provide and expensive out of proportion to its use. If demanded, a complicated procedure involving an operator would have to be used. As the telephone system has developed so has the pressure to make all facilities automatic, and the solution of class of service problems by a combination of opportunism and arbitrary decision could not be practised in perpetuity as later developments will make clear.

The quantity of exchange lines which may be connected to an exchange is related not only to the quantity of available numbers but also to how they may be used. With step-by-step selection and an invariable relationship between directory and equipment numbers, p.b.x. groups of lines identified by the directory numbers of the first lines of the groups causes the directory numbers corresponding to the equipment numbers of the other lines to be unusable except for special services within the groups, and equipment numbers left spare for possible future growth of p.b.x. groups but not taken up cause further wastage of directory numbers. Each of the parties on party lines having an individual directory number by which it can be called, results in more directory numbers being used than lines connected and there are other losses due to service and test numbers and to unfilled one-hundred-line blocks. As a consequence, the usual four decimal digit exchange numbers available to an exchange of the type being described generally permit less than nine thousand subscriber lines being connected to the exchange.

If two sets of bank contacts and wipers are mounted on one mechanism, the wipers are operated mechanically in parallel and can be controlled electrically by one processor to produce the equivalent of two one-hundred-line final selectors or

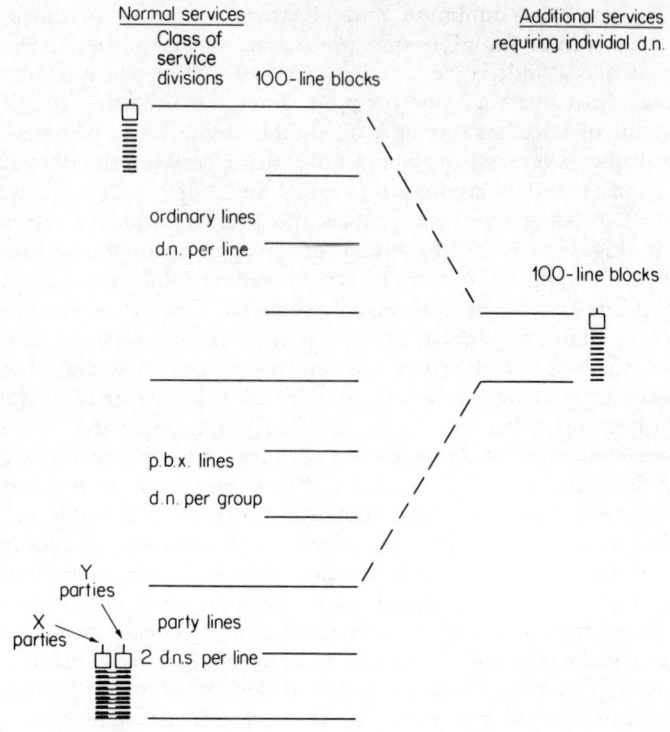

Figure 4.3 Exchange line terminating call divisions and blocks

of one two-hundred-line group selector with ten levels of twenty contacts in each level. In a group selector the direct control applies only to the selection of the level to be used, a line in a selected level being selected by hunting as yet to be described. The objective of increasing the size of exchange switches is to reduce the quantity of switches and therefore of processors required in the interests of economy.

The data storage and processing for terminating calls distinguished by directory number addresses may be used in principle, though with some differences in detail. for originating calls distinguished by the equipment numbers of the bank contacts of the line finders or the wipers of the pre-selectors to which the lines are connected. For both line finders and pre-selectors, arcs 1 and 2 are essential for message circuit switching and arc 3 for the free-busy data. Arcs 4 provide data incidental to processing, namely for the line finders lines calling but unconnected and for the pre-selectors the home positions of the wipers. Other arcs to store other data of the lines and states of lines of Table 4.1 may be provided and are shown in Figure 4.1 but not being essential, they are seldom found in public exchanges consistent with the practice for terminating calls. The data storage and processing must use other means of which separation of the lines into divisions as for terminating calls is one. As will be realized from Table 4.1, the originating call data are not systematically related to those for terminating calls, hence the divisions for the two types of call are separate and independent.

Referring to Table 4.1, the directory number d.n. is required from the e.n.c. address mostly for accounting purposes. For metered calls the charges are stored on message registers using the e.n.c. addresses but with labels giving the directory numbers which, read as the meters are read manually, provide the necessary e.n.c. to d.n. translation. Calls which are charged but not metered have to be handled by operators who have to ask the callers to tell them the directory numbers of the telephones from which they are speaking, for entry as data on the tickets for the calls. In effect the subscribers are assisting if not actually performing the e.n.c. to d.n. translation.

The classes of service appropriate to exchange lines have to be provided either by equipments individual to the lines which with few exceptions is too expensive, or by time shared common processors prA provided in variety sufficient to cover all the classes of service. When exchange lines call they have to be connected to a prA processor of the appropriate kind. In step-by-step systems this means collecting the exchange lines and the processors into divisions each satisfying one class of service, and if the connection between the lines and the processors is by line finders, further division into blocks as for selectors is necessary. Division for originated calls encounters more difficulty than for terminating calls for two reasons. First, more classes of service and combinations of classes are required and second, the equivalent of more than one directory number is not possible, every line having one and only one e.n.c. equipment number, that of its connection to a pre-selector or to line finder bank contacts. A division to serve each combination of class of service is thus required and the services and their possible combinations being numerous, the divisions tend to be small and uneconomic. Referring to Table 4.1, any of the ordinary or p.b.x., coin-box or disabled subscriber lines and each party of two-party lines can theoretically have the unlimited, l.d.b. or i.b. class of service as well, which is eighteen combinations, nine for the party lines and nine for the others, without including possible other services such as press-button dialling as in section 3.4.1 and which developed later. This situation required the combinations of services to be reduced by arbitrary decision that not all lines, party lines in particular, could have any one of the unlimited, l.d.b. and i.b. classes of service in addition to a main class. The private meter class of service when it came to be required was and still is applied individually to exchange lines, otherwise it would multiply the combinations of classes of service by a factor of nearly two, being less than two because not all classes, coin-box for example, can make effective use of private meters.

Referring again to Table 4.1 and to the states of lines, X or Y party of a party line calling has to be identified when the call button is pressed and dial tone not given until then. Figure 4.4 shows one way in which the requirements can be satisfied. The party line division of processors prA is itself divided into two, one half responding to X party and the other to Y party calls. The line equipment for a party line has two line relays, LX and LY, both connected to the exchange battery so that neither operates to the initial calling loop but one is operated when a call button is pressed. The one relay which operates causes a pre-selector or line finder to connect to a processor prA in the required sub-division, then the LA and B relays to operate and thus to operate the K relay in the line circuit. The LA relay holds first to the earth on one wire of the exchange line and subsequently to the loop. Subsequent meter pulses operate the meter MRX or MRY according to the calling party. Thus the prA processors are not different from those of ordinary line service: the class of service discrimination is obtained by adapting the structure to the requirements.

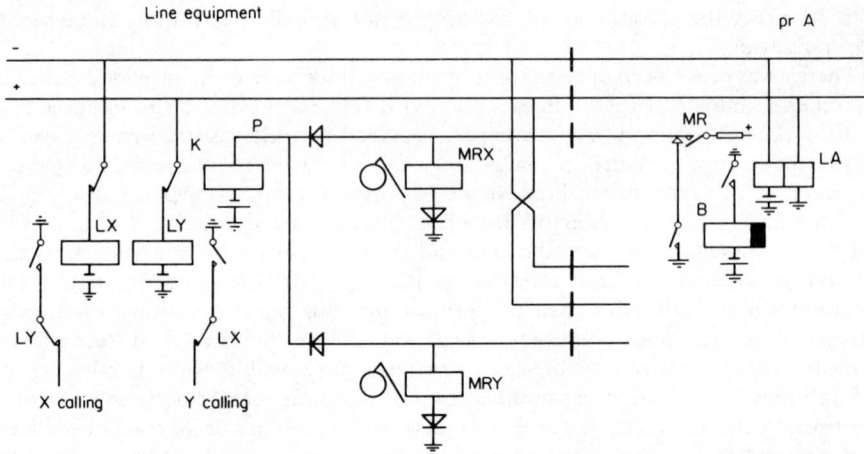

Figure 4.4   Line equipment for two-party line

The o.b. outgoing calls barred state of line occurs as period data, preventing calls from being originated but not from being received, commonly preliminary to t.o.s. for non-payment of the account. Observation state of line means that all calls made over a line have to be observed from their origin to completion by an operator trained to do that kind of work. Ordinary service observation s.o. is a continuous activity for the collection of service statistics, the observations being made on exchange lines selected at random. Special service observation s.s.o. is made on lines selected for some reason such as complaints by the subscribers of bad service or over-charging. The data storage and processing for meter state of line has already been described.

Examination of Figure 4.1 shows no provision for any originated line or state of line data storage other than meter record and what might be accommodated on additional data arcs and wipers if provided, which is rare. Group class of service by groups of processors prA each suited to one combination of classes of service is supplemented by maintenance staff operations as for terminating calls. The o.b.

state of line is very simply applied manually by the insertion of a wedge under the armature of the L relay to prevent its operation and the observation states of line require special processors to be connected into the lines to be observed, in the same way that interception processors are connected.

Step-by-step operated systems were the first automatic systems. Initially the only services required and provided were basic dialling and connection and the more complicated facilities grew piece-meal as the telephone services expanded. New developments were satisfied rationally if possible but by opportunism if not, using the structure in effect to store some of the line data and denying the public some services which it probably did not know about and therefore did not press for. With the advantage of hindsight, the limitations and disadvantages of such methods are clear but they were pursued and were at least acceptable for a surprisingly long time before a fundamental change of technique was patently needed. Cross-bar systems made some advance as will be seen, but major change had to await the advent of electronic devices in large scale production. The problems nevertheless remain much the same and group class of service and others of the original solutions are useful to and can be traced in modern systems.

4.2.3 Selection and Trunking

The designation 'step-by-step' refers to every dial pulse causing the wipers of a selector to take one step, being one level or contact in a level, towards the final destination of a call being set up. The example of Figure 4.1 illustrates some important principles of exchange switch and call control but is limited to one block of one hundred lines in one division. Generally more than one division and more than one block per division are needed together with access to service circuits and to junctions as indicated in the generalized manual exchange of Figure 2.12.

With the system of Figure 4.1 limited to one hundred lines, the quantity of final selectors is the same as that of line finders or trunks from pre-selectors and each prA processor could be amalgamated with the prB processor to which it would be directly connected, to form one processor: the quantity of such processors would be dependent on the total traffic originating on all the lines. Generally, however, many one-hundred-line blocks within divisions according to the terminating classes of service are needed, the quantities of final selectors per block being sufficient to carry the terminating traffics of the lines in the blocks at some specified grade of service. In Figure 4.5 two blocks of final selectors are represented, each block by the symbol of Figure 4.3, namely a square box for the processors and ten parallel lines for the ten levels of bank contacts. Line finders being used for the originating traffic, enough blocks are provided to accommodate the exchange lines in originating class of service divisions and the quantities of line finders per block are adjusted to the originated traffics of the blocks. The exchange lines are distributed over the final selector banks by jumper wiring between the two sides of the m.d.f. and independently over the line finder banks via the i.d.f. Using pre-selectors in divisions instead of line finders, the quantities of trunks to selectors in the divisions are determined by the traffics originated in the divisions. The quantities can be

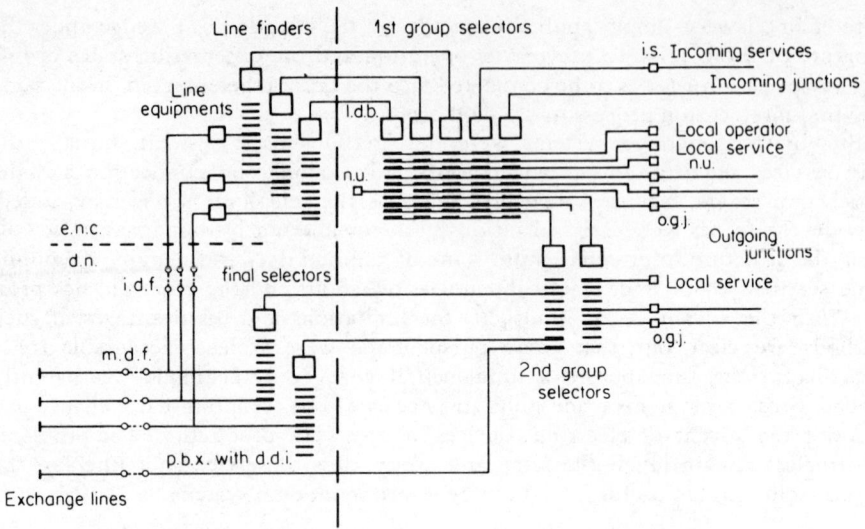

Figure 4.5 Step-by-step exchange trunking

readily matched to the quantities of pre-selectors, one per exchange line, by a system of connection which is called grading, an example of which is shown in Figure 4.6. Assuming pre-selectors with a home and ten working contacts in the banks, and twenty-two trunks to be connected to the banks, the pre-selectors are divided into four groups each producing about the same quantity of traffic, and the banks of each group are connected in multiple. In Figure 4.6 each row of ten short lines represents the ten sets of multipled contacts of a group of pre-selectors. The trunks are connected to the multiples, some trunks to a set of multipled contacts in one pre-selector group, some to multipled contacts in two groups or more, to form a pattern progressing from trunks individual to groups to trunks common to all the groups. A simple rule for the design of gradings is that the product of the number of groups and the availability — the availability being the number of trunks to which each pre-selector has access, which is the number of contacts in the bank — should be about twice the number of trunks to be connected, and the connection pattern should give the smoothest progression from individuals to commons that uses up the available trunks. Full availability is the term used when every switch has access to every trunk, and limited availability when switches have access to only a limited number of the total trunks. Grading is one form of limited availability. Limited availability means that sometimes switches will fail to find a free trunk even though at least one exists, which in turn means that to realize a given grade of service, the quantity of trunks which would be sufficient for full availability, must be increased if the availability is limited.

In Figure 4.1, the line finders or pre-selectors in one division and one block are

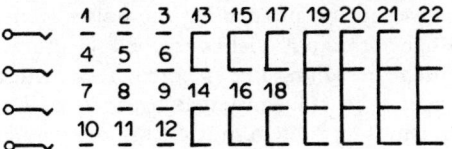

Figure 4.6   Four-group grading

connected by loop trunks to the final selectors in one division and one block. With many divisions and blocks variable connection is required between the originating and terminating switches. Effectively the solution is to cut the trunks at one of two places indicated in the figure by broken lines, one line horizontal and the other vertical, and to introduce group selectors between the two sets of ends thus produced, as indicated in the trunking diagram of Figure 4.5 in which the vertical broken line corresponds to whichever broken line of Figure 4.1 is the one which is cut. If it is the vertical line, the part of the prA processor which falls on the selector side of the cut is transferred to the prB processor. The relative advantages of the two arrangements have been argued in the past but are not now of interest. Both can be found in systems in existence but the differences being in details and very little in final result, it is not useful to pursue both and only that corresponding to the vertical line will be considered.

In Figure 4.5 the exchange lines are shown connected to line finders in three divisions or blocks and the final selectors in two blocks but many more of both are to be imagined. Each line finder switch and each incoming junction is terminated on a first group selector. Incoming services not able to use ordinary incoming junctions use dedicated junctions terminated on first selectors and including processors i.s. suited to the service. The wipers of a group selector are stepped under the control of incoming dial pulses of one digit to a level of contacts corresponding to the digit dialled, and then under the control of its own processor the wipers are hunted over the contacts of the level until a sct which is not already engaged is encountered. The wipers coming to rest on the free contacts, the trunk terminating on the wipers is connected to the trunk connected to the contacts. If the hunting tests all the contacts without finding a free trunk, the selector processor sends busy tone back to the caller. If the level is not in use and leads to nowhere, the dialling must be in error and circuits sending n.u. tone back to the caller will be found on the contacts. The bank contacts of the working levels of the selectors are multipled and graded to suit the numbers of trunks to be connected. The trunks may lead to second group or to final selectors, second group selectors being shown in Figure 4.5, or to outgoing junctions each with a circuit processor o.g.j., or to a service. The second group selectors may be followed by third group selectors giving access to group or to final selectors, services and junctions, and so on, each exchange being planned and constructed to suit the exchange lines, junctions and services which it has to interconnect and the numbers which will be dialled to make connections through the exchange. Each digit which is dialled finds a selector ready to receive and be stepped by the pulses transmitted directly from the dial contacts, and hence the term dial controlled selection. The selector may be in the exchange to which the calling line is connected, or in another exchange to which the dial pulses have to be transmitted as signals over the intervening junction or junctions in tandem, and is a problem in itself where many junctions or very long junctions are involved. Four kinds of connections are recognized by the designations 'local' for exchange line to exchange line connections on the same exchange, 'outgoing junction' for exchange line to outgoing junction and 'incoming junction' for incoming junction to exchange line connections, and 'junction tandem' for incoming junction to outgoing junction connections.

The line finders are in divisions to give different classes of service by different processors as described in section 4.2.2, and also by differences in the trunking as illustrated in Figure 4.5 by the l.d.b. class of service. The l.d.b. class of service first group selectors are barred access to the level 5 group of junctions over which all long distance traffic is routed, receiving n.u. tone instead. This is an example of a basically control class of service being realized as a structure class of service.

A method commonly used to provide direct dialling-in facilities to p.a.b.x. stations is shown in Figure 4.5 by exchange lines connected to a level of a second group selector, the last two digits of the numbers dialled over the level operating final selectors at the p.b.x. to connect to extensions or the attendant's console. The exchange lines are usually operated in separated groups for traffic to the public exchange and from the public exchange with d.d.i. Because the p.b.x. with d.d.i. class of service was introduced long after step-by-step exchanges started to be installed, it had to comply with already existing conditions of directory numbering and quantity of digits. The method described absorbs public exchange directory numbers for the p.b.x. extensions, thus reducing the exchange line capacities of the public exchanges and, with other drawbacks as well, is not a satisfactory general solution to the problem.

If in Figure 4.1 the loop trunk is cut along the vertical broken line, which is very commonly the case, the group selectors have to provide a three-wire connection for the message and signal transmission — and + wires and the P-wire data. Figure 4.7 represents a group selector with −, + and P wires incoming to the wipers and from banks with three contacts per set. A loop incoming on the wiper side causes the A relay to operate, which communicated to the processor prGS individual to the selector, results in the B relay operating and busying the P-wire against seizure by other connections. The dial break pulses which follow step the level selecting ratchet motor DM and the processor hunts the wipers over the selected level, testing the P contacts for free or engaged. If no free outlet exists, the processor sends busy tone back to the caller by means not shown in the diagram. If and when a free outlet is found, the processor stops hunting the wipers and operates relay H which holds to the P-wire earth until the end of the connection. The B relay releases after a delay to give the B relay in the next selector time to operate and earth the P-wire. The selector thus provides a straight-through connection of three wires, from the trunk connected to its wipers and thence to the trunk connected to the bank contacts on which the wipers stand, and the processor takes no further part in the

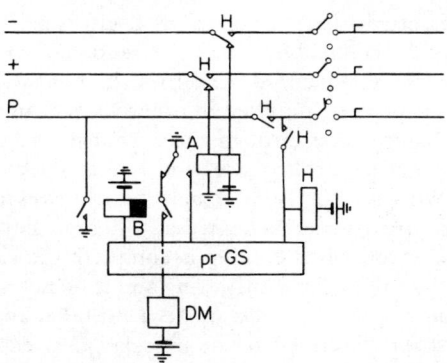

Figure 4.7   Group selector

connection processing. Any number of group selectors in series may be included in the connection between the terminal switches and processors, each group selector controlled by one dialled digit to select a level of bank contacts and by its own processor to hunt the wipers to a free trunk on the level. At every stage of the establishment of the connection, the switches already connected are held connected by earth potential applied to the P-wire by the last switch in the chain, which is finally the final selector. If junctions are included in the connection, then either the junction must have three wires, one being a P-wire for holding the connection, or some means of operating with only two wires must be used. Junctions with three wires are sometimes but rarely used in practice. For reasons of cost, two wires is the almost universally used arrangement which, with the problems it entails, is described in section 4.2.4.

In some systems the connections are held by earth potential applied not by the last processor but by the first in the chain, namely the prA processor. In general these are systems with the cut between the prA and prB processors made along the horizontal broken line of Figure 4.1 and which has already been said to produce differences of detail but not of principle or result.

Figure 4.5 is a trunking diagram representing the groups of selectors which are needed and their interconnections to satisfy the connections to be made through an exchange, but it does not convey the quantities of equipments required nor their interconnections in detail. The quantities of selectors and of trunks connected through graded multiples to their banks are determined by the traffic to be carried, and the quantities and the connections have to be designed individually for each exchange. This work which is part of equipment engineering involves a considerable amount of planning and design effort per exchange.

A feature of step-by-step dial control operation is that any number of calls may be in process of being connected simultaneously without mutual interference. The only danger is that of double connection due to two calls simultaneously finding the same outlet free and both seizing it. The probability of simultaneously testing within the few milliseconds that it takes is itself small, and by circuit and equipment design double connections may be rendered impossible. The mutual independence of different connections simplifies the control in some ways but as will be seen later, the control is simplified in other ways if only one connection can be in the process of being made through the exchange at any one time. Another feature of step-by-step dial controlled selection is that selection is complete when dialling is complete. With no operator to help him, a subscriber needs a positive result of every series of digits which he may dial, whether they are valid numbers or not. With direct dial control, when dialling is complete the caller should hear ring, busy or n.u. tone or the called party answering and if he hears nothing he knows at once that either he or the machine has made a mistake, and he must dial again. Later developments to be described result in delay between the completions of dialling and connection and some difficulty in knowing how long to wait before assuming that something has gone wrong. On the other hand, the direct dial operation means that the switches must be capable of working at the speed of the dial pulses, commonly ten pulses per second, and to have completed all group selector rotary hunting and selection processing in the time, called the inter-digital pause and sometimes the inter-train pause, between the end of one train of pulses and the beginning of the next. As a result there is a limit to the mass and robustness of the construction of the switches. In addition ratchet motors are generally more difficult to make and make reliable than other kinds of motors. Nevertheless satisfactory designs giving reasonably reliable service are possible, with the merit of low cost per line for exchanges of all sizes from very small to very large, so long as the services to be rendered are relatively simple and the range of territory over which direct dialling is practised is limited. Departure from the principle of step-by-step dial operation was forced as networks grew in size and complexity, not only because the systems become difficult for the subscribers to operate but also because with dial controlled selection the junction plant has to be designed to suit the numbering and becomes unnecessarily expensive.

Once the stage of short distance dialling is passed, the subscribers have to compose the numbers which they need to dial to reach specific destinations, using published directory lists of codes defining exchanges and which have to be prefixed to the exchange numbers of the stations on the exchanges. This becomes such a problem and difficulty when subscriber dialled calls may be made nationally, and still more so for world-wide dialling, that help must be given by the exchange processors to the subscribers, to simplify the problem for them, as described in section 4.5. Then there is the problem of charging which becomes progressively more difficult as the range of subscriber dialled connections extends and finally becomes world-wide. For message register stored charges the quantity of units to be recorded for each call is a function of the distance and the duration of the call after it is answered. The distance has to be derived from the digits dialled and the duration measured by a clock device and the total quantity of equipment involved is very considerable if concentrated in one processor capable of dealing with all possible calls which may be dialled and which in Figure 4.1 would be the processor prA. The cost and complexity are much reduced by locating the metering processing not where the traffic enters but where it leaves the exchange, that is in the final selector, service and outgoing junction processors. Each of these processors generating meter pulses as if the call in which it becomes included originated in the same exchange, the metering is effective if the call is in fact originated by a local exchange line. If not, the metering at that exchange is ineffective but that is of no consequence. The important point is that for most calls the metering processing is very simple, because the majority of calls are local to the same exchange or to the local area over junctions which carry traffic of only one charge rate. Only on a relatively small quantity of junctions carrying the longer distance traffic is the metering processing complicated and expensive: for these processors o.g.j. in Figure 4.5 to detect, store and process the exchange code digits dialled over the junctions to determine the rate of charge appropriate to each call could be impractical without some simplification if the range of the calls connected included the whole of a national network and possibly some international calls as well. The first stage of simplification is exchange codes systematically related to the geographical locations of the exchanges, and the second tariff structures which generously relate charges to distances, which together make possible the determination of charge rates from just a few of the initial digits which are dialled. The timing of effective calls to determine charges from the charge rates is relatively simple. Space does not permit of systematic numbering systems being described, nor of tariff structure problems in detail.

Charging by bulk metering is acceptable by the subscribers of a network and the administration, up to some limit above which it is desirable to bring into the connections operators or ticketing machines to record the details of the calls made. In this there are some problems, notably those of identifying and recording the exchange lines from which calls are being made, but on the other hand the treatment of the most difficult calls in this way has the effect of simplifying the metering of the less difficult calls. The acceptable limit to bulk metering varies considerably between countries. Generally the limit is not reached with dial controlled selection exchanges, because of technical and other difficulties, but with later systems.

4.2.4 Message, Signal and Data Transmission

A generalization of step-by-step dial control systems, corresponding to that of Figure 2.12 for manual systems, is given in Figure 4.8. The common processors and interface become zero because all the common processing has been transferred to circuit processors. Despite the very different exchange systems which they represent, Figures 2.12 and 4.8 are not different in basic principles and corresponding features can be found in each.

Referring to Figure 4.8, in conformity with Figure 4.1 and its description, the exchange lines providing message and signal transmission terminate in the exchange on line equipments with message, signal and data transmission facilities through exchange switches to trunks with processors prA for originating and prB for

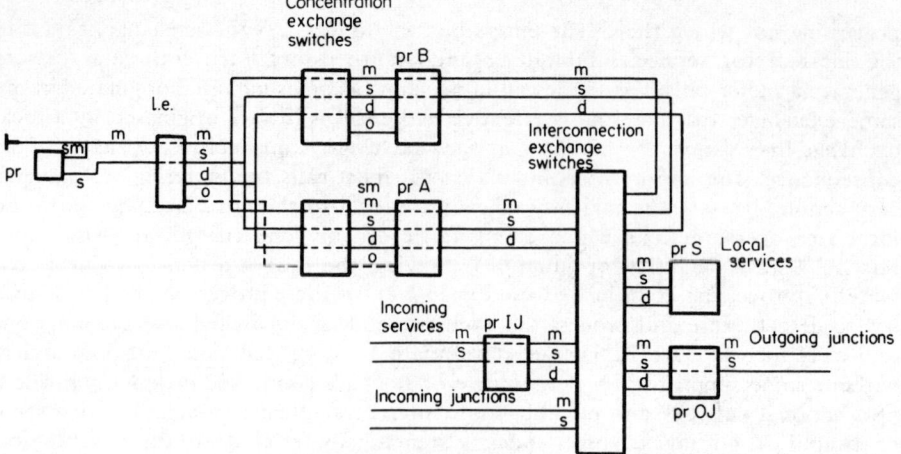

Figure 4.8 Generalized step-by-step exchange

terminating traffic. The exchange switches between the exchange lines and the processors prB are controlled by the processors, and those between the exchange lines and the processors prA are either line finders controlled by those processors or pre-selectors controlled by the line equipment processors, dashed lines in the figure indicating the alternatives. Incoming ordinary and service junctions have associated m and s transmission circuits and are terminated with processors prIJ when needed. Outgoing junctions are similar with processors prOJ, and local services are terminated with processors prS. The exchange switches provide associated m, s and d transmission paths for the circuits and the trunks which they switch, the interconnecting exchange switches being group selectors each controlled by its own processor prGS. There is message m and signal s transmission between exchanges, by no data d transmission. The processors of one kind prA, prB and so forth, are not all the same but are in divisions each providing a class of service to selected subscribers or, in the case of the prS processors, a particular service to most if not all of the subscribers.

Economy requires that bothway transmission of messages, signals and data should be achieved with the minimum of wire conductors through the switches within the exchanges and through the junctions between exchanges. In practice, two wires for junctions and three wires within the exchanges are used as in Figure 4.7 with some problems which are examined with the help of Figure 4.9. That figure shows exchange lines, incoming junctions and outgoing junctions which can be interconnected by exchange switches and trunks. Calls originating on exchange lines are connected via a trunk with a prA circuit processor to a trunk with a prB circuit processor and thence to another exchange line for a local call, or to an outgoing junction with a prOJ circuit processor for a call to another exchange. A call from another exchange incoming on an incoming junction is connected via a trunk with a prB circuit processor to an exchange line for a terminating call or to an outgoing junction for a transit call. In addition to the message, signal and data transmission requirements, the exchange lines have to be supplied with line current to operate the station equipments.

Two wires through the exchange suffice for message transmission so long as the transmission is two-wire. This condition is true of all exchange lines and short junctions. It is not always true of longer junctions which may use four-wire transmission over metallic conductors or over f.d.m. or p.c.m. multiplex channels. When step-by-step exchanges first came to be used, however, all transmission was two-wire over two metallic conductors and provision was made for that kind of transmission alone and it still applies to local area switching, that is to say that exchanges at the lowest hierarchic levels provide and accept only the conditions represented in Figure 4.9. Junctions using transmission of any other kind have to be equipped at both ends with transducing equipments to match the junctions transmissions to the two-wire transmission required by the exchanges.

Referring to Figure 4.1 and to loop trunks cut along the vertical broken line, the exchange line relays LA and LB become located in the prB processor, as shown in Figure 4.9, which is satisfactory for local calls for the supply of d.c. line currents to the lines, and for loop signalling to the processor and thus, among other operations, for holding connections by earth potential over the P-wire. Incoming junction calls are also satisfactory in the same way: the loop signals control group selectors and a final selector with processor prB, the junction not requiring a processor prIJ. For exchange lines which are calling and are connected to outgoing junctions the processor prOJ has to provide the equivalent of the LA relay which is shown in the figure as relay A. The relay receives calling line loop signals and supplies d.c. current to the line, and the processor holds and releases the connection by applying and removing earth potential to and from the P-wire. Without the capacitors shown as blocking d.c. loop transmission through the prOJ processor, calling exchange line loop signals would operate the A relay of the processor in parallel with the junction terminated in the distant exchange in the A relays of group selectors and the LA relay of a final selector, and the requirements of calling line loop control of the connection would be satisfied if the equipments responded to the line currents available. The vagaries of junction transmission of loop signals and of numbers of junctions in tandem, each shunted by an A relay, makes such a simple arrangement impractical, for which reason the d.c. loop signals are blocked by capacitors in the prOJ processors and repeated by contacts of the A relays to loops individual to the junctions, as shown in Figure 4.9. A calling line is then able to establish a connection to another exchange line or to a service, in its own or any other exchange, without restriction of the number of group selectors or junctions in the path, and to hold and release the connection despite the lack of a third wire in the junctions. A processor in each exchange holds and releases the switches in its own exchange, namely a prB processor in the last exchange and a prOJ processor in any other exchange.

The called line loop signals have to be transmitted to the originating exchange in order to control the metering for the call. As described with reference to Figure 4.1, the LB relay operating in the prB final selector processor causes the M relay to be operated to add one unit to the meter reading of the calling line if on the same exchange. If not, the loop signal has to be transmitted over the junction concerned to the distant exchange, which is shown in Figure 4.9 as accomplished by contacts on the LB relay which reverse the polarity of the current supplied to the junction line by the LA relay. At the other end of the junction, in a prOJ processor, is a relay D polarized by rectifiers so that the relay operates when the junction current is reversed: contacts on the D relay reverse the polarity of the current supplied by the A relay to the message wires of the circuit connected to the incoming side of the prOJ processor, and if that circuit is an incoming junction, it will be terminated in the distant exchange by a prOJ processor with D relay which is operated and so on. In this way the called line loop signals are transmitted back to the originating exchange. In each exchange the operation of the D relay not only reverses the polarity of the incoming loop current: it also causes the prOJ processor to operate a meter relay M, as in Figure 4.1, to transmit metering pulses on the P-wire connection through the exchange and appropriate to the charge for the call if it has originated at that exchange. Only at the originating exchange are the meter pulses effective, by operating the message register of the calling line the appropriate number of times which the prOJ processor has to determine from the digits dialled to establish the connection. If a long distance call is made, the originating exchange may not have the means of determining the rate at which the call should be charged if successful, in which case the charge is determined at an exchange of higher hierarchic level and meter pulses generated and sent from that exchange to operate the meter at the originating exchange. To do so the junction incoming to the higher level exchange includes a processor prIJ with means of determining charge rates from dialled digits received, and of generating meter pulses corresponding to the charge rates and effective durations of the connections made. The meter pulses are transmitted back to the originating exchange, each pulse as a brief polarity reversal such as the operation of a D relay produces. At the originating exchange, each operation of the D relay by the incoming meter pulses adds one unit to the meter record of the calling line. The operation is designated metering over junctions and by its use the charges for long distance subscriber dialled calls may be metered automatically at even the smallest of exchanges which are able to afford very little equipment. A high level exchange with many dependents for metering over junctions requires much semi-permanent data storage for meter pulse charging if the rates of charge are different for the different dependents. Usually the charges are the same for any exchange in the area served by the higher level exchanges in order to simplify the charge determining processing.

By the means described the basic message, signal and data transmission and exchange line d.c. current feeding requirements are satisfied, but with some disadvantages, difficulties and limitations which it is important to evaluate.

Signal transmission through the processors prB and prOJ of Figure 4.9 is required to be interrupted so that the signals received on one side may be relayed on the other without interference between the two sides. The signal transmission is, however, associated with message transmission, and message transmission must at all times be continuous, a requirement satisfied by the series capacitors which together with the relays on each side are termed a transmission bridge. Ideally bridges are high pass filters allowing transmission of messages with very small attenuation and offering very high attenuation not only to the d.c. steady states of the signals but also to the transient disturbances associated with changes of state. The inductor I in the prOJ processors is necessary to message transmission which would otherwise be short-circuited when relay D is not operated. In practice the bridges are far from ideal as filters, with the consequence that each produces significant message attenuation and attenuation distortion and interference to transmission on both message and signal paths. The sum of all the message transmission effects of several junctions in tandem, each with a transmission bridge, is serious impairment of transmission already near to the tolerable limit because of the transmission distance. The effect on signal transmission is time delay and time distortion — the distortion causing difficulties with dial pulsing. When a change of signal occurs, the transient transmission on the message path through the bridge causes noise on the message transmission and also interference with signal transmission repeated on the other side of the bridge. The interference is made negligible by the addition of passive components to limit the rate of change of signal currents but which also reduce the rate of signalling to about two reversals per second and increases the time delay in signal transmission. The calling line loop signals for calling, dialling and clearing cause interference at times when conversation is not normally to be expected. The called line answering and clearing signals occur when the calling party may be listening but noise could be tolerated to some extent. Metering over junctions produces meter pulses during conversation and it is for this reason more than any other that audible interference between signal and message transmissions has to be prevented. To do so not only limits the rate of meter pulsing but also increases the delay between the answering of a call by the called subscriber and the arrival of the answer signal at the originating exchange and this causes some operating difficulties where, as occurs on many long distance calls, the message transmission path is not completed until the called subscriber has answered.

For junctions which do not adequately transmit the d.c. signals of Figure 4.9, and many of the longer junctions do not transmit d.c. at all, the signals have to be transduced to forms which the junctions will transmit efficiently. The transducing equipments, which can be very expensive, have to be located in terminal processors, incoming junctions then having to have processors prIJ. The signals may also have to be pulse coded. Transmission over many of the longer junctions is limited to frequencies necessary to speech transmission. For those, pulse coded signals at frequencies within the speech range have to be used and consequently are termed voice frequency or v.f. signals. In the present context they are v.f. signal-messages.

The use of a processor in each exchange to hold and release the switches making a connection through the exchange enables junctions of two wires instead of three to be used but introduces some timing difficulties. In one exchange all the switches held by earth potential, applied by the one processor involved, release simultaneously when the earth is removed. With more than one exchange in a connection, all the switches in all the exchanges release nominally together in response to the

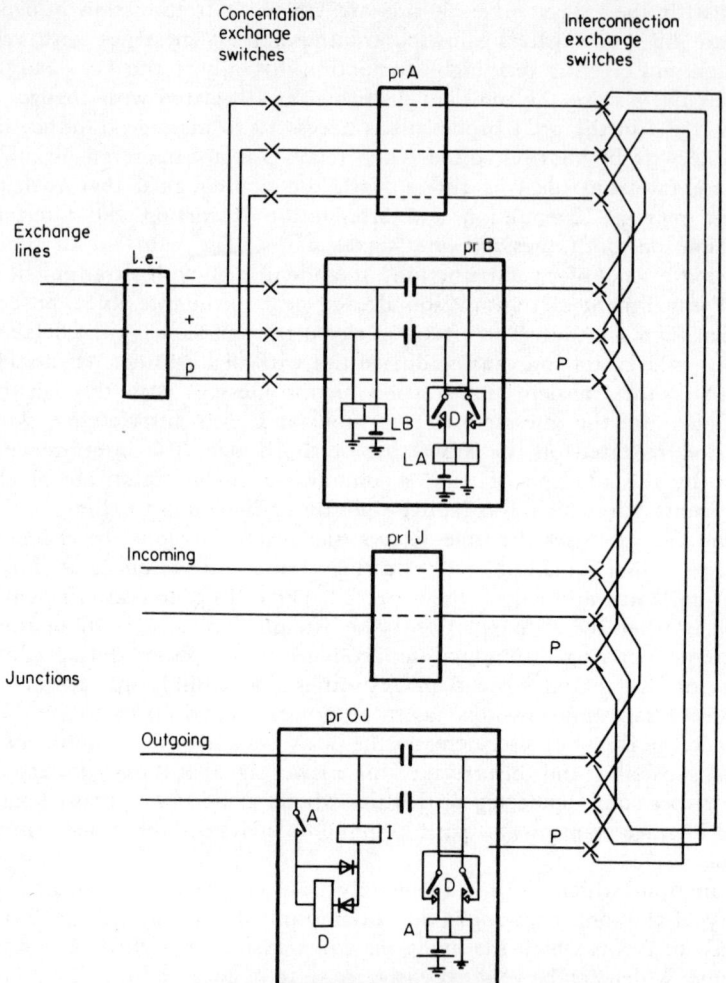

Figure 4.9   Message, signal and data transmission

calling line clearing signal, but not in fact because of signal transmission times and differences in the response times of the various processors. As a consequence it is possible for any one of the junctions to be released at one end before it is free at the other, and if taken into use in that condition for another call, that call will fail to be completed satisfactorily. To reduce the risk of this occurring the processor circuitry is made more complicated than it would otherwise need to be. A completely satisfactory solution requires the transmission of more data than is practical under the conditions of Figure 4.9.

In the ways described the basic and unavoidable processes of call connection, supervision, metering and release are accomplished through a number of exchanges in series. Other means different in detail but similar in principle and producing the same results are known, but no way of increasing for general use the two-state signals in each direction of signalling to more than two states, except for the use of one wire with earth return of which opportunist use is made but is not a generally usable state. The important point is that, for practical and economic reasons, the exchange control system has to make do within exchanges with two wires for signal and signal-message transmissions and one wire for data transmissions and between exchanges with only two wires. Theoretically any quantity of information can be transmitted by such means in both directions, as in telegraph services, but the associated speech transmission renders impractical more than has been described because of the low tolerance of speech to noise interference. These conditions having existed from the beginning and persisting, have had the result that some features which would be of advantage, some to the subscribers but many to the administration and to maintenance, have not been used in exchanges and are only now being considered for systems with data link junction signalling analogous to the order wires of manual systems, and made possible by electronic processors. For example, coin-box coin deposition and collection controlled by operators in centralized service centres, and exchange line ringing controlled by operators, are not generally possible with transmission facilities limited to those of Figures 4.8 and 4.9 because of the difficulties and the expense.

4.2.5 Observations

In the course of time, step-by-step systems came to be regarded as too slow in operation, and because the wipers and banks were limited to base metals to stand up to the rubbing wear, the circuit switching became suspect. From the economic and design points of view, of greater importance was the very limited data storage and transmission facilities available to the exchange equipments. The consequences were manyfold. There was some undesirable restriction of the facilities and classes of service which could be offered to the public. The processing had to be divided between many different designs of processor made appropriately available by the arrangement of the structure which thus was made more complicated and expensive. The exchange engineering force had to be used to perform operations which could not be performed automatically, thus raising cost and continuous staffing problems. The multiplicity of different processors which had to be used not only increased the cost of design and manufacture but also involved much equipment engineering to design and install individual exchanges: if change to the processing was required for new services or other reason, either changes had to be made to already manufactured and working processors in great quantity or new processors designed, manufactured and substituted. It is interesting to trace how the course of later developments was directed to mitigating or overcoming these defects.

Step-by-step dial control using two-motion switches was the first solution to the problem of a machine operated telephone system. The simplicity of operation and maintenance, and the economy over the whole range of exchange sizes, caused the system to be widely adopted and it is still in evidence in practice after more than 80 years. The influence of manual practices can be seen in the equating of directory and equipment numbers with its advantages and disadvantages, and in the storage of exchange line and state of line data, and in some aspects of signalling and control. Whereas a manual system tends to a near-minimal structure because the operator is such a good processor, the first machine systems comprised maximum structure exchanges because machine processors were so undeveloped. As machine processing has developed it has understandably been accompanied by a transfer of processing from the structure back to common processors akin to the original operators.

A succession of developments have been aimed at mitigating or eliminating some disadvantage of the rigid connection between the digits dialled and the selection processing, or of the switch construction limited to operating directly from dialled pulses, or at improving the reliability of operation or extending the services offered to the public or reducing costs. Developments have often followed more than one direction at a time and overlapped in time so that it is difficult to pick out a distinctive chronological order. Therefore they are treated in subsequent sections in an analytical order useful to understanding but only broadly chronological.

4.3 Delayed Dial Controlled Selection

The construction of exchange switches operated directly by dial pulses is constrained by the speed of dialling to relatively light ratchet motor drives, to small numbers of contacts in the levels of the banks, and to low pressures of the wipers on the bank contacts which in conjunction with base metals used for the wipers and bank contacts does not ensure good and reliable electrical contact under all the conditions of use. Continuous and uniform motion of the wipers by rotary motor drive are necessary conditions for more robust and powerful switches but are not directly compatible with operation under dial pulse control directly. Two general solutions emerged. One chronologically the first uses relatively slow moving switches mentioned in section 3.5, of the uniselector or two-motion type, driven by clutched motors operated by continuously rotating common shafting, in systems to be described in section 4.4. The second, the subject of this section, uses delayed but still effectively dial control of very fast operating switches of the uniselector type, with a hundred sets of contacts in the banks and wipers driven at about two hundred contacts per second by a rotary motor controlled by an auxiliary ratchet motor switch. The essential operating elements are shown in Figure 4.10 which is to be substituted for a corresponding part of Figure 4.1, the part which includes the final selector arcs 3 and 4 and driving magnet DM. The s wire relaying the calling line loop signals to the prB processor is not connected to the magnet DM for the direct dial control of the switch, but temporarily via contacts of a relay Z which operates when the selector is first seized, to the driving magnet DM of an auxiliary ratchet motor switch forming part of what is termed a marker, and which is common to a number of selectors. The marker has its own processor prMK. The ratchet motor is driven by the dial pulses of two directory number digits to bring its wipers on to one of a hundred positions. In fact, two ratchet motors are used, one for each digit: the one set by the first digit has a wiper and a bank of ten contacts, each contact being connected to a wiper, one of ten, of the switch set by the second digit. With ten contacts per wiper, the second switch has one hundred contacts which are the marker bank contacts MK in Figure 4.10 and they are connected contact to contact with the hundred contacts of the arcs 4 of the rotary motor driven switches of the selectors controlled by the marker. Immediately the second digit is finished, the prMK processor operates its relay SS to operate relay ST in the connected selector to start the selector switch wipers moving over the bank. If, as the wipers move, a circuit is completed from earth, through the marker MK arc and the selector arc 4 to the T relay in the selector, that relay will operate to the exchange battery via the K relay in the exchange line circuit, unless the arc 3 wiper is held at earth potential because the line is already busy. The T relay operating stops the switch on the line contacts and thus the line is selected according to the directory number digits which have operated the marker. Groups of lines all with the same directory number have their arc 4 contacts all connected to the one directory number contact of the MK arc, provision being made for variable connections to be made manually, such connections being indicated by the dashed line in Figure 4.10 and commoning arc 4 contacts. During the rotation of the selector switch all the lines in the group are tested and the revolution completed only if they are all engaged. Thus the lines in a group must be located within a one-hundred-line block but within the block they do not have to have consecutive equipment numbers with the attendant difficulties and disadvantages. If the selector switch makes one complete revolution without being stopped on a contact, the prMK processor causes the prB processor to stop the rotation and to send back busy tone to the caller.

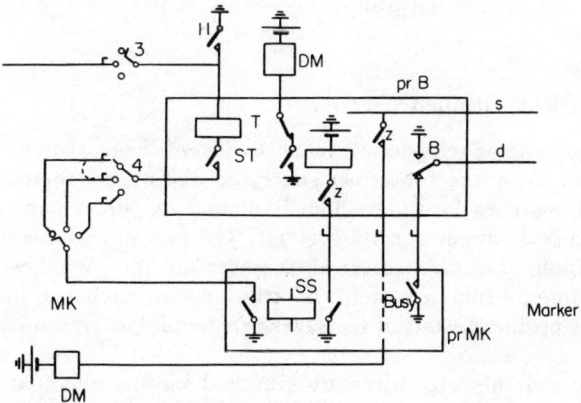

Figure 4.10 Delayed dial controlled selection

Group selectors operate in the same principle. The marker switch having only ten contacts and responding to only one digit, the ten contacts of the MK bank are each connected to a group of arc 4 contacts, the groups of contacts corresponding to the levels of two-motion group selector switches but with the difference and the advantage that the groups do not have all to be the same size. The selector switch had to be capable of at least one revolution of the wipers within the inter-digital pause, a requirement which compels a very high speed of rotation of the selector wipers and which in turn compels very fast operation of the relay T when the wipers are on a marked and free outlet, and very fast and accurate stopping of the rotation, all of which is possible with robust construction and high contact pressures to ensure good contact. In one system the message circuit wipers do not touch the bank contacts during rotation but are moved into contact when the wipers are stationary, by which means rubbing is avoided and silver can be used for the contacts to improve the contacting. The marker switches of all systems are step-by-step dial control switches no different in their operation from the switches of systems of that kind but different in their constructions in that, requiring only ten or one hundred contacts and not one hundred sets of contacts, they are smaller and can be designed to be more reliable and durable.

For every exchange switch to be able to operate independently of every other would require every switch to have its own marker and wiring between its arc 4 and the MK arc of the marker switch, which would be costly and inefficient for processes completed in a few seconds at the beginnings of calls which average over 100 seconds in duration. Clearly there is need for some form of time sharing, but there is insufficient time in the inter-digital pause for the connection of markers to selectors by normal pre-selection or line-finding by which processors prA are connected to calling lines. The exchange switches are therefore divided into groups of three or more which share one marker between them. When any one in the group is using the marker, none of the others may be seized for a new call, which the prMK processor ensures by applying earth potential to the data d wires of all the otherwise free selectors in the group.

4.4 Reverted Dial Controlled Selection

The development of telephone exchange systems using switches which were not directly controlled by the subscriber generated dial pulses, thereby to gain for the design of the switches latitude which is denied by direct control, proceeded in parallel with that of direct control systems. The two developments differ in detail consequent upon the mechanical differences in the switches, but in general principles of step-by-step selection and trunking, in exchange line data and data storage and its problems, and in message, signal and data transmission, there are no differences.

Briefly, the switches are driven by clutched motors operated by continuously rotating motor driven shafting as described in section 3.5, the digits dialled by the subscribers are stored in registers and the registers set the switches by step-by-step selection. The switches are not stepped under the control of dial pulses but as they move they generate a pulse for the equivalent of each step of selection. The pulses thus generated by a selector are reverted, that is sent back through the switches, to the register controlling the selection, the register counting the pulses until they correspond in number with those of a received dialled digit, when the register sends a stop datum impulse to the switch to cease moving. Another switch in the connection then operates similarly until selection is complete through an exchange, in the same way that step-by-step dial control systems operate but with some detail differences.

Figure 4.11 shows part of a trunking diagram for a basic system, variants of the basic system being described later. The trunking diagram is essentially the same as that of Figure 4.5 but with the addition of registers and register finders. When an exchange line originates a call, connection is made to the line by an exchange line finder which then gains connection to a register via a register finder. The register provides dial tone to the calling subscriber and receives and stores the digits which he dials, using relays or uniselectors as temporary stores. When the register is ready to set the first selector, it sends a start datum through the register finder to the selector to operate the level selection clutch motor. As the selection mechanism moves, it reverts a pulse to the register for each level reached until the required level is attained, whereupon the register sends a stop datum to the selector and the motor declutches. If the selector is a group selector, it operates the wiper carriage rotation motor immediately after level selection and the selected level of wipers making contact with the bank contacts, they find a free trunk to the next stage of selection as described for direct dial selection switches. Subsequent group selectors, if any, are controlled in the same way, by the register sending operational data and the selectors sending reverted pulses through all the switches in series. When a final selector is reached, the second motion also takes place under the control of the register to select a line with the stored directory number, and then under the control of the switch processor if the line is busy but one of a group all with the same directory number. Thus the selection is step-by-step as for direct dial controlled selection systems, it is still decimal because it is directly related to the digits dialled by the subscribers, but the switches do not necessarily have to have ten levels of ten contacts. The storage of the dialled digits in the register before selection commences gives the opportunity for some different numbers of steps by the switches. The five-hundred-line switch of Figure 3.7b provides a good example. Each final selector has capacity for five hundred exchange lines in twenty-five levels, angularly displaced planes actually, of twenty contacts. So the register has to examine the last three digits of the exchange number and translate them into one digit of steps up to twenty-five and one digit of steps up to twenty. Similarly the group selector requires one digit of up to twenty-five steps. The twenty-five group selector levels may each be used for access to a five-hundred-line group of exchange lines or to a group of junctions or service lines. Thus one group and one final selector stage may together suffice for an exchange of up to ten thousand lines and for which the step-by-step dial control systems require two group selector and a final selector stage. The registers have to provide translations from decimal digits

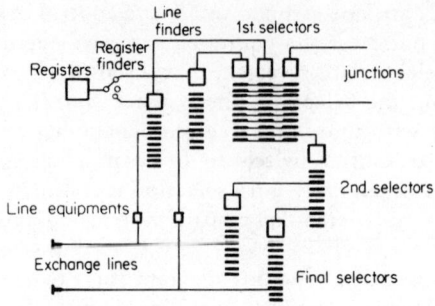

Figure 4.11 Exchange with reverted dial controlled selection

received to selector reverted pulses to be received, which translations being systematic and not random, are neither difficult nor complicated.

The register in the originating exchange controls switches in distant exchanges over junctions which are capable of transmitting the necessary control data with adequately small time delays. Each register is a processor large in comparison with the exchange switch circuit processors prA. A register when connected to a circuit processor is effectively in parallel with and part of the circuit processor. In other words, because the register part of the prA processor is needed for only a small fraction of the average duration of calls connected, and at a predictable position in the program of operations, the first being a condition for economy and the second for the possibility of space division time sharing, the registers are space division time shared among the exchange line line finder trunks.

The two-motion switch systems which have been described in this section are best suited to large exchanges. An all-uniselector system which is economical for the smaller exchanges is similar in its trunking to the delayed dial controlled selection systems of section 4.3.

4.5 Register-translator Controlled Selection

Going back to the problem of subscribers composing the numbers which they need to dial to make required connections in large networks and ultimately to anywhere in the world, if the selection processing is directly related to the subscriber dialled digits, the subscribers must dial digits to suit the selection processing. Although the exchange numbers are invariable, the digits required to reach particular exchanges vary according to the origins of the calls. In the example of Figure 4.5, a subscriber on the exchange needs to dial only the exchange number to connect to a station on the same exchange but for every call incoming over a junction the exchange number must have been preceded by the exchange code: and for a call which is routed through several exchanges, the number which is dialled has to include an exchange code for each of the exchanges on the route. Under these circumstances, directories of exchange codes would have to be issued separately for every exchange to reach every other exchange. The exchange codes would vary from no digits for calls on the same exchange through one, two and so on perhaps to ten or more digits. Even to dial his own station from not far away from it, a subscriber would need to consult a local directory of exchange codes. What is needed from the subscribers' point of view is invariable exchange codes to be dialled from anywhere to reach particular exchanges and to be followed by the exchange numbers to call particular stations on the exchanges. One penalty of this arrangement is that for some calls the subscribers have to dial more than the minimum necessary number of digits in order that on other calls they may dial fewer digits than are actually needed and used for the switches. It is possible to achieve the stated objective by choice of numbering in conjunction with suitable exchange and junction plants but it becomes very difficult and expensive in plant to do so for large networks. The usual practice is to divide the network into areas called linked numbering areas within which the exchange codes are invariable and achieved by a combination of numbering and arrangement of plant, to link the areas together for longer distance calls by prefixing the linked numbers by invariable area codes systematically chosen and to control calls between the areas by the use of register-translators.

The problem is basically one of addressing telephone stations and is the same as postal addressing which is composed by the letter sender according to the distance to be travelled and interpreted by the postal sorter according to the progress of the letter toward its final destination. To make a telephone call, the caller must compose a suitable number, starting with the linked numbering area number which is preceded if necessary by an area code, and that if necessary by a state or country code and so on. The interpretation of the address as connection proceeds is the function of the register-translators using arbitrary number translation. Systematic translation as described for the five-hundred-line system of the previous section, produces selection which is no different in principle from dial controlled selection. Arbitrary number translation means that given any number within the range of valid numbers it can be translated to any other number to operate exchange switches.

A further complication in the early days of automatic exchanges arose out of what would now be called the human factor in the problem. Dialling telephone numbers was an activity new to most people and tests showed that errors, measured statistically as digits in error per hundred complete numbers dialled, seldom occurred with four digit numbers, occurred more often with five digit, significantly with six digit numbers, and with seven digits the error rate became a serious difficulty. Subscribers being familiar from manual exchanges with stations being defined by an exchange name followed by a number, suggested the idea that subscriber dialling accuracy might be assisted by specifying a name and a number to be dialled instead of all decimal-digit numbers. Lettered dials as in Figure 3.2 were the result. The subscribers dialled the first two or three letters of the exchange name and followed them with the exchange number, and a register-translator made the necessary translation of the exchange name into digits. This practice persisted for many years until increasing complexity in the systems made it impractical and all-decimal digit numbers imperative. It is a reflection of the increased dialling ability of the present members of the public, by now most of them brought up from infancy to use a telephone dial, that decimal numbers of up to eighteen digits are admitted for international calls and with which the subscribers are expected to cope successfully.

Figure 4.12 shows the modifications caused to the message, signal and data transmissions of Figure 4.9. The paths are interrupted in the processor prA and diverted to a register-translator, which is a signal and data processor. The calling subscriber's loop signals are received by a relay LAA and the connection held by the register-translator. The digits dialled are stored and when sufficient for the address to be interpreted have been received, they are translated to other digits which, had the subscriber dialled them directly to control the exchange switches, would have extended the connection towards, if not actually to, the terminal exchange. The digits are transmitted by the register-translator relay A and followed by the exchange number digits unchanged. The register-translator then releases itself and switches the transmission to the through condition by sending operational data to operate the relay BH in the prA processor. Equipment which interrupts through transmission for its diversion to temporarily connected other equipment will be called a diversion bridge. If the dialled digits do not correspond to a working exchange code, the register-translator sends data to the prA processor for it to send n.u. tone to the caller. Economy clearly requires the register-translator processing to be time shared among, and not individual to, the prA processors, the same as the

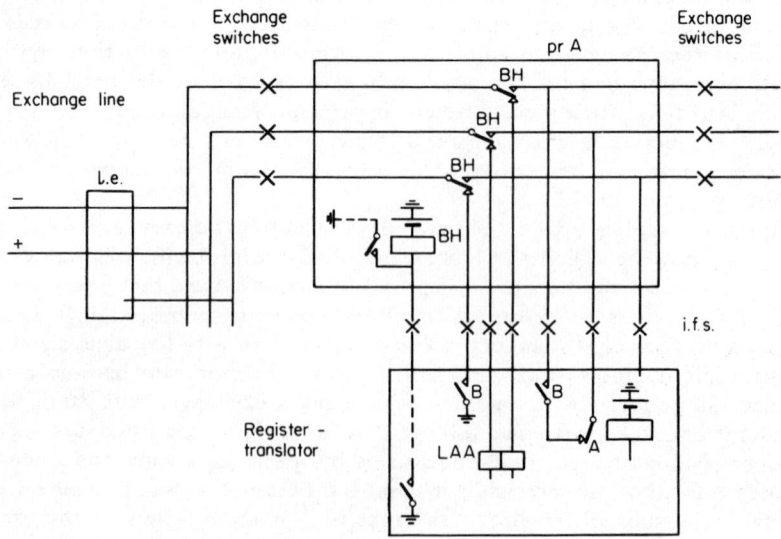

Figure 4.12   Diversion bridge for register-translator

registers of Figure 4.11 and as indicated in Figure 4.12 by the crosses representing interface switches in the connections between the prA processors and the register-translators. Clearly the reverted dial control systems described in conjunction with Figure 4.11 already have the elements described and merely require the registers to be substituted by register-translators.

The general principle of translation processing using electro-mechanical equipment is illustrated by the simple example of Figure 4.13 which assumes sliding contact switches for digit storage and the translation of directory numbers comprising exchange codes of two digits followed by exchange numbers of four digits. The basic requirement is for a set of contacts for every two-digit number, that is one hundred sets of contacts, with means of making the contacts effective when the number is dialled. In Figure 4.13 a switch sw1 with six wipers is shown, with one hundred sets of contacts 00 to 99 onto any one of which the wipers may be located under the control of two dialled digits in the same way as the wipers of a one-hundred-contact final selector are controlled. In fact two-motion dial controlled and one-hundred-contact uniselector delayed dial controlled switches are commonly employed. Two exchange code digits are stored on switch sw1 and four exchange number digits on four other switches sw2 to sw5, each with one wiper and a home and ten working contacts in the bank. When the register-translator is taken into use the switches sw1 to sw5 are set to positions corresponding to the first two and next four digits dialled by the caller, by means not shown in the figure, and the positions of the switches control digits transmitted to the exchange switches by relay A of Figure 4.12. Digit sending uses a digit pulse generator which pulses the A relay and simultaneously the driving magnet of a uniselector sw6 the wiper of which takes one step for each pulse until it encounters earth potential on the bank contact on which it rests, when it stops the pulse generation. The register-translator processor then drives the switch sw6 to its home position and an interval of time later appropriate to the inter-digital pause required by the exchange switches, it allows pulse generation and the stepping of the uniselector sw6 to restart. The switch sw6 has a home and ten contact positions, its ten bank contacts are multipled with those of the switches sw2 to sw5 and also connected to rows of jumper wire terminals. Jumper wires may be connected between the sw1 switch bank contacts and any of the jumper wire terminals, which is the means of achieving random number translation.

The digit to be sent is controlled by a digit switch, the uniselector sw7 in Figure 4.13. It is pulsed one step when digit sending is to commence and thus earths its first bank contact and also one of the bank contacts of the switch sw6 via the wiper 1 of switch sw1, the bank contact on which the wiper is standing and the jumper wire from the bank contact. Pulsing of the A relay commences, the relay contact sending break pulses to the exchange switches as if they were pulses from a subscriber's dial, until stopped by the wiper of switch sw6 finding the earthed bank contact. During the following inter-digital pause, the switch sw7 is made to take another step, so that when pulsing out recommences, it is the second translated digit which is sent, and so on. Up to six translated digits can be sent but less are usually required and arranged by jumpering the first unwanted digit to a digit cut

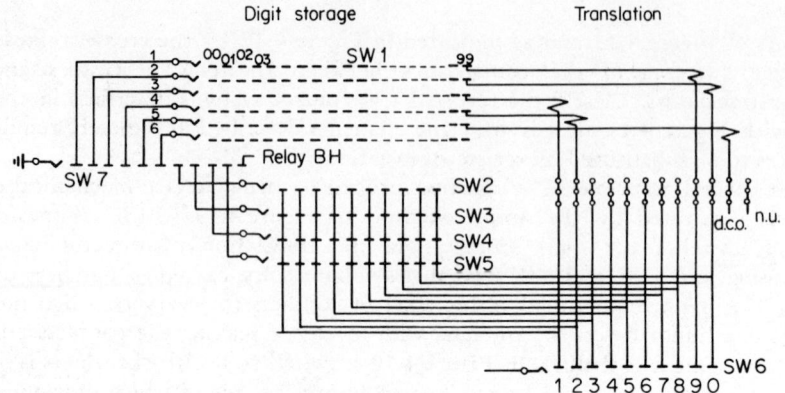

Figure 4.13   Digit storage and arbitrary translation

off, d.c.o., terminal which when earthed, causes the digit switch to step immediately to its seventh bank contact and thus to send the first digit of the stored exchange number. The figure shows the exchange code digits 99 to be translated to 9012. Exchange code sw1 bank contacts for codes not in use are jumpered to an n.u. tone terminal which when earthed causes the register-translator to send a data instruction to the prA processor to connect n.u. tone to the calling circuit. When all the exchange number digits have been sent, the switch sw7 steps onto and earths its eleventh bank contact which sends to the prA processor the operational datum to operate the relay BH and thus to release the register-translator.

Temporarily while the register-translator is connected and until the BH relay in the prA processor, Figure 4.12, is operated, the three wires outgoing through the exchange switches are used for data transmissions and only when the BH relay is operated do they assume their normal role of message, signal and data transmission. With the BH relay unoperated, the transmission from the exchange line and through the prA processor is suitable to the operation of step-by-step dial and delayed dial control selectors. Hence the register-translators may be preceded by group selectors to increase to three, four or any desired quantity, the exchange code digits which may be translated. Figure 4.14, which is similar to that of Figure 4.11 for a reverted dial control system, uses direct dial or delayed dial control switches. The exchange line line finders have pre-selectors which are the switches shown in Figure 4.12 by crosses in the trunks between the prA processors and the register-translators, but the pre-selectors have access to group selectors which are called A-digit switches because they are operated by the first dialled digit which originally was the first or A digit of an exchange name. When a subscriber calls he is connected to an A-digit switch which provides the dial tone and accepts the first digit. Ten groups of register-translators connected to the levels of the switch extend the translatable exchange codes to one thousand.

The use and development of registers and register-translators began when the range of subscriber dialled calls was very limited, all other calls having to be handled

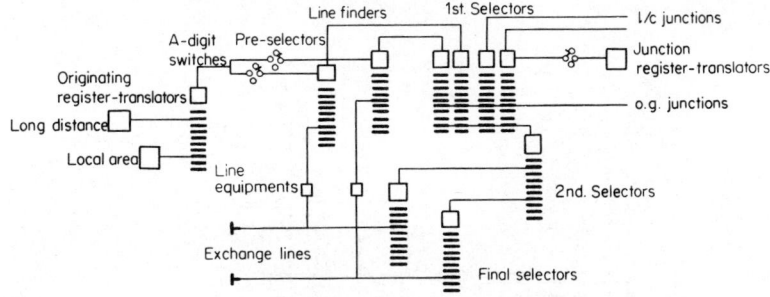

Figure 4.14   Exchange with register-translator controlled selection

by operators. The design of the first such equipment was kept to the simplest possible by such means as making all the diallable exchanges have exchange codes with the same number of digits. As the range of connections and services available to the subscribers by dialling increased, so did the problems of control. In particular it is not possible to maintain uniformity in the number of exchange code digits and it becomes uneconomic and difficult to provide, in every exchange, translations for every other exchange in a constantly increasing area which eventually becomes the whole national network at least. Systems of the kind represented in Figure 4.14 solve the first problem by the levels of the A-digit switches selecting register-translators translating different numbers of directory number digits, as shown in the diagram by local area and long-distance register-translators, or by register-translators all the same but with different numbers of ranks of exchange code switches between them and the pre-selectors. These are opportunist solutions dependent on subscriber rotary dial direct control of selection and they fail with press-button dialling and with reverted dial control systems. Register-translators in series is the general solution to the problem of translations for every exchange in networks of any size and eventually for every exchange in the world, and it is analogous to the postman who delivers letters to their final destinations in his local area and those which he is unable to deliver he sends bearing their addresses to another postman nearer to the final destination. The solution is dependent on systematic numbering for the exchange codes. Some incoming junctions are shown in Figure 4.14 with access via their terminal processors, which have diversion bridges for the purpose, to junction register-translators by pre-selectors. The digits received on one side of a bridge are stored in the register, and translated to appropriate digits which are sent out over the other side of the bridge to further the connection. Originating register-translators which receive dialled digits from subscribers complete local area calls, that is local calls and calls terminating in nearby exchanges. Longer distance calls are routed toward their destinations over junctions with access to junction register-translators which take over the further routing.

Figure 4.15 is Figure 4.8 developed for register-translator control of step-by-step selection exchanges. A subscriber originating a call dials into an originating

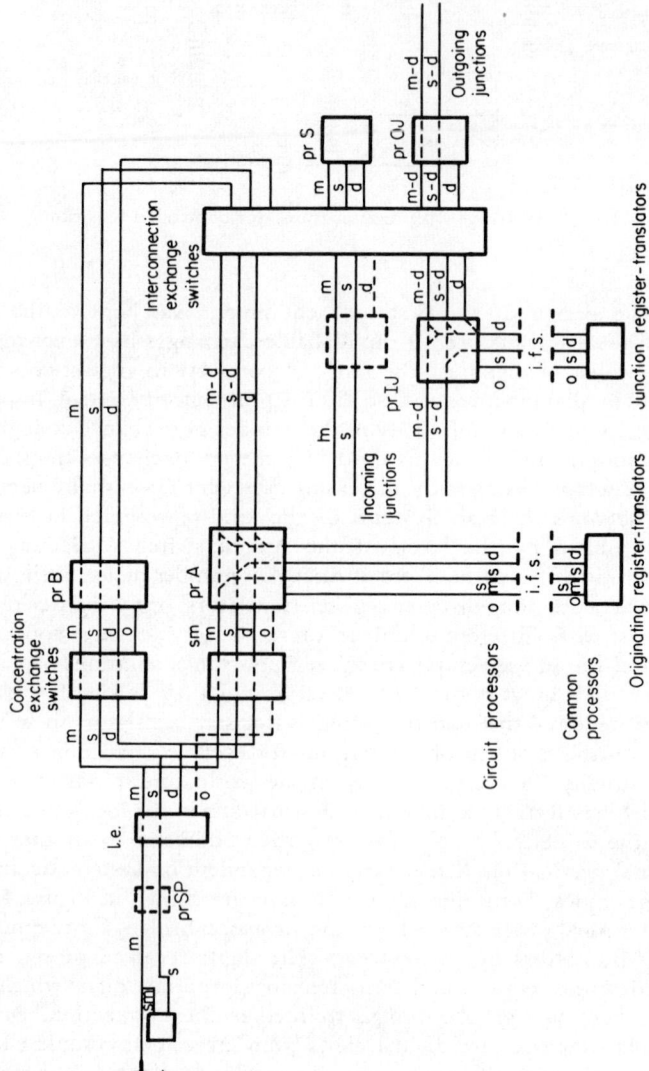

Figure 4.15 Generalized register-translator controlled step-by-step exchange

register-translator which sends out digits to extend the connection to another line on the same exchange, to a local service possibly a manual board, or to an outgoing junction which at its far end may be terminated in a service processor or a manual board, but more often on exchange switches in another exchange through which the connection is extended to an exchange line or a service, or over yet another junction until the call is completed. An incoming junction may terminate directly on the exchange switches or include a processor prIJ which does not give access to a junction register-translator, in both cases the exchange switches then being stepped by impulses over the junction from a distant register-translator: or an incoming junction may terminate on a processor prIJ with access to junction register-translators which take over incoming call processing on instructions received over incoming junctions from distant register-translators. When an outgoing junction is first seized for a connection and a loop signal has been sent to call the distant end and to connect a junction register-translator as in Figure 4.15, the transmission paths are as shown in Figure 4.16. Within the exchanges there are m, s and d paths through the switches but interrupted by the diversion bridges. Between the two exchanges is the junction providing m and s transmission but isolated from such transmissions by the diversion bridges. In this condition the junction is available for data transmission between the registers. Over this path the called register informs the calling register when it is ready to receive data and it may give other information. The calling register transmits enough address, full or partial, for the called register-translator to advance the call by sending out digits to extend the connection through the exchange to terminate there or to continue over another junction. If the call involves another junction, the called register-translator may release and thus establish through connection of the prIJ processor for the calling register-translator to control further connection operations: or the calling register-translator may send all the call connection data to the called register-translator for it to take over call connection to completion which may involve yet another register-translator. As each register-translator completes its function it restores the through transmission of the diversion bridge to which it is connected and releases.

The data transmission over a junction between diversion bridges may use the s path two-state conditions of loop in one direction and polarity in the other as

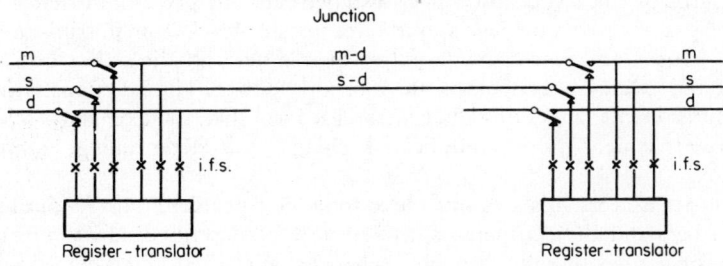

Figure 4.16   Transmission paths between register-translators

described with reference to Figure 4.9, or to reduce the time taken for transmission, high speed transmission over the m path may be used. Data transmission over the message path usually takes the form of discrete audio frequencies providing five or six mutually independent states which are sent in combinations two at a time to represent items of data similar to press-button sending described with reference to Table 3.2. The time advantage of such high speed data sending is limited by the subsequent selection if it is step-by-step, but the sending means is useful over long distance circuits which will not transmit d.c. necessary to s signals. It is useful and used for its time saving in cross-bar and later systems which provide rapid selection and connection.

To indicate that data may be sent over junctions between exchanges, on the message and signal paths but not simultaneously with message and signal transmissions, the junction transmission paths are designated m—d and s—d, and data over these paths are designated data messages and data signals respectively.

Arbitrary number translation and the transmission of data between registers are important and powerful parts of communication system control when fully exploited. Data transmission between registers can include class of service, charge rate and some special instructions if needed, in addition to the address of the wanted line, although step-by-step systems are not able to take advantage of this facility because they do not have the means of using data other than the address of a wanted line. The fact that translation and data transmission between registers are available only during connection processing is a disadvantage not limited to step-by-step systems. One means of surmounting the difficulty not easily applied to step-by-step systems but used in cross-bar and other systems is to transmit during call connection processing data then available but not immediately useful, the data being stored in the receiving exchange for processing at a later stage of the call. Another possibility making use of the fact that the translation digits are arbitrary, is to choose digits which not only route the calls but also convey data relevant to them. This device is used by step-by-step systems to simplify the metering processing. For example and referring to Figure 4.14, for subscribers originating traffic it can be arranged that the metering rates applied by the first selectors to the calls passing through them are determined by the levels to which the selectors are set by the registers. The digits of a register translation being arbitrary, the first digits of the translations are chosen to suit the charging rates for the calls which are directed through the levels. For longer distance calls charged at a higher rate than is provided by the fixed rate levels, one level usually level O or possibly more than one level is used to receive meter pulses over junctions, the pulses operating the meters of the calling subscribers as previously described. The junctions terminate at their far ends on exchanges of higher hierarchic level than the exchanges of origin of the calls, at which more complicated charge rate determining equipment is available.

Register-translators in series introduce some new problems and features for both the subscribers and the equipments, as will now be described. Many are related to the processing times of calls through exchanges or to the limited data transmission facilities. As processing and data transmission facilities have improved with the passage of time, so many of the problems and features have become less important.

Register-translators in series extend the range of subscriber dialled calls almost indefinitely but impose on the subscribers the necessity of dialling many digits for the longer distance calls and of composing the numbers to be dialled, using exchange code directories and other data and a set of rules which, incidentally must be easy to apply or difficulties and mistakes will occur too frequently. When the digits have been dialled there is a delay before connection is completed and the subscriber knows it to be completed because he hears a tone or the called party speaking. With slow selection, long distances and particularly if the subscriber uses press-button dialling, the delay can become so long that the subscriber may be in doubt whether to go on waiting or to clear and start again because something has gone wrong. If the subscriber dials incorrectly he may receive n.u. tone but not necessarily and in particular if he has dialled too few digits for a translation to have been made, he will receive nothing because the register-translator is waiting for more digits to come in. For that reason and also to remind subscribers who delay dialling that they are holding common equipment which other subscribers may be waiting to use, n.u. tone is returned to the caller after the register-translator has been held for a predetermined time either from the beginning or since the last dialled digit to be received.

The register-translators translate the exchange code digits which they receive to digits to be sent out and be followed by the exchange number digits unchanged. They then release to restore normal message, signal and data transmission through the exchange and without which the caller is unable to hear the result of his dialling, and if release is unnecessarily delayed, the traffic and therefore the quantities of register-translators required are increased. The problem of determining when all the digits to be dialled have been received by a register-translator so that it may release can be a difficult one if the quantities of digits to be dialled by the subscribers for calls to different destinations vary unpredictably. For that reason directory numbering is commonly arranged so that as far as possible, the quantities of digits are constant for connections of given types. The register-translators have to be able to recognize the types and to count the number of digits received to know when to release. This is usually possible for local area calls but not always for longer distance calls, and reasons for exceptional quantities of digits to be allowed are constantly being advanced and cannot always be resisted. If the quantity of digits is not known for certain, the usual procedure is for the register-translator to count up to what is known to be the minimum possible quantity and thereafter to release after a given interval without a digit being received, or to release when the call is answered if that anticipates the expiry of the waiting period. The last exchange always and the penultimate exchange often knows when all the digits have been received and if a register with the knowledge sends back a release datum to all the registers previously involved in a connection, to release them, the release difficulties disappear, but such a solution was not practical for the early exchange systems.

4.6 Conclusion

With register-translation added to step-by-step selection, large networks including the national system of the U.K. have been successfully operated for many years. The main advantage is cost closely related to size of exchange and to traffic over the whole range of sizes of exchanges in a national network, including the growth periods of exchanges starting from small fractions of their final sizes. The disadvantages became more pronounced as telecommunication spread from national to world-wide dimensions. They include a slow rate of establishment of connections amounting to tens of seconds from the start of dialling to completion of calls set up over very long distances, and a very limited data storage and processing ability resulting in the provision of services by opportunist methods which become increasingly difficult to find, and costly, as the scale and complication of the services increase. An appreciation of these characteristics is useful to an understanding of the later developments which are aimed at reducing the disadvantages and in doing so, introduce other problems notably of economics.

Chapter Five
CROSS-BAR EXCHANGES

5.1 Switches

As a means of finding suitable paths through automatic exchanges, step-by-step selection together with group hunting were inventions, obvious now but probably brilliant at the time, which became less attractive as time passed and telephone needs increased. Not only path search but also simultaneous path search and the addressing of line information stores were all achieved by step-by-step selection with a minimum of processing. The basis of the path search is the identity of equipment and directory number which for exchange numbers became more difficult to maintain with the increasing demands for p.b.x.s, party lines, night service numbers and all those circumstances which result in more than one exchange line and equipment number per directory number and more than one directory number per exchange line. It was shown in the previous chapter that the difficulties of maintaining the identity of directory and equipment numbers for exchange codes increased even more rapidly with the extension of the range of subscriber dialling until the link became non-existent for calls over a medium distance and arbitrary number translation a necessity. Not surprisingly, the idea that the identity did not have to be maintained even for exchange numbers began to be pursued. Means of providing flexibility between directory and equipment numbers having started, it has become more easily and completely applied as apparatus and systems have developed, yet in some recently developed systems some systematic association between the two can still be traced, as a means of economy in the storage of directory number to equipment number translation and the inverse translation of equipment to directory number.

Twenty years and more of apparatus development showed how to make relays reliable by the choice of magnetic structure, by the use of precious metal contacts and adequate contact pressures produced without too much mechanical rubbing, and by twinning the contacts — which means for every contact two contacts in parallel and side by side on the same springs — and by the use of quench circuits to reduce electrical erosion of contacts as they are separated. Sliding contact switches were also improved but the sliding motion seems to set a limit to the improvement possible.

The initial attraction of relays as exchange switches and which caused them to be developed later into cross-bar switches was the prospect of greater reliability of operation, and random number translation having shown that equipment and directory numbers need not be directly related, relay and cross-bar systems started on a new line of advance in exchange design. Exchange switches made up of relays were developed very early for the small exchanges of the p.a.b.x. type and were cheaper than the sliding wiper switch exchanges for exchange line quantities in tens rather than hundreds. The cross-bar operation improved the switch economics and with further advances in apparatus and system design and manufacture, cost eventually became comparable with those of sliding contact systems for public exchanges of all sizes and during the 1930s large scale installation commenced. The economy of automated manufacture which became a prominent feature of industry in the 1950s benefited cross-bar systems more than the older ones with the result that from that time onward, a world-wide preference for cross-bar systems developed.

Cross-bar switches are co-ordinate relay contact switches as in Figure 3.7(c), but without an individual magnet for each set of cross-points. There is instead, as indicated in Figure 5.1, a select magnet DM per row and a bridge magnet DM per column of contacts. Any one set of contacts is closed by operating first the appropriate select magnet, which moves an armature to select a row of contact sets, and then the bridge magnet which has an armature common to all the sets in a column but closes the contacts of only those sets for which a select magnet has been operated prior to the movement of the bridge magnet. The bridge magnet having operated, the select magnet or magnets may be released for the operation of other sets of contacts. When a bridge magnet is not operated, all the contacts of its column of contacts are open and none is affected by the movement of any select magnet. When a bridge magnet is operated, the contacts in the column which are closed and those which are open stay closed and open respectively no matter how the select magnets may subsequently move. Only the operation first of a select magnet and then of a bridge magnet is able to change the conditions of contacts by closing those at the co-ordinate point of the horizontal select and vertical bridge magnets. If more than one select magnet is operated the bridge magnet will close contacts corresponding to each, and thereby sets of contacts in the same column may be operated simultaneously or one set at a time. In some switches the bridge magnet coil currents do not have to be maintained to keep the armatures operated, this being achieved by mechanical or remanent magnetism latching. Latched armatures are released by the momentary operation of a magnet which according to the design may be the select or the bridge magnet. Compared with individual relay operation, cross-bar switches require fewer but larger and more powerful magnets, the operation being slower in consequence and the energy required to operate them greater. The releasing speed is comparable.

Figure 5.1 shows only one wire of the trunks switched by the cross-bar switch. Usually the trunks consist of two message wires m which may of course carry signal s and signal-message sm data, and a data wire d used as a P-wire, but up to six wires in total are known in practice. The horizontal trunks have in addition an operational data wire o terminated on a select magnet and the vertical trunks not only an operational data wire o terminated on a bridge magnet but also a data wire

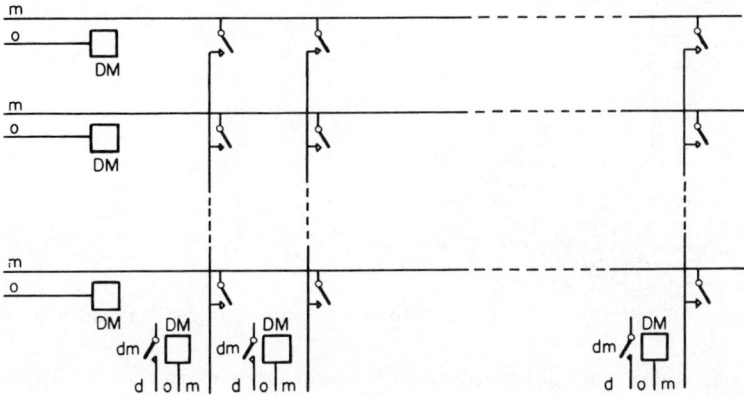

Figure 5.1   Cross-bar switch

or wires d to contacts dm controlled by the bridge magnet, the contacts being makes or breaks and by which the state of the trunk, free or engaged, is indicated. Because the sets of contacts in rectangular array are multipled to horizontal and to vertical trunks, each set may be considered to be located at the point where a horizontal trunk and a vertical trunk cross, and the sets are for that reason commonly referred to as cross-points.

Cross-bar switches with ten rows of ten sets of contacts in ten columns may be used as the basis of a step-by-step dial control system analogous to two-motion switch systems, but they are rarely used in this way and then only to achieve economy of control for small exchanges. Merely because decimal numbers are convenient, ten rows and ten or twenty columns of contacts occur in practice but the quantities of rows and columns being arbitrary, other quantities commonly occur and up to seventy-four rows and columns in excess of twenty are in use. Where large numbers of rows in excess of twenty are used, they are produced with the aid of the circuit device shown in Figure 5.2. The cross-bar switches have in addition to rows of cross-point contacts, rows of auxiliary contacts selected and operated by select and bridge magnets as for normal cross-points. The cross-point sets of contacts are in fact each two or three sets of cross-points but when all are closed only one set is effective. This is because each set is in series with the auxiliary contacts common to the column of contacts. Prior to the operation of a bridge magnet, in addition to a cross-point select magnet, one of the auxiliary row select magnets is operated to determine which of the cross-point sets shall be effective. Figure 5.2 shows the contacts of a one-wire path through a switch which has two rows of auxiliary contacts c and d selecting one or other of the two sets of cross-points a1 or a2, b1 or b2 and so on, depending on which cross-point select magnet is operated. Three paths in parallel at each cross-point is common practice, and up to three rows of auxiliary contacts.

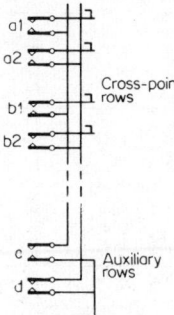

Figure 5.2   Auxiliary contact switching

What has so far been described are the salient features of all cross-bar switches. In detailed characteristics of quantities of rows and columns of contacts, and of mechanical construction, there is considerable variation in practice, as there also is in the systems based on those switches. In the following sections a generalization of all the systems is described and illustrated by examples which are typical of but not necessarily followed in all systems in practice.

5.2 Trunking

Of the several ways in which the cross-bar switches are connected together to form exchanges of various sizes, one way shown in Figure 5.3 uses switching stages of three kinds. One kind concentrates the exchange line traffic to originating or.j. and to terminating t.j. circuit processors, called junctors, another kind interconnects or.j. and incoming junction junctors i.j. with outgoing junction o.j., service s.j. and terminating t.j. junctors, and the third connects originating and incoming junctors with originating o.r. and incoming junction i.c.r. registers respectively. Figure 5.3 is typical of trunking used in practice although because of drawing space limitations, the switch sizes and the quantities of trunks shown are much smaller than practice requires, and connections through large exchanges need to use more switches in series than in Figure 5.3. The symbol used for the cross-bar switches is horizontal and vertical trunks, it being understood that there is a cross-point at each intersection. Another characteristic readily observed in Figure 5.3 is that the patterns of interconnections between the switching stages are regular and easily defined, which is intentional so as to simplify the path search and selection through the switches when calls come to be connected. An exchange line when it originates a new call is connected through its A and a B switch to an originating junctor or.j. suited to its class of service and thence via a switch F to an originating register o.r. which receives, stores and processes the subscriber dialled directory number digits which define the connection wanted. A calling signal on an incoming junction actuates the incoming junctor i.j. to connect via the switch G to a junction register i.c.r. which receives digits from a register connected to the other end of the junction. When either kind of register has received sufficient digits to define a connection through the exchange switches, which depending on the destination of the call may be all the directory number digits or only some of them, the register actuates a translator-maker-connector switch to gain connection to a translator-maker. The register then repeats to the translator-marker the digits which it has in store and at the same time sends a datum through the F or G switch which is connecting it to a junctor, to the junctor to instruct it to mark the point from which selection is to start. If a connection can be made, there being free trunks and switches able to do so, the connection is made by path search and connection processors associated with the cross-bar switches not individually as in step-by-step systems but in groups. The connection is made via the C, D and E switches to a service junctor s.j., or to an outgoing junctor o.j. and junction, or to a terminating junctor t.j. suited to the class of service of the called station, and thence via B and A switches to an exchange line of the called station. If no effective connection can be made, the register connects busy or n.u. tone, as appropriate, back to the caller. For calls terminating in the exchange on an exchange or service line or tone circuit, the register is immediately released. For calls outgoing on junctions, the register sends a calling signal over the junction to prepare the distant end to receive the call by the finding and connection of an incoming junction register, i.c.r.. The distant register when connected sends a proceed-to-send-the-address datum which received by the calling register, causes that register to transmit to the called register the

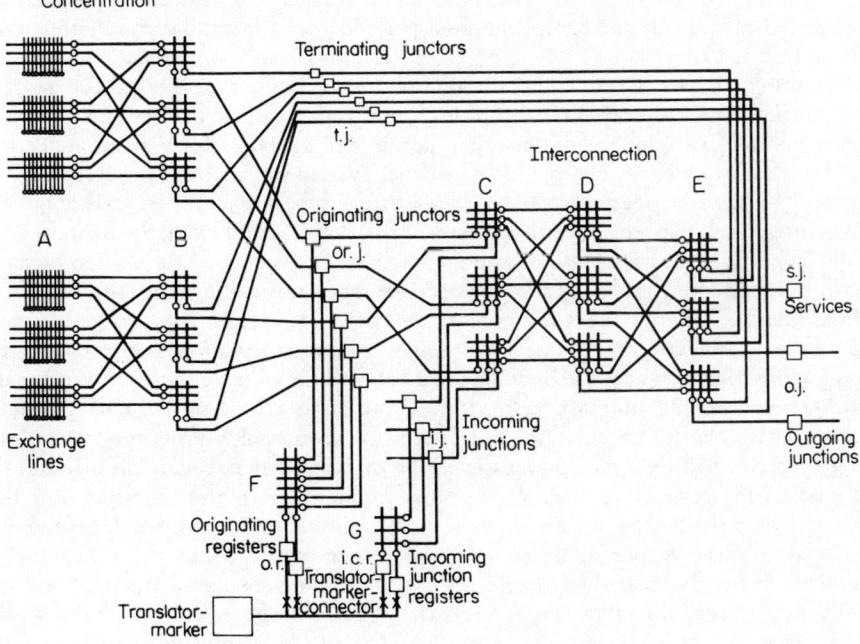

Figure 5.3 Typical cross-bar exchange trunking

information which it needs to advance or to complete the connection. The information is of necessity in the form of decimal digits and being data to be stored before processing, it may be sent at high speed, ten digits per second using multi-frequency coded digits being common compared with ten pulses per second for loop pulse dialled digits. When all data have been sent the calling register releases.

A connection when completed through the exchange and released from the register which established it, contains only the circuit processors, that is the junctors, necessary to the supervision and control of the call up to its release, together with the cross-bar switches which make the connection by bridge magnets operated and held, if not of the latching type, by parallel connection to a P-wire continuous through all the switches and earthed by one of the processors. The release of a connection held in this way through an exchange needs no more than the removal of the earth from the P-wire. On established calls the junctors receive, relay and process signals and data and exercise all control, applying ringing, tones and metering pulses and finally releasing the connections, in the same ways that circuit processors perform those functions in the step-by-step systems of chapter 4. Cross-bar systems are the same as step-by-step systems with register-translator common processors as symbolized in Figure 4.15, except for a different kind of exchange switch which requires a different kind of trunking and control. The principles of the trunking and the operation of cross-bar and all systems other than step-by-step, the trunking of which is determined by the selection process, are simple and obvious as the foregoing description demonstrates, but their application may not be.

Trunking alternative to that of Figure 5.3 is chosen generally for reasons additional to the necessary traffic and interconnection requirements. The system of Figure 5.4 achieves some economy of junctors and switches for a small increase in the control processing compared with the system of Figure 5.3. Referring to Figure 5.4, the connection of an originating register o.r. to an exchange line originating a call uses the concentration and interconnection exchange switches and does not include a junctor. When the register has received all of the dialled digits, it controls the making of a second connection from the calling line, this time to a called line or circuit, the first connection being released. The figure indicates temporary connections at different times to an originating register and the connections finally made, one being a local call and including a local junctor l.j. and another an outgoing junction call with, in the connection, an originating junctor or.j. suited to the junction. After connection to an outgoing junction, the register has to interchange data with a register in the distant exchange which is difficult for the originating register, it having no direct connection to the junction. For that reason, when the outgoing junctor is seized it connects through sender-connector switches to a free sender to which the register communicates all the data which it has in store for the call, then it releases and leaves the sender to continue the call connection processing. Data communication between the register and the sender takes place not through the exchange switches but over bus-wires to which the register connects itself as it begins call connection processing and to which the

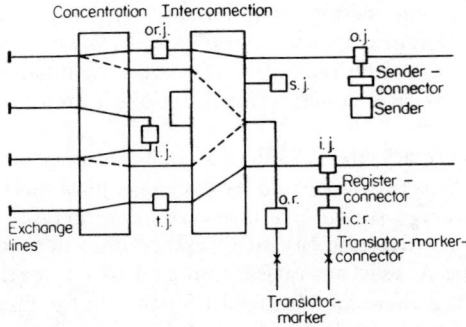

Figure 5.4   Variation of exchange trunking

sender connects itself when it is selected for connection. As will be clear from later descriptions, no other equipments can be connected at the same time to the bus-wires, over which data transmission can thus take place without interference by other connections. Other connections which are made include terminating calls from an incoming junction and junctor i.j. to an exchange line via a terminating junctor t.j., and transit calls from an incoming junctor i.j. to an outgoing junctor o.j. via a transit loop trunk. Incoming junctions and junctors i.j. have register-connector switches for the temporary association of registers, as in the system of Figure 5.3. The trunking of Figure 5.4 is of interest not only as one of the variations from that of Figure 5.3 but also as illustration of the fact common to them all, that whatever the reason for the variation, which may be to gain some economic advantage or simplification of control or even for a company to make its product different from that of other companies, the scope for a reduction in the quantities of exchange switches and junctors and for overall cost reduction compared with the basic system of Figure 5.3, is very limited.

5.3 Construction and Operation

Switch and exchange manufacture require that switches be of very few and possibly of only one standard size, and that exchanges of different sizes and traffic loadings be constructed by the assembly and interconnection of sufficient quantities of the standard units. To be sufficient the trunking, that is the switches and their interconnections, has to provide enough paths in parallel through the exchange to carry the traffic at the grade of service specified by the administration, and enough switches in series to enable any one of the transmission circuits terminated on the exchange, namely exchange lines, service lines and junctions, that might need at any time to be connected together or to internal equipments such as registers, to be so connected. As a general principle, switches are assembled in ranks, each rank being designed to carry at the grade of service specified the traffic expected to be offered to it, and sufficient ranks are connected in series to satisfy the circuit interconnection requirements. In Figure 5.3 the exchange switches to carry the message traffic are in ranks A, B, C, D and E, and ranks F and G carry the register traffics. The ranks F and G are functionally interface switches but in cross-bar systems are constructionally no different from the exchange switches.

In the system of Figure 5.3, the A rank of switches has exchange lines connected to the vertical trunks. The traffic carried by exchange lines varies very widely from line to line and the average traffic per line varies considerably from exchange to exchange. In consequence the quantity of exchange lines needed to load fully the horizontal trunks of the A switches varies from exchange to exchange, for example from twenty to seventy if there are ten trunks for the traffic. One way of satisfying this requirement in practice is to have standard units of the order of twenty verticals and to use for each exchange an appropriate quantity of units in series multiple, horizontal trunk to horizontal trunk, to make up for each A switch the quantity of vertical trunks required. It is part of traffic engineering to determine the quantity when the exchange is designed and subsequently to allocate the exchange lines to the switches so that the lines per switch and the traffic per switch are very nearly the same for all switches despite the fact that the traffic carried by the lines individually varies so widely. Having loaded the A to B trunks as much as possible, the B switches provide further concentration to the junctors between the B and C switches, but little further concentration is possible and the remaining ranks of switches are needed to satisfy the line interconnection requirements. In very small exchanges, switch ranks A and B are sufficient by themselves and in very large exchanges more ranks of switches than are shown in Figure 5.3 may be required. A single cross-bar switch being an inconveniently small unit with which to construct most exchanges, it is common practice to manufacture larger units often called frames because each unit comprises a frame on which is mounted a quantity of switches. The switches are in two and sometimes three ranks, the ranks are interconnected by a pattern of trunks standard for all frames, and included with the frames is much of the common processor equipment needed to control the operations of the switches. The frames are standard factory production units very suited to quantity and economical manufacture. Exchanges of different sizes and traffic loadings as they occur in practice are built up of frames in parallel to form ranks and the ranks are connected in series on site by wiring between the frames, the wiring between the ranks on the units already being in place. By this method of construction, the design and installation of exchanges are simplified and cheapened as well as their manufacture.

The horizontal and vertical trunks of cross-bar switches being interchangeable, in practice the switches are found connected sometimes one way and at other times the other way round without departing from the general principles of the trunking of Figure 5.3. In particular, where the quantity of horizontal trunks is or may be made large by the method of Figure 5.2, an invariable quantity of bridge magnets per switch and their vertical trunks are connected to the trunks to the B switches and the exchange lines to the horizontal trunks, and the quantity of exchange lines per switch is varied by a factor of 1, 2 or 3 by the use of no rows or two or three rows of auxiliary contacts.

The total quantity of equipment required for and therefore the cost of the exchange switches, their interconnection and control, varies with the standard size or sizes of exchange switch is used. For some size not necessarily the same for all ranks of switches, the quantity of cross-points or the cost of the switches and their interconnection or the cost of the switches, their interconnection and control, is a minimum for a given exchange. Sizes thus determined can be different for the different minima and more so for the different exchanges which in a large network commonly vary in size over a range of more than ten to one. The designer has to choose one or two exchange switch sizes to suit best the application for which he is designing, in which he is assisted by the fact that none of the minima is very sharply defined so that substantial departures from minimum conditions have only small effects on the overall result. Designers have also made use of this fact to use larger or smaller switches to suit some objective in which they are interested other than one of those described.

In that transmission circuits terminated on the exchanges are interconnected by metal contact switches in both step-by-step and cross-bar systems, the subscribers are aware of the differences between them only in respect of the speed of connection and the reliability of operation. There are, however, important traffic and engineering differences. One concerns the trunking: as seen from Figures 5.3 and 5.4, in cross-bar exchanges the exchange line originating and terminating traffics are not separated at the entry of the lines to the exchange, as they are in step-by-step exchanges, but after concentration. In consequence, the segregation of exchange lines into divisions for the identification of class of service which is practised in step-by-step systems with increasing difficulty as the range of services offered increases, is impractical with cross-bar systems but nor is it essential. Instead a more convenient and economical method is made possible because of another fundamental difference, that of the means of storage, retrieval and transmission of data within the exchanges. In step-by-step systems as was described in Chapter 4, exchange line and other data are stored on the banks of the exchange switches but in quantities which are small because of practical limitations: the data being stored on the banks means that they are available to control processors only after the selection processing is completed and by transmission over the signal and data paths of the connections made. Cross-bar systems store data by analogous means and in quantities which are somewhat greater, but as the data transmission is independent of the exchange switch connections, the data are available to processors before any control operations or processing takes place. That this is possible and the means by which it is accomplished will be made clear in the more detailed descriptions of system operation given in section 5.4. Here the effects on exchange construction and operation are outlined.

Because the class of service of an exchange line is known before it is connected through the concentration switches, lines of different classes of service may be mixed on the A switches, Figure 5.3, and selection controlled so that the lines become connected to appropriate processors as path search and connection proceeds. Originating and terminating junctors suited to ordinary, party line, coin-box or other class of service, in quantities proportional to the traffics of those classes, are distributed evenly over the outlets from the B switches. When a connection comes to be made, selection is limited to the type of processor indicated by the class of service of the line being connected and of the class for originating and terminating calls separately. At the same time additional data required for the call may be transmitted to and stored in any of the processors which become included in the connection as it is made. For example the kind of ringing to be applied to a line may be communicated to a terminating junctor as it is connected. Also for example, the long distance barred l.d.b. class of originating call service is stored in the register as it is connected and used later to cause n.u. tone to be sent to the caller if he dials a long distance call, the class of service being thus used as control data: Figure 4.5 shows it having to be used as structure class of service data in a step-by-step system. The trunking and the provision of equipment thus become more flexible and able to provide more and better facilities than in step-by-step systems limited to the segregation of lines and switches into divisions each of one class or combination of classes of service, and with an economic advantage in addition.

The connection and data problems of traffic originating and terminating on junctions could be treated in the same way as those of exchange lines if the need existed but the requirements are less varied and effectively limited to the kind of signal and data transmission available over the junctions and the charging rates for calls routed over the junctions. The charging rates are not junction but call data dependent, on the destinations of calls; which being determined by the directory numbers dialled and stored in registers, can be identified by the register data and used to extract from a store the charge rates and transmit them to processors concerned to control the charging of successful calls. The signal and data transmission problem is usually solved by standardizing one method to be used on all junctions. The method chosen is one which readily suits the majority of junctions, and those junctions unable to use the method are individually equipped with transducing equipment to adapt them to the standard system. In these ways class of service problems are generally avoided for junctions.

Despite the intrinsically greater power and flexibility of cross-bar control systems, the services and facilities offered are not markedly in advance of those of step-by-step systems because of the limitations of electro-mechanical processing equipments and of data transmission over junctions as described with reference to Figures 4.9 and 4.16. A marked increase had to await the advent of electronic equipment and processing.

5.4 Path Search and Connection

5.4.1 Generalized Operation

The remaining difference between cross-bar and step-by-step systems concerns the part of call connection processing known as path search and connection and which comprises the finding and connecting of suitable and free paths through the exchange switches. Each step-by-step switch has an individual processor to control it, in Figure 4.15 the prA or line equipment processor for line finding or pre-selection and the prB and prGS processors for selection; but cross-bar switches have processors, called markers, common to groups of switches for one kind of path search and connection used for both line finding and selection. For selection, step-by-step path search and connection starts from a point, a first or incoming junction selector, and proceeds step-by-step through the exchange switches to a destination circuit out of the exchange, and different calls may search for and connect paths through the exchange switches simultaneously without mutual interference. In cross-bar systems, path search and connection is end-to-end, between a defined starting point and one or more defined destination points. The start and destination points being defined by electrical conditions called marks, a marker finds and makes a connection between the start point and a free one of the destination points if there is one free and with a free path between it and the start point, but only one call can be in process of search and connection at one time through the same switches and using the same marker. However, because the rate at which connections can be made is much less than the rate at which they originate in all but the smallest of exchanges, it must be possible for more than one connection to be in process of connection at any one time, to satisfy which requirement the exchange switches and their common control equipments are arranged in sections through any one of which only one connection can be made at a time but different connections may be simultaneously using different sections.

In Figure 5.3 and the generalized system of Figure 5.5, a call datum received over an incoming junction into its incoming junctor marks the junctor as the start point of a connection to an incoming junction register i.c.r., the free ones of which are continuously marked as destination points. A marker immediately responds and controls switches which connect the junctor to a register if a free one exists with a free path to the junctor. Figure 5.5 does not show the switches and markers which are similar to those which are shown for the concentration and interconnection switches. Where calls which originate simultaneously are liable to mutual interference, circuit means has to be used to ensure that the calls are connected one after another and cannot mutually interfere. One circuit means is relays connected as in Figure 5.6 to form a one-only selector o.o.s. with inputs and corresponding outputs, and with only one output corresponding to one of no matter how many inputs may be activated at one time. The circuit can be seen to be analogous to a sliding wiper switch moving from contact to contact and stopping on the first marked contact that it encounters.

Selection in the reverse direction, from one free register to what would usually be only one junction which was calling, would require a greater provision of switches for the same grade of service, because of the reduction of choice limited to one free register.

The starting point for an exchange line originated call is the line equipment of the exchange line, and the terminating points are the originating junctors or.j. to which it has access and provide the class of service appropriate to the line: the originating junctor which becomes connected is the starting point for a connection to an originating register, the free originating registers being the destinations. In Figure 5.3 the exchange lines are in two thirty-line sections. Within one section

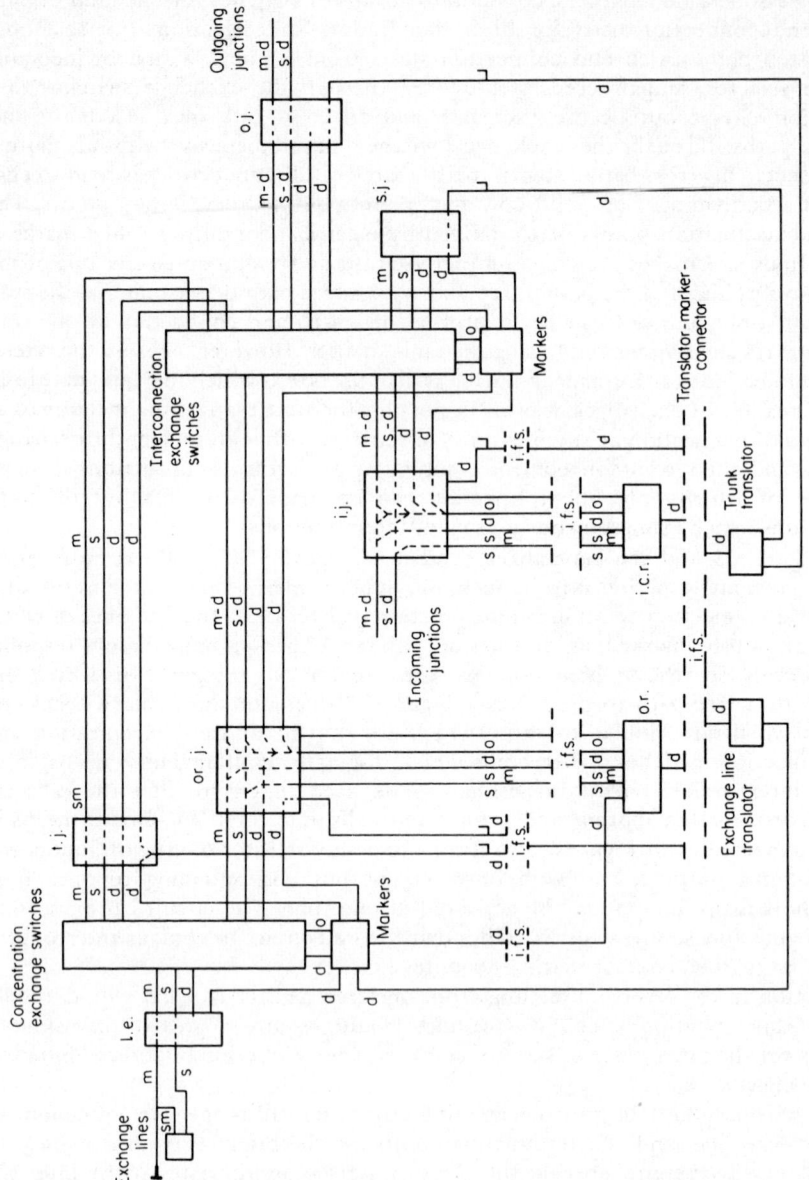

Figure 5.5 Generalized cross-bar exchange

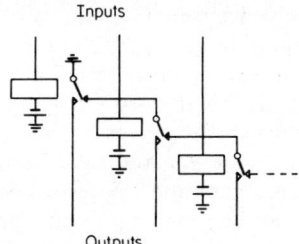

Figure 5.6 Relay o.o.s.

only one line can be in process of connection to an originating junctor, or terminating junctor, at any one time, but connections in different sections may proceed independently and simultaneously.

When a register has received enough digits to make a connection through the exchange, it has to mark the or.j. or i.j. junctor to which it is connected, as the start of the path search and connection, which it does with a datum d transmission to the junctor. It also has to mark the destination points according to the digits received, which means the translation of the directory number digits received to the equipment number addresses of the junctors of a group of junctions to another exchange, of a group of service circuits or of a group of exchange lines which may be only one line. The translation includes the communication of the addresses to the markers concerned, which markers then make the required connection if it can be made. The quantity of equipment required for translation is considerable and the time required for translation short. Translators time shared among the registers is thus both a possibility and an economic necessity, and is shown in Figures 5.3 and 5.5.

The average duration of calls originating from exchange lines and engaging or.j. junctors being usually about 120 seconds, within that period the junctor needs a register o.r. for about 20 seconds on average, so that by time sharing the registers they are only about one sixth as numerous as the junctors. For junction originated calls, the average i.j. and i.c.r. holding times are about 150 and 5 seconds respectively, giving a ratio of about 30 to 1. The time of translation, marking and connection through the switches is of the order of one second which justifies the second stage of time sharing, shown in Figure 5.5 by a second interface between the registers and the translators. One per second is the rate at which exchange line and junction calls originate and are connected through the switches of a lightly loaded exchange. For such and smaller exchanges, one translator would be sufficient but with a not negligible risk of failure the consequences of which could be disastrous, complete break-down of service in fact. Hence there is a security problem which compels at least two translators in every exchange. At the other extreme, large heavily loaded exchanges require connections to be made at the rate of twenty per second and more, which means many translators and problems of mutual interference between connections of necessity in course of simultaneous establishment through the exchange. Security problems, which also affect the markers common to many switches, are solved by duplication. Mutual interference is avoided by sectionalization of the equipment, exemplified by the exchange lines in Figure 5.3, which allows connections to be made simultaneously in different sections, and by lock-out circuits which suspend the processing of a connection that would interfere with a connection for which the processing has already commenced.

Referring to Figure 5.5, the total processing can be seen to be divided between circuit processors, common processors which are the registers space division time shared among the or.j. and i.j. circuit processors, and second common processors which are the translator-markers space division time shared among the registers. Although the division of the processing is thus broadly decided on a functional basis, the detailed division has to be decided by the designer on economic and operational factors, for which reason it varies in detail between systems in practice. The point at which a register has received sufficient data, that is directory number digits, to be able to make a connection through the exchange switches, depends on the destination of the connection, which makes it difficult for the registers to avoid performing at least the first parts of the translations themselves. The requirement is for a register to receive sufficient digits, then obtain connection to a translator as the first step in an attempt to make a connection through the switches. The only way that a register is able to determine that it has received sufficient digits without translating them is by counting them as digits. Some service numbers, the emergency number for one, never have more than three digits, and in a national network the local exchange numbers may have from three to seven digits and junction calls to be sufficiently recognized may need, depending on their final destinations, a quantity of digits from one to five. Therefore merely counting a predetermined quantity of digits to be received is no solution and some form of translation is necessary to determine when a call may be connected through the exchange switches. A method which avoids translation in the registers themselves is for them to count the digits received and when the number reaches the minimum that occurs for any translation, to communicate the received digits to a translator. If the translator processing shows that the received digits can in fact be translated, an attempt is made to establish a connection through the exchange switches, but if not, the translator informs the register which releases the translator and connects to a translator again later after the receipt of further digits. Continuing in this way, a translation is eventually encountered and the call being processed is brought to a positive conclusion. The advantages of this method are that no constraints are imposed on the directory numbering in respect of quantity of digits necessary or which may be used, and the registers are uniform throughout the network and not individual to exchanges. The translators have to be designed individually for each exchange, but this is unavoidable independently of the part played by the registers in translation. The method is well suited to electronic registers and translators not limited by operating speed and durability, but both speed and durability limit its use in electro-mechanical cross-bar systems to exceptional circumstances. In consequence, a general arrangement, assumed for Figure 5.5, is translation in the registers to detect local exchange calls. A local call being detected, further digits received are counted by the register up to the number appropriate to local calls, and the register then connects to an exchange line translator for the connection to be made. Otherwise, the register connects to a so-called trunk translator after receiving digits which by translation it knows to relate to a service call or which by counting the digits it knows to include most other calls. Only the exceptional calls cause the translator to inform the register to come back again after the receipt of further digits. Where, because of traffic and the consequent rate at which calls must be connected, more than one translator of each kind is required, division of the translation between two or more different kinds of translator has economic advantages.

A translator given translatable digits received and stored by the register, produces the addresses of corresponding destination circuits and communicates the addresses to the markers as electrical condition marks on marking wires. The electrical condition most convenient to use is earth potential applied to mark. The marks may be applied directly to the markers, as shown in Figure 5.5 from the exchange line translators to the markers for the exchange lines, the markers having to determine the states of the lines, free or engaged, or indirectly, as shown from the trunk translators to the service and junction junctors and thence to the markers only if the junctors are free.

Referring to Figure 5.3, an exchange line originating a call may be connected to an originating junctor which has no free path to a free register o.r. even though some other junctor which might have been chosen does have a free path to a free register. Commonly for that reason circuit means are used to make available to calling lines only junctors which if selected can obtain connection to a register.

In Figure 5.3 only one call at a time can be in process of connection through the C,D and E switches but larger exchanges for which simultaneous path search and connection is essential are trunked differently, with more ranks of switches. The final ranks being made up of sections similar to the A—B sections of Figure 5.3, their outputs are connected to terminating, service and outgoing junctors distributed as groups as evenly as possible over the various sections, the terminating junctors being in class of service groups, the service junctors in service groups and the outgoing junctors in groups associated with groups of junctions to other exchanges. Between those sections and the originating and incoming junctors are other ranks of switches also in sections. Path searches and connections may proceed simultaneously and independently provided that no two connections use the same section at the same time; a condition which is ensured as already mentioned by circuit lock out such that when a section is in use for one call it is locked out of use by any other call. A path search which encounters a locked out section will either choose another section if it can or wait until the lock out ceases. On the other hand to reduce abortive searches, no section is taken into use if it has no free outlet to the kind of destination circuit required for the connection which is to be made.

A register having received sufficient digits for a connection through the exchange to be made, and having obtained access to an exchange line or trunk translator as appropriate, transmits to the translator some or all of the digits which it has received from an exchange line or from a register in another exchange. It also transmits a datum to mark, as the start of the connection to be made, the originating or incoming junctor to which the register is connected through the register connector interface switches. The translator operates markers to mark the destination circuits for selection, and common processors associated with the exchange switches then make a connection from the starting circuit to a free one of the destination circuits if there is a free one with a free path from the starting circuit. In the case of terminating calls to exchange lines, the connection has to be made in at least two stages: in Figure 5.3 from a destination exchange line to a terminating junctor and from the terminating junctor to the originating or incoming junctor or in the inverse of that order. Not any free terminating junctor can be selected to make the first stage of the connection. Not only must it provide the class of service required by the called line, but the marking is arranged so that when the first half of the connection is made, it is made to a terminating junctor to which it is known that the second half can be completed. To confine the search to one exchange line section, all the exchange lines in a group with one directory number are connected to one exchange line section unless the group is large, when it is divided between different sections, and a section with at least one free line is selected for terminating call path search. The terminating junctors available to one exchange line section are distributed over the interconnection sections if there is more than one, as groups of outgoing junctors and service junctors are so connected. Finally, originating and terminating calls within the same exchange line concentration section cannot be in the course of connection simultaneously: so one in operation must lock the other out.

5.4.2 Exemplary System

Some idea of the practical implementation of the processes described is conveyed by the circuitry of Figure 5.7 part of which is expanded in Figures 5.8 and 5.9, and the following description. As was the case with step-by-step system circuits, the cross-bar system circuits which are given here and later do not relate to any particular system in practice but are designed to illustrate the problems involved and typical ways of solving them.

In Figure 5.7 the circuit convention for cross-bar switches is relay contacts multipled on both sides, as shown for the A and B switches between the exchange lines and the originating or.j. and terminating t.j. junctors which should be related without difficulty to the trunking diagram of Figure 5.3. Three-wire connections are made through the switches, as for step-by-step systems, one wire being a P-wire which is connected to earth potential to busy and hold connections. The exchange line circuit processors have L relays, dm bridge magnet contacts equivalent to K relay contacts and message registers, MR, as in step-by-step systems and in Figure 4.1. The switches are in sections, in Figure 5.3 in sections of thirty lines, each section with its own common marker and equipment for line selection and connection. The vertical trunks occupying the same position in the different switches of one rank being designated a file, each exchange line within a section is specified by an A switch and file address according to the vertical trunk of the switch to which the line is connected. The address is given to the section marker as marks on file and switch wires, the marks being applied for originating calls by the contacts of the line relay L when operated by a calling loop, and for terminating calls the marks are applied by an exchange line translator given the directory number of the line by a register. Figure 5.8 shows the A switches of one section comprising three switches each with three horizontal trunks to B switches and three vertical trunks to exchange lines each with an L calling relay. The P-wires of the trunks are shown with connections between them which can be made by the switch cross-points indicated by crosses. The message transmission circuit wires of the trunks are not shown as they do not take part in the path searching and selection. Marking wires for the horizontal trunks are shown marked o for operational data and they terminate on select magnets as shown in Figure 5.1. In fact, as will be seen from Figures 5.7, 5.8 and 5.9, both ends of the A–B trunk o wires terminate in select magnets, one end in the A switch and one in the B switch. The P-wires of the A switch vertical trunks are also the operational o wires, also as seen in the diagrams. A cross-point is operated at the intersection of a horizontal trunk and a vertical trunk by a mark first on the o wire of the horizontal trunk to operate the select magnet, followed by a mark on the vertical o wire to operate the bridge magnet, as was described with reference to Figure 5.1. For the B switches the P-wires and the o wires of the vertical trunks are separate. In Figures 5.8 and 5.9 the o wires are shown but not the operate magnets.

The file and switch co-ordinates which are the equipment number of an exchange line within a section, are indicated to the section marker by a mark earth on a file wire common to all the lines in a file and mark on a switch wire common to all the lines on a switch. Because of the use of common wires there is ambiguity in the addresses of more lines than one marked simultaneously if more than one file common is marked at the same time as more than one switch common. For that reason circuit means ensure for both originating and terminating calls that lines marked more than one at a time are all in the same file and thus only one file wire is marked even though more than one switch wire may be. A line originating a call operates its L relay and the two contacts of the relay mark the file and switch common wires defining the equipment number of the line, but to be effective the file marking must operate a file relay FAx, Figure 5.7, via a one-only selector o.o.s.l.. In figure 5.8 file relays FAA, FAB and FAC are shown. A file relay connects the switch marking contacts of all the L relays in the file to the switch wires ay which are wires a1, a2 and a3 in Figure 5.8. For terminating calls an exchange line translator given the directory number of a line or lines to be marked, it translates the number to equipment numbers and marks one file terminal e.n.f. in Figure 5.7, the mark being earth potential which operates the file relay, and it marks one switch terminal e.n.s. for each line. That only one file terminal is marked is ensured by allocating all lines with one directory number to one file in the section. If one file of lines is too few for the group, the lines are allocated to files in different sections; and if there are too few sections for all the lines of a group to be accommodated by one file in each, the group is divided into sub-groups, each sub-group is connected as if it were a group and the sub-group to be used when a terminating connection is to be made is determined by one-only selection. At the same time that an exchange line marker marks lines in a section, it operates a lock-out relay TC which prevents originating call connections despite L relays being operated.

The operation of a file relay also connects the P-wires of all the lines in the file to P relays in the marker, one P relay for each and which operates if the line is already connected through the switch. A break contact on the P relay disconnects the line 'switch' marking from the switch wire ay, to prevent a line already engaged from being marked for connection. This is necessary to terminating calls because the translator applies marks for lines irrespective of whether they are engaged or not. It is normally not needed for originating calls because the L relay is disconnected and unable to operate when the line is connected but under some fault conditions it is useful to be able to remove a persistent mark by making a connection to the line.

A switch having a connectable exchange line being marked on its ay wire, is required to mark all the free trunks from that switch to a B switch, for which purpose a marking wire mk and relay TK in series are provided for each trunk. The switch wires ay are connected through decoupling rectifiers to the mk wires, each switch wire being connected to the mk wires of all the trunks to which the switch has access, which reference to the trunking diagram Figure 5.3 will make clear. An earth mark on a switch wire produces current in the mk wires and operation of the Tk relays of all the trunks to which the switch has access except those trunks which are already engaged. Current for an mk wire and TK relay is obtained from the exchange battery via a resistor which is shortcircuited to earth by earth potential on the P-wire of the trunk, the P-wire being earthed when the trunk is busy. Hence when a switch or switches are marked, TK relays for all the usable A—B trunks to those switches are operated and contacts on the relays are used to mark the B switches on which they terminate, each B switch having a marking wire bz similar to the ay marking wires of the A switches. In Figure 5.8 there are nine relays TK1 to TK9 corresponding to the nine trunks between the A and the B switches, connected through resistors to the exchange battery indicated by the alternative symbol of an arrow head to simplify the drawing. There are bz wires b1, b2 and b3 earthed by TK contacts and continued to marking wires MK4 for the switches B1, B2 and B3 respectively.

The vertical trunks of the B switches terminate in junctors providing the various classes of service required for originating and terminating calls. For each connection which is made the bz switch marking has to be confined to trunks providing the class of service appropriate to the call. Distinction between originating and terminating junctors is made by TC relay contacts in series with the switch marking wires. With the TC relay unoperated, a mark on a switch wire bz marks all the originating junctors or.j. of the switch, by applying earth to the MK4 wires of the junctors. If a junctor is already in use, its MK4 wire is disconnected by a B relay contact but otherwise the earth on the MK4 wire marks, via decoupling rectifiers, file and switch wires in the same way that A switch lines are marked. At that point all the circuits and trunks which could possibly be used for an originating connection are marked, namely, a file of exchange lines with individual lines in the file distinguished by a switch mark ay, A–B trunks with TK relays operated, B switches with their switch wires bz marked and originating junctors indicating their B switch and file positions. The common marker processing now has to select and connect a unique path containing a junctor of the class of service appropriate to the exchange line.

Class of service selection is facilitated by imposing constraints on the trunking, firstly that all the lines in one A switch file of one section of switches shall have the same class of service. This requirement is compatible with all the lines with one directory number also having to be in the same file because all the lines have the same class of service with few exceptions which can be taken care of by other means. Hence the common file marking also defines the class of service. The second imposition is that all the originating junctors in a file of the B switches shall be of the same class of service, which reduces the class of service selection of junctors to a selection of a file. The circuitry is further simplified by the use of a basic class of service which applies to more exchange lines than does any other class and which is given is the absence of specific class of service data. Referring to Figure 5.7, the basic class of service applies so long as relay BSO is not operated, and selection starts from the B switches by the selection of a file by a one-only selector o.o.s.2 and the operation of a file relay FBw. In Figure 5.9 file relays FBA, FBB and FBC are shown in detail, the first two being associated with basic class of service junctors and to which selection is limited, when the BSO relay is unoperated, by BSO relay contacts in the input leads to the one-only selector. The file wire of an A switch file of exchange lines requiring other than basic service is jumpered via a decoupling diode, as shown in Figure 5.7, to an originating class of service relay CSp of which there is one for every class of service other than the basic class. A class of service relay when operated operates the BSO relay to remove the basic class from selection and substitutes the class of service required by the calling line, by means of contacts, the contact CSA in Figure 5.9, in the inputs to the one-only selector o.o.s.2.

The B switch selection being limited to the files providing the required class of service and the file relay FBw being operated, the B switches on which the call may terminate are marked by earth connections to switch wires sbz, one for each switch, and the switch wires are connected through TK relay contacts to A–B trunk marking wires mtk. In consequence, all the A–B trunks which might be used are marked and one is selected by the one-only selector o.o.s.3. The output from the selector defines the trunk to be used and therefore, because the A and B switches are already selected, an exchange line in a A switch and a junctor in a B switch are also defined, and selection is complete. The earth output from the one-only selector makes the connection by operating directly the select magnets for the trunk, being magnets in the selected A and B switches, and operating, via decoupling diodes, common switch wires and file relay contacts, the bridge magnets of the selected exchange line and the selected junctor. Before the marking can be released, the connection which has been made must be held from the originating junctor by an earth potential on the P-wire through the switches. This occurs when the A relay in the junctor is operated by the loop on the calling exchange line and the B relay is operated by an A relay contact. In practice the circuitry is more complicated than is shown to ensure that the select magnets have time to operate before the bridge magnets, that the bridge magnets are held by the marker long enough for the A and B relays in the junctor to operate and take over the holding of the connection, and that all the operations associated with one file shall have been completed before those for a calling line in another file may commence. Lines calling but unconnected and existing simultaneously in the same section of the exchange are a rarity normally but many occur during and after peaks of traffic which cause all junctors or registers to be in use and calling lines have to wait for equipment to become free.

The selection and connection of exchange lines for terminating calls is very similar to that for originating calls but with the difference in the example being considered that it is made in two operations as previously described, to ensure the selection of a terminating junctor on a free path from the start to the destination of the connection. Detailed description follows later in the continuing account of originating connections. In practice, originating connections are also controlled to ensure that an originating junctor selected for connection to an exchange line can also be connected over a free path to a free register.

When the B relay in the originating junctor operates, it extends a mark MK1 to indicate the starting point of path search and connection to a free one of the originating registers o.r., all of which are continuously marked as possible destination points except when locked-out by terminating calls. When a register is connected, transmission over eight wires is available between the junctor and the register. The three wires of the connection through the exchange switches are diverted to the register over wires r2, r3 and r4, via contacts of the relay DB, the break in the transmission being bridged by transmission from the register over wires r5, r6 and r7. When the connection through the switches is complete between the junctor and a register, it is held by the earth connection over the r2 wire, and current from the register over wires r3 and r4 operates relay I to remove the marking from MK1. The I relay locks to one of its own contacts until the junctor is released.

Originated call data required by the register at some later stage of the call, such as long distance barred class of service needed during call connection processing, are generated by the mark which operates the A switch file relay, Figure 5.7, and transmitted, as data d in Figure 5.5, to all the registers connected to a section of switches through which the connection is made, and accepted and stored by the selected register as it becomes connected. The transmission needs interface switches to steer the data through the sections of switches involved in making the connection, but being concerned with sections and not with individual trunks as are the interface switches between the registers and the junctors, the switches are different and shown separately. As the switches are also small in quantity, the paths through them can use many wires in parallel for the transmission of many data.

A register having been connected, it receives dialled digits from the calling line. The digits are stored and when sufficient to define a connection through the C, D and E switches to a junction, to a service junctor or to an exchange line on the same exchange, the register starts the call connection processing by gaining connection to a translator-maker to which it communicates the received dialled digits, and at the same time earths the wire r8 to provide the MK2 mark datum which defines the start of the connection. The translator-marker marks all the possible terminating points of the connection as will be described in section 5.5. The path search and connection equipments of the switches between the start and the terminating points select and make a connection if one can be made. The register having looped the wires r5 and r6, senses when a connection is complete from current flow in the loop and it holds the connection forward of the junctor by earth potential on the r7 wire until the relay DB is operated and through transmission established through the junctor. The relay DB is operated by an o datum from the register when its processing is completed. Earth potential on the wire r1 operates the relay DB which in addition to establishing through transmission of the message and signal paths, removes the holding earth from wire r2 so that the register and the switches connecting the junctor to the register are released.

The circuit operations involved in call connection processing to connect a calling originating junctor to an exchange line on the same exchange can be gained from the diagrams of Figures 5.7, 5.8 and 5.9 and will now be described.

When for a terminating call a line translator is given the directory number of a station to be called, it marks a directory number terminal d.n. corresponding to the number, as will be described in detail with reference to Figures 5.10 and 5.11. The terminal has jumper wire connections to equipment number terminals e.n. which define the file and switch addresses within a section of exchange switches wherein the lines with that directory number are located. In the system of Figure 5.7 two equipment number terminals per line, e.n.f. and e.n.s., are required to define the file and switch co-ordinates separately, as previously described together with the operation of a TC relay to make terminating junctors available instead of originating junctors. Terminating junctors are of different types to suit the exchange line classes of service for terminating calls: typically ordinary, coin-box and p.b.x. without d.d.i. are included in a basic class, with party line and p.b.x. with d.d.i. as additional classes. The different types of terminating junctor are distributed over the B switches as for and intermingled with the originating junctors, and the junctor individual marking wires MK3 are disconnected by B relay contacts as are the marking wires MK4 of the originating junctors, if the junctors are already engaged. There are class of service relays BST and CSq corresponding to the relays BSO and CSp for originating calls, but with contacts individual to terminating junctors and not to files of junctors as for originating calls. The difference occurs because the MK3 marking has to define not only a B switch to which a free called line has a free path, the MK3 mark corresponding to an MK4 mark for a calling line, but also an individual terminating junctor of the switch. The junctor is a possible terminating point for a path search and connection operation first to be carried out through the C, D, E switches and starting from an originating junctor or incoming junctor depending on the origin of the calling line. The terminating junctor must be of the class of service required by the called line or lines and, as for originating junctors, a basic class of service is assumed in the absence of a specific class being defined by the operation of a CSq relay. Any CSq relay which is operated operates the relay BST to suspend the basic class of service as well as substituting a defined class of service. An exchange line has both originating and terminating classes of service corresponding to its one equipment number e.n.: these classes are defined separately by relays CSp and CSq because the originating and terminating classes of service are used in many different combinations by the exchange lines of an exchange.

Referring to Figures 5.3 and 5.7, the MK2 marking of an originating junctor defines the start of a connection to be made in the same way as the L relay of an exchange line defines the start of a connection from an A switch to an originating junctor. The marking of the exchange lines to one of which a terminating call is to be made causes all the free terminating junctors of the required class of service and having a free path to a free called line, to be marked as terminating points for the connection through the C, D and E switches. If a connection can be made, it is made first from the originating junctor to one of the marked terminating junctors and then from the terminating junctor to a called line using, for each of the two processes, the same path search and connection techniques as were described for the originating call made through the A and B switches. In more detail and referring to Figure 5.7, a line marker e.n.f. file marking operates a file relay FAx and the switch marking on terminals e.n.s. marks the line or lines which may be used in the file. P relays are operated for lines already engaged so that only free lines are marked as far as the ay wires. The operation of the TC relay removes the originating junctors and substitutes the terminating junctors in the path search and connection processing which then proceeds as described for originating calls to the point of marking the usable junctors at their MK3 wires, the equivalent of the MK4 wires including class of service selection. The marks on the MK3 wires of free terminating junctors co-operate with the MK2 marks from the originating junctor to make a connection through the C, D and E switches, the selected and connected terminating junctor t.j. then receiving current in the message wires to operate its relay A. It also receives earth potential on the P-wire through the C, D, E switches. The A relay operates the B relay, a contact of which supplies B switch file and switch data to the A and B switch marker which in turn, already receiving A switch file and switch data, completes the connection to an exchange line. The completion is detected in the t.j. junctor by the presence of earth potential on the P-wire through the A and B switches, the H relay operating to extend to the whole connection the P-wire hold from the or.j. junctor, or from the i.j. junctor if the call was incoming over a junction, and to remove the file and switch markings.

A terminating call may have a main class of service to gain connection to a terminating junctor of the appropriate kind, and another class of service to control some feature of the junctor operation. For example, terminating junctors may normally apply loop ringing, such ringing being suited to non-party lines and to X parties of party lines: but a junctor receiving as it is connected data denoting Y party class of service, will apply ringing suited to Y parties of party lines. The translator provides the additional class of service, to mark via interface switches wires to which the relays DR in the junctors are commoned, using battery potential marking. If such a relay is marked in a junctor which becomes selected and

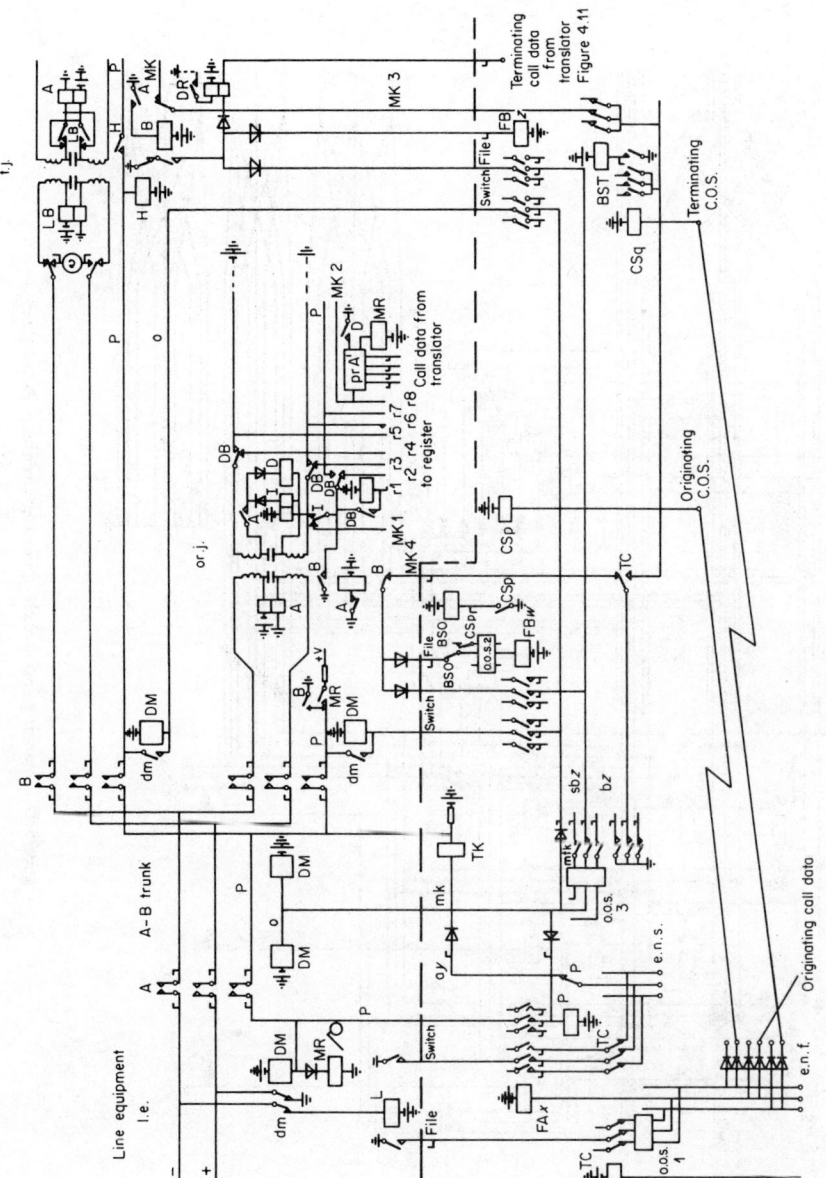

Figure 5.7 Marker path search and connection

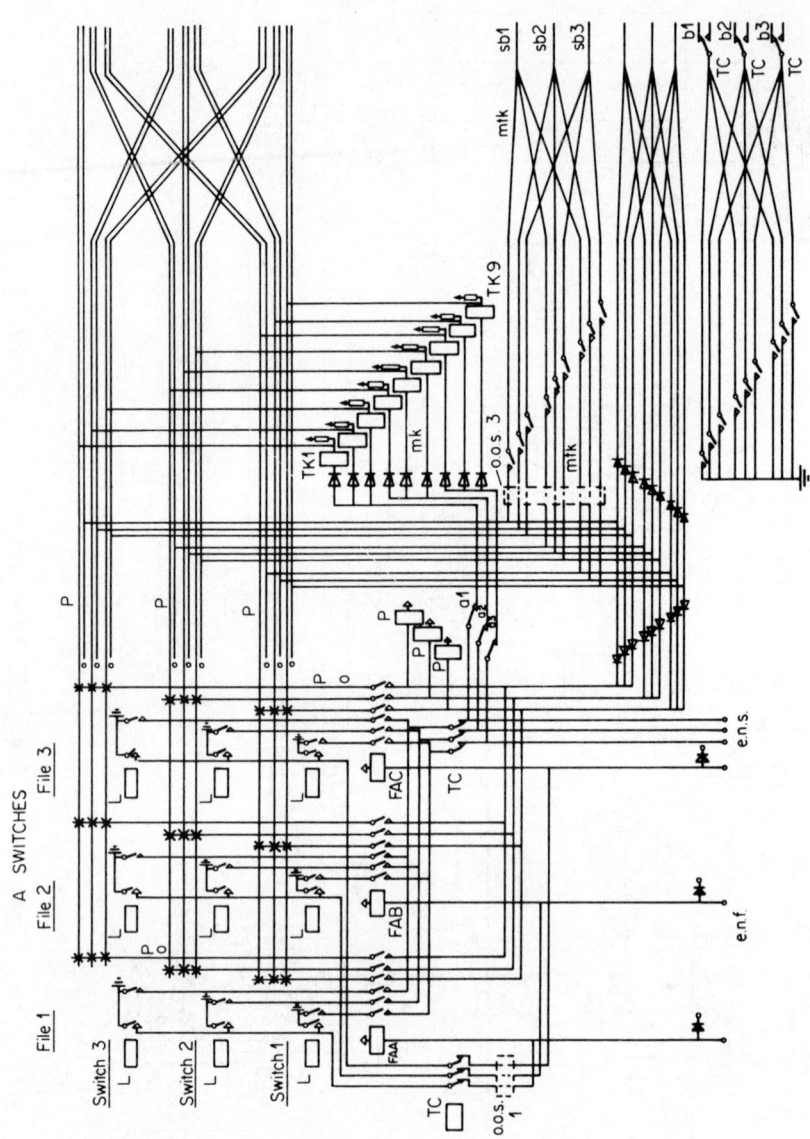

Figure 5.8 Marking and switch operating circuits — A switches

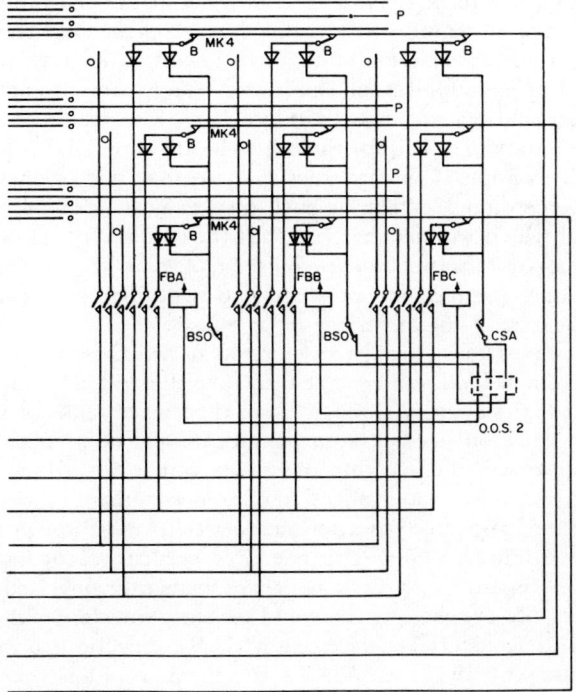

Figure 5.9 Marking and switch operating circuits — B switches

connected for the call, it will be seen from Figure 5.7 that the DR relay is operated as the junctor is marked, to store the class of service temporarily to affect the processing for that call. The call data previously mentioned as communicated to a register as an originating line is connected to the register, is communicated in the same way over bus-wires common to all the registers and accepted into the register which becomes connected for the call. Call data dependent on the destination of the call may may also be required notably charge rate dependent on the final destination. At the time of call connection processing, the originating junctor or incoming junctor involved is uniquely marked within its section by earth potential on its MK2 wire and this enables the junctor to accept data communicated, as shown in Figure 5.5, from the translator through the translator-marker connector interface switches to wires common to all the junctors which might be concerned in the connection. The processor of the originating junctor of Figure 5.7 is shown with a part prA' as a black box which includes means for accepting call data over common wires when the MK2 marking wire is earthed. The charge rate received in this way as the call is connected causes, when the call is answered and the D relay operated, the MR relay to be operated periodically at the charge rate to add each time one unit to the meter MR of the calling line. Exchange line translator-markers commonly do not supply charge rate data, the lowest charge rate being assumed in the absence of data communicated. The lowest charge rate also applies usually to calls to near neighbouring exchanges, with no charge rate data needed. It is possible to have a charge rate stored and for the call to be answered, but for the charge rate not to apply, for example if the call goes to a freefone line or unexpectedly to an operator for interception. Charging in such cases may be avoided by with-holding the answer signal which would otherwise start the metering, even though the call is answered and message transmission established. A new system would preferably have some different arrangement amounting to two kinds of answer, one starting and the other not starting the metering.

Markers for connections through three ranks of switches, the C, D, E ranks of Figure 5.3 for example, follow the same principles illustrated by Figure 5.7 but are more complicated. Connection through more than three ranks of switches in one stage of marker selection becomes too complicated to be practical. If more than three ranks are involved, the selection and connection is divided into two and three rank stages, as for example the connection of an originating or incoming junctor of Figure 5.3 to an exchange line for a terminating call, first through the C, D and E ranks and then B and A ranks. The needs of security of operation also cause complications. In Figure 5.7, a fault which permanently operated an L relay or earthed one of its file or switch wires would seriously interfere with connections in the section until it or its effects were removed. A connection to the line removes the switch marking by the operation of the P relay, but if the file marking is persistent, it prevents all files of lower order in the one-only selection of o.o.s.l from being connected. For that reason a second one-only selector with the selection order reversed from that of the first is required with means of using the more suitable one as necessary. The effects of faults in the one-only selectors and relays of the marker vary from minor to disastrous, leading to the need for some if not complete duplication of the marker. The translators are similarly vulnerable and are usually partially if not completely duplicated. Being equipment common to much other equipment, the duplication and other complications of the markers is not a very significant part of the total cost.

Further complication arises from the fact that not all path search and connection processing is successful even without faults. No connection may be possible because all the destination circuits are engaged or there is no free path to a free destination circuit, or the number dialled does not correspond with a working directory number. The reasons for failure have to be determined and appropriate action taken. Usually busy or n.u. tone has to be sent to the caller, or if all the junctions on one route are busy, the call may have to be directed over an alternative route. The processing is a question of data communication between and programs within the units of equipment involved. Tones may be sent from the junctor or the register, or the marker may mark service junctors for connection, the junctors supplying the tone or whatever processing is required. Alternative routing may be a program in the translator or the register. The precise arrangements depend on the system and the circumstances.

5.5 Translation and Translators

A register having decided that it has received sufficient digits to start call connection processing and the kind of connection to be made, it connects to a free one of the appropriate kind of translator. If none to which it has access is free, the register waits until one is free, the order of waiting time being one second. When connected to the translator, the register transmits to it the appropriate digits at high speed, usually in coded form one digit at a time serially over a few wires or the digits in parallel over many wires. The transmission takes place through the translator-marker connector which may be constructed of relays, or of cross-bar mechanisms if the translators are numerous. The translator decodes the digits to an identification of the exchange line or lines, or of the service lines or outgoing junctions, to a free one of which is to be connected the junctor to which the register making the connection is itself at that moment connected through the interface switch.

5.5.1 Exchange Line Translators

Figure 5.11 indicates one form of exchange line translator, for exchange lines having numbers comprising four decimal digits and divided into blocks of five hundred numbers. The quantity of exchange lines served by five hundred exchange numbers may be more or less than five hundred, depending on the relative quantities of numbers allocated to groups of more than one line, mainly p.a.b.x. groups, and of lines requiring more than one number, mainly party lines and lines having special services, such as night service, in addition to normal services. The lines are a number group which is contained within one concentration section of the exchange, within which only one connection may be allowed to be in the process of path search and connection at any one time. For that reason the translation and marking within one number group may be treated as one process with no clearly defined separation between the translation and marking, and the total equipment is called translator-marker. Equipment which is termed translator performs only translation but generally only selected stages of translation, which is true of the translator of Figure 5.11, the remainder being found in the markers.

The principles of translation are simple but not the drawing, explanation and understanding of a necessarily symbolical diagram of a full-scale translator which Figure 5.11 represents. For that reason a translator miniaturized so that it can be drawn in circuit detail to illustrate the principles, is given as Figure 5.10. The translator is for exchange lines having numbers comprising three ternary digits and divided into blocks of nine numbers, there being a total of 3^3 or twenty-seven numbers and three blocks. The diagram shows two translators which may be in use simultaneously and independently of one another provided they do not both want to use the same block. A register having received three digits called N for nines, T for threes and U for units, connects to a translator to which it sends the digits. The translator stores the digits and uses them to mark one wire in each of three sets of three wires. The mark for the N digit operates one of three relays N0, N1 or N2 depending on the digit being 0, 1 or 2, but only if the corresponding relay located in the other translator is not already operated. A lock-out circuit not shown in the diagram is therefore necessary to prevent two N relays corresponding to the same digit in different translators from being operated simultaneously. If the N digit is x and the Nx relay is operated, the T digit operates one of three relayss Tx0, Tx1 or Tx2: and the second digit being y, the Txy relay is operated and the third digit z marks one wire xyz of twenty-seven wires. The two translators can operate independently in this way subject only to the N relay lock-out. It will be seen that the contacts form what are called trees with the trunks of the trees at the entry to a translator, then branches which lead to more branches. Each translator has access to three trees ending in relays (one tree via relay N0 ends in relays T00, T01 and T02) and three ending in marking wires (for one tree the wires 000, 001... 022 as can be seen from the diagram) and the topmost branches of the trees are common to both translators.

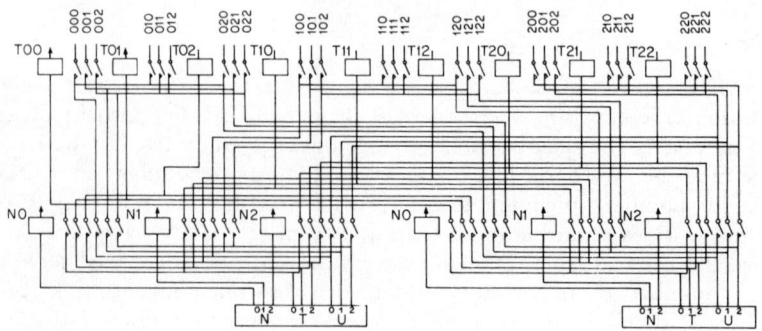

Figure 5.10   Miniature translator

More than one translator is necessary for security reasons and more than two may be necessary for traffic reasons, as previously explained. The trees for the separate translators have to be commoned at some point. If they are commoned at the marking wire outputs, each translator has to have a full complement of relays but insecurity, measured as the chance that not one translator is able to operate satisfactorily because of faults, is a minimum. With the topmost branches commoned as in Figure 5.10 there is an economy of relays at the price of some loss of security. If four or more digits have to be translated further economy of relays is possible by commoning the trees at two levels of branches, which is a feature of the translator of Figure 5.11. The design of translators is thus a question of satisfying traffic, security and economic problems.

Figure 5.11 shows one of a quantity of translators provided according to the traffic to be carried but an even number so that they can be paired for security. A register connected through the translator-marker to a translator transmits to the translator the exchange number digits which it has received and which the translator decodes to marks on one of ten wires for each of the thousands Th, hundreds H, tens T and units U digits. A wire is marked when it has earth potential connected to it. The H wires 6 to 0 are connected through decoupling rectifiers to a relay Z which is operated when any of the H digits 6 to 0 is marked. Combination of the Th marking wires with Z relay contacts gives access to twenty demi-thousands relays dThz, one for each five-hundred-number group. The working numbers on an exchange may not need all of the number groups, the relays for those not required being substituted by equipment which will cause n.u. tone to be sent to the caller if a number in any of the groups has been dialled. Because not more than one call may be in process of connection at any one time within a five-hundred-number group, the operation of a dThz relay in one translator must lock out operation of the corresponding dThz relays in every other translator. For this purpose, the battery supply for the same number group relays in the different translators is available through the contact of a relay HSz common to them all, and the contact is operated when any one of the dThz relays is operated, to prevent the operation of any other. The relay which operates locks to one of its own contacts and the same contact supplies current to operate and hold the

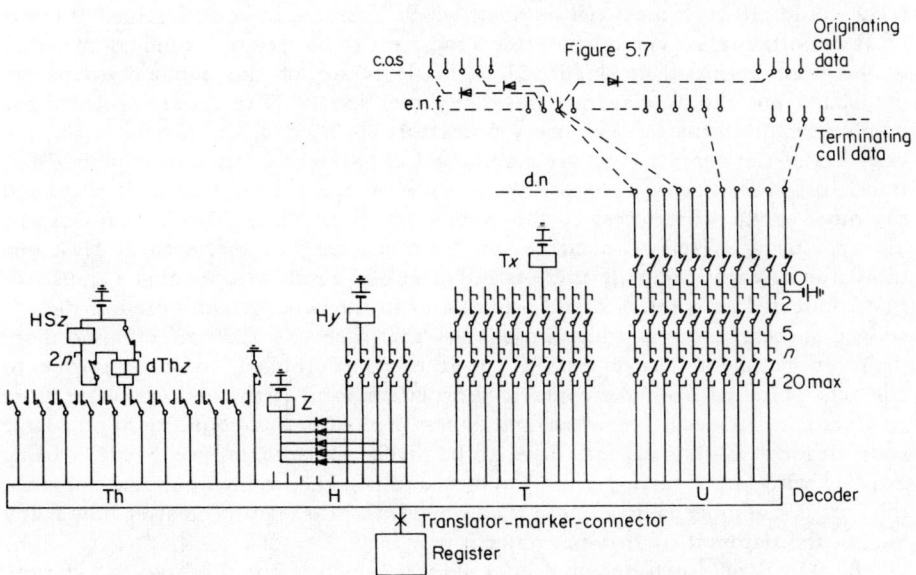

Figure 5.11   Exchange line translator

common relay. Another contact on the dThz relay in closing informs the register that call connection processing may proceed. The operating speed of the common relay HS is so much higher than that of the number group relays that even if two number group relays try to operate simultaneously, only one will succeed. Also the reliability of the HS relay must be beyond question as its failure would affect five hundred numbers and something more complicated than a single relay would be used in practice.

An operated number group relay in a translator connects the appropriate five H wires and the T and U wires of the translator to the number group equipment which comprises five H relays and 50 T relays. Each H relay has a set of ten contacts to connect the T wires from the translator to ten T relays, each with ten contacts connected on one side to directory number terminals d.n. and multipled on the other side with the contacts of the other nine relays of the group, the multipled contacts being connected through a second set of ten contacts of the H relay to the U wires from the translator.

By means of the equipment described, an exchange line translator given by a register a working number in a five-hundred-number group not already in use, causes a directory number terminal d.n. individual to the number to be marked by earth potential. The d.n. terminals are cross-connected by jumper wires to the equipment number terminals of the exchange lines corresponding to that directory number. The equipment number terminals are those shown in Figure 5.7. Figure 5.11 shows only the e.n.f. equipment number terminals: a duplicate set of d.n. terminals cross-connected to the e.n.s. terminals is also necessary. A faulty T relay would affect at most ten numbers which is assumed to be tolerable but faults on H and other relays could have effects too great to be accepted. Some duplication is therefore essential. In Figure 5.11 the H relays of the number groups are duplicated but the T relays are not. Only two sets of H relays are provided per number group no matter how many translators are needed, and to achieve this the branches of the contact trees are commoned at two levels. An even number $2n$ of translators is used in two groups, one group with access to one set of H relays and the other group with access to the second set of H relays. The T relays can be operated using H relays in either set. Each register has access to at least one translator in each group. If there is no successful result of operating a translator first connected, it is released and a translator in the other group is used. If there is still no success, there is a high probability that there is a fault affecting no more than ten numbers. Disconnections are the main problem, most being due to contacts which fail to make contact. Short circuits and earth connections are more troublesome but occur less often, and do not prevent proper connections on a large scale although some calls may have to be dialled more than once. Security being satisfied with one T relay per ten numbers and two H relays per hundred numbers, the cost per number is little more than one tenth of a relay no matter how many lines or translators there may be, which is very little.

The translator, when operated by a number which is not allocated to a station, marks terminals which result in n.u. tone connected to the calling line. A working number causes file and switch equipment number terminals to be marked. Unlike step-by-step systems, the exchange lines of a group do not each absorb one directory number but only one number for the group. A translator given the number for a group causes one file and more than one exchange switch to be marked. More exchange lines in a group than switches in a file, which in most exchanges occurs for only a small proportion of the exchange lines and an even smaller proportion of the stations and directory numbers, is accommodated by special circuitry. The lines are connected in files preferably, for traffic loading reasons, in different switch sections and when the group directory number occurs, one of the files is selected for use. Exchange lines with two directory numbers result in two d.n. terminals being jumpered to one e.n. terminal, without difficulty if none other than the three terminals is involved, but if any of the three terminals is also jumpered to yet another terminal, rectifiers or other means of decoupling are needed to prevent interference between the various markings.

Party and some other lines require two directory numbers for each line but p.b.x. stations have only one directory number however many lines they may have. As a consequence, more lines can be accommodated on cross-bar exchanges than on step-by-step for the same directory number capacities. Ten thousand exchange number cross-bar exchanges without party lines and in business areas with many p.b.x. stations commonly have over twelve thousand connected lines.

Translator directory number d.n. terminals individually connected to marker equipment number e.n. terminals theoretically provides complete flexibility between directory and equipment numbers but in practice complete flexibility is less valuable than the simplification and reduction in the cost of processing which limitations to flexibility can realize. Exchange lines in one section of concentrating switches being limited to directory numbers in one number block together with lock-out which limits a number block to one call at a time, solves in a simple way the problems of ensuring only one connection at a time in a section of switches, while allowing connections in different sections to be made simultaneously without mutual interference. The lines in a file being limited to the same class or classes of service, called group class of service, and the same originating call data, simplifies the data storage by enabling items of stored information to serve for more than one line. A still further limitation in Figure 5.11 is that the terminating call data shall be the same for the ten directory numbers pertaining to one T relay in the translator, for which purpose an additional contact is provided on each relay and a jumper terminal for the ten numbers. With these limitations sufficient data storage capacity is realized with the very limited facilities available. The consequences include administrative effort to fit the lines and their requirements into the available facilities, some wastage of numbers or of equipment because the fits cannot always be exact and some difficulties when the data of connected lines change. In these respects the translator and marker connections have some resemblance to step-by-step final selector wiper bank connections which limit the equipment and directory number relations and supply line data storage and retrieval, with similar problems and difficulties which were described as being due to those characteristics. The quantity of data which can be stored is, however, somewhat greater in cross-bar systems than in step-by-step, and the difficulties somewhat less.

5.5.2 Trunk Translators

Trunk translators are similar to exchange line translators in having output terminals marked one at a time according to dialled digits supplied as input, but differ in many other respects. The output terminals denote junction groups and service groups of circuits, some groups comprising large or very large quantities of circuits, a hundred or more in some exchanges, the translator producing data for the circuits of the group, their equipment numbers in particular, and for the calls as they are connected over the circuits. In particular the rates of charge for calls over junctions from an originating exchange are related to their final destinations which may not be the exchanges on which the junctions terminate but exchanges needing two or more junctions to reach. Translation having to be continued until both route out and rate are known, the quantity of digits required for successful translation is not uniform for all calls but may be one, two, three and possibly up to five decimal digits. For junction groups carrying only terminal traffic to one exchange, one translation suffices. More than one translation for one group of junctions generally implies that the junctions carry traffic not only to the exchange on which the junctions terminate but also to other exchanges by transit switching.

In the trunk translators of Figures 5.12 and 5.13, digits supplied by a register are decoded to marks on wires, ten wires for each digit, as for line translators. In the universal general case translator of Figure 5.12, the ten wires for the first digit are taken to terminals 0 to 9 and also through the contacts of ten relays each with ten contacts and operated by the decoded second digit, to provide connections to marking terminals 00 to 99. Similarly the 00 to 99 wires taken through the contacts of ten relays each with one hundred contacts and operated by the decoded third digit give connections to terminals 000 to 999. This process may be continued as far as it is practical which is unlikely to be more than three digits. It is a general solution applicable to any exchange, but being wasteful of terminals and relays many of which would be unused in most exchanges, it becomes economic to design translators individually for each exchange, of which Figure 5.13 is an example.

Figure 5.13 indicates a one-digit translation for 0 as the first digit, all calls starting with that digit being long distance calls which are routed out over one group of junctions to an exchange of higher level at which the charge determination is accomplished and there recorded or meter pulses generated and sent back over the junctions. There are two-digit translations for the numbers 10 to 19, these being various manual operator services, and for the numbers 90 to 99, these being other services including 99 for emergencies. Three digit translations for the numbers 200 to 899 are for calls controlled by the exchange including the charge determination and recording. Not all of the translations are needed, many of them in blocks being contracted to two digits and some to one digit for which one translation suffices for the whole block. The relays and contacts required to combine the digits are not shown in detail but as 'black boxes'. Shown in Figure 5.13 are route relays, each with contacts to terminals from which jumpers can be run to define the equipment number addresses of lines in a service or junction group and to define data needed for processing calls connected to the group. If more than one switching section

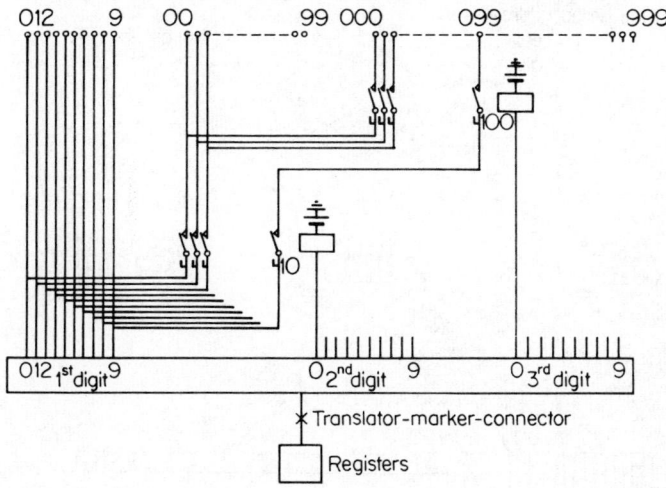

Figure 5.12 Universal trunk translator

exists to permit simultaneous path searches and connections, the lines of a group are divided into sub-groups each connected to a different section. The vertical files in the sections are normally restricted to lines all in one group or sub-group which if too large for one file in each section may occupy more than one file in some sections. If there are many groups or sub-groups filling only half a file or less, some files may be sub-divided to accommodate two or more groups or sub-groups. Except for the sub-divided files, the addresses to be provided by the translator comprise only section and file for each group and sub-group. With more than one section, as previously described, when call connection processing is commenced, a section is chosen for use only if it contains at least one free line in the group of lines involved, thus making all the lines of the group available for every connection to the group.

The route relay operating coils are jumpered to outlets of the translator. Some route relays are operated by more than one outlet where the same group of lines with the same charge or other call data are the same for more than one series of translatable directory number digits. Calls routed out of the exchange over the same junctions but requiring different call data because of different destinations defined by different directory numbers, use individual route relays operated by the directory numbers, the route relays defining the same junctions but different call data mostly defining the charge rates for the calls. The call data are stored in junctors selected and connected by the line marking, as described with reference to Figure 5.7 for data stored in the processor prA$'$. Directory numbers not in use are jumpered to a route relay which causes n.u. tone to be connected.

The translation and the call data may be modified by class of service data such as

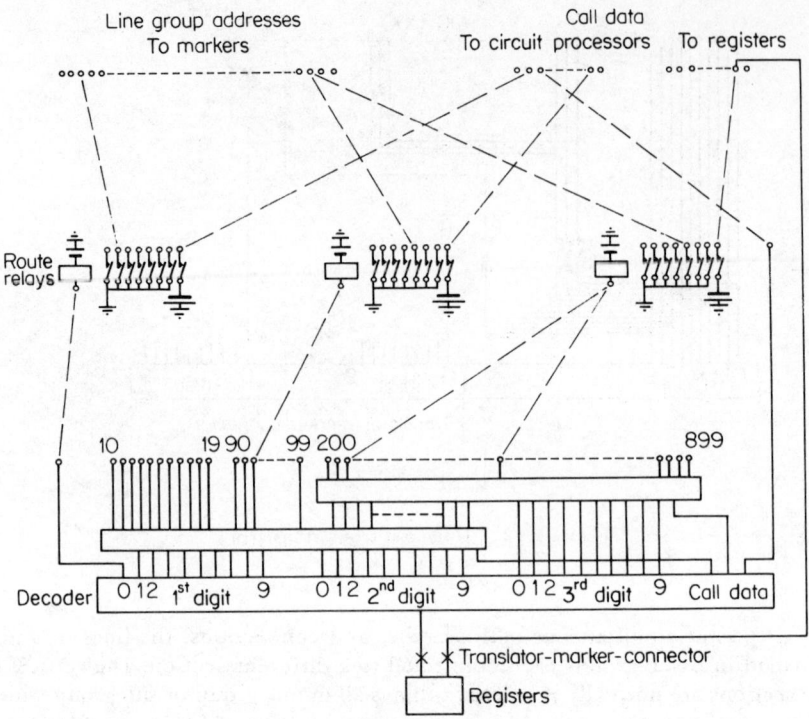

Figure 5.13 Trunk translator individual to exchange

l.d.b. stored in the register as the register is first connected. The stored data are communicated to the translator and decoded at the same time as the received dialled digits are communicated and decoded. The data may affect the route relay which is operated and they may include data to be stored in the circuit processor which becomes connected. It can also happen that the translation produces call data required by the register and for which a path is shown in Figure 5.13 from the contacts of route relays via the translator-marker-connector. Commonly the data are numbers defining how many dialled digits are to be received before the register may release.

At least two translators are provided in each exchange, each with its own route relays, to provide adequate security, with every register having access to at least two translators, for the same reason. Lock-out means ensure that no two translators are used to route calls over the same switching section simultaneously.

In these ways, trunk translation is solved by means which are relatively simple provided that the exchange codes are systematically arranged relative to the locations of the exchanges and not random, which would require every one of thousands of codes to be identified individually.

The registers themselves are left with the problem of when to start call connection processing, which type of translator to connect to and when connected, which of its stored digits to transmit to the translator, and after connection to a junction, what data to send over the junction to advance the call. The solutions to the problems include some amount of translation by the registers themselves of the received and stored digits, in conjunction with a count of the number of digits received. For the translator of Figure 5.13, in order to determine when to connect to a trunk translator, the register must combine the first digits of 1 and 9 with a count of two and other first digits with a count of perhaps five digits in total, as well as recognizing local calls which will need the register to be connected to an exchange line translator instead of a trunk translator. The register transmits to a trunk translator to which it becomes connected as many digits up to three as it has received. Data required to continue the processing after connection to an outgoing junction include how many digits in total have to be received, and which of them have to be transmitted over a connected junciton to further the progress of the call. The data are given as call data from the route relay as in Figure 5.13. Instead of the register itself recognizing the code of its own exchange so as to connect to an exchange line translator for a terminating call, it can first connect to a trunk translator whatever digits it receives, and leave that translator to recognize the local code and tell the register to use an exchange line translator instead.

Translation from a directory number to a connection to be made and data to be transmitted is one of the most important elements in the control of telecommunication systems. The problems are basic but the solutions are not unique even in one system, as seen from the present example where the translation is divided, with an arbitrary element in the division, between the registers, the translators and the markers.

5.6 Data Storage and Processing

It should now be clear that, at least so long as only electro-mechanical equipment is available, the problems of designing a structure with a given type of exchange switch (step-by-step or cross-bar) to interconnect transmission circuits terminated on the exchange switches, and the control of the switches to make the connections, are less difficult and more satisfactorily solved than are those of data storage, retrieval, movement and processing which decide the connections to be made and their supervision when made and up to release. The difficulties of data storage and processing are partly those of the techniques used but mostly of their cost which imposes demands for the utmost economy in the use of both. The difficulties of data retrieval and movement are partly economic but mostly technical in respect of the transmission of data within exchanges and between exchanges. The final solution in practice to all of these problems is limitation of the services and facilities offered to the public to what is within the capacity of equipment at a cost which the public is prepared to pay. The extent of the limitation and its consequences are often not fully realized but should be for new system design. Because the data problem was at the time more difficult than the switching problem and the more urgent, it is perhaps not surprising that electronic components and techniques developed by and for an industry, the computer industry, not primarily concerned with switching, were and have been applied almost exclusively to the data and not to the switching part of exchanges. Electronic components and techniques reduce the difficulties of data transmission and processing to the extent of making possible a marked increase in the capacity of exchanges with respect to services and facilities offered to the public, and available to administration and maintenance, at moderate cost. Nevertheless, new systems despite their advantages must compete economically with the existing systems, new exchanges must work in series with exchanges already installed in the network and new equipments work in parallel with existing equipments in exchanges still growing by being extended with new equipments. Nor are the basic requirements and problems of services offered to the public changed by change of equipment technology. For all of these reasons it is important to current exchange systems that past systems and technologies should be known and understood not in detail but as they have here been treated, by analysis and generalization. The equipment and circuit descriptions given here have been chosen to illustrate the analyses and generalizations and to make them more easily understood, for which purpose systems which could have existed have been more useful than systems which have actually been put into practice.

The means of data storage, retrieval and processing are different in step-by-step cross-bar systems but the objectives and accomplishments are not very different, with the result that there are no significant differences in the services and facilities offered, not in the quantity of help required from the maintenance staff in period and temporary data storage or by operators in processing calls which are out of the ordinary. Exchange processing is divided between four main operations serially performed, namely

(a) detection of a calling exchange line or incoming junction and its connection to a register;
(b) transmission to and storage in the register of digital data defining the connection required;
(c) call connection initiated by the register to connect the circuit already connected to the register to a suitable other circuit over which, if it is a junction to another exchange, selected data to further the advance of the call have to be transmitted;
(d) supervision of established calls up to and including release and including charge accounting.

The processors concerned with path searching and connecting through the exchange switches are regarded as part of the exchange switches. The circuit processors are mainly concerned with supervision and contain relatively small quantities of data stores and of wired logic. The common processors comprising register-translators in step-by-step systems and the registers, translators and translation parts of the markers in cross-bar systems, are large and complicated relative to the circuit processors, and they execute call connection processing for which considerable amounts are required of semi-permanent and temporary data storage and processing logic. The contribution of each kind of equipment to the total cost of an exchange is influenced by the quantities required, with the result that the exchange switches make the largest contribution and the common processors the least.

The cross-bar system of division of the common processing between registers and second common processors, namely the translators, results in most of the logic and temporary data storage being located in the registers and most of the semi-permanent storage in the translators with economic advantage and somewhat greater data storage capacity than that of step-by-step systems. The registers of step-by-step systems may be divided in the same way and are so divided in some systems but the facilities for data transmission being limited to the paths through the exchange switches is a handicap relative to those of cross-bar systems in which data may be communicated, to various units to be used in a connection, as the connection is established and over paths not dependent on the exchange switches. As a consequence cross-bar systems are able to transfer more of the total processing from circuit to common processors than is possible in step-by-step systems.

Some of the details of cross-bar processing are now examined together with some possibilities which have a bearing on systems to follow cross-bar and have problems to solve in much the same or in analogous ways.

5.6.1 Supervision

Consider the circuit processor junctors or.j., i.j. and t.j. of figure 5.3, the processes with which they are concerned and the distribution of the processes among them: collectively their main functions are ringing called lines, metering effective calls and releasing connections when they are no longer required.

With regard to ringing, no ringing is required for exchange lines to p.a.b.x.s with direct dialling-in. Such lines require a seizing signal, usually a loop, to which the p.a.b.x. responds with proceed-to-send data signals to the public exchange to cause it to send decimal digits defining an extension to be connected without the intervention of the p.a.b.x. attendant. Such lines are identified by class of service marking for an appropriate type of terminating junctor to be included in terminating connections to them. All other lines require loop ringing except two-party lines for which the artifice of step-by-step systems in reversing the − and + wires of a line to ring Y parties with normal loop ringing is not possible in systems other than step-by-step. The two parties connected to one line being identified for a terminating call by directory numbers individual to each, the exchange line translator has to mark one line when given either of the numbers and depending on which number, provide the data to control the ringing. The data may be either class of service or call data. Class of service if used selects one of two kinds of terminating junctor, one suited to ordinary and X party ringing and one suited to Y party ringing. A call datum, if used, controls the kind of ringing sent out by junctors of one kind. All ringing processing is simplified by the assumption that a line is to be rung as soon as it is connected and that the ringing is tripped when the called subscriber answers and is not thereafter applied again during that call. There are circumstances in which an operator may wish to connect to an exchange line and to hold it ready for ringing later when she has completed some other operations, called delayed ringing: or having rung a line and the call having been answered and cleared, she may wish to re-call the subscriber by re-ringing the line: or having answered a call incoming from a subscriber who she is still holding although not on the line, having cleared, the operator may wish to re-call by ringing called ring-back. For calls to and from p.a.b.x.s with through dialling and with d.d.i., the ringing facilities and holding would be required of the p.a.b.x.s instead of the public exchanges. The facilities available to operators would also be useful to the maintenance staff. The difficulties of incorporating such facilities into exchange systems are not so much data storage or processing as data transmission, of the operator instructions to the terminating junctor or p.a.b.x. to ring or to hold. Because of the minor importance of the facilities and their difficulties of provision, none is found except rarely in present practice but all should be incorporated into a new system.

Time sharing of ringing equipment is a possible means of economy. Universal junctors with through transmission normally would on seizure for a new call, be connected by a diversion bridge and through interface switches, as in Figure 4.12, to a space division time shared common processor set by called line data to apply the required kind of ringing as and when required and to release when the called line answers. Re-ring and ring back come at any time after a called and a calling line respectively have cleared and they are dependent on an operational data transmission, from the operator or other person controlling the call, to a terminating or originating junctor respectively to instruct it to ring the line or to gain connection to a common space division time shared processor which rings the line. Time sharing is economical when a substantial quantity of processing can be transferred to a common processor, which is not the case just for ringing. Alternatively, on the model of Figure 2.5, an exchange line may be rung by a junctor not part of the ultimate connection and held only until the called line answers, when the ringing junctor is released and the connection from the calling to the called line is completed through a junctor without ringing facilities and simplified to that extent. While the called line is being rung, ring tone has to be sent back to the caller usually by a ring tone junctor connected to the incoming side at the same time as the ringing junctor is ringing the called side of the connection being made. The quantities of ringing and ring tone junctors required are for each about one sixth that of the through-connection junctors and the processing is greatly increased, all of which taken together make the method unattractive economically for electro-mechanical systems. It is commonly used in electronically controlled systems, the conditions being different, but for first time ringing, not for re-ring or ring back.

In cross-bar systems the processing for metering does not have to be related to the trunking as in step-by-step systems: data for charge determination are supplied by the translator common processors and transmitted to the circuit processors as call data as described for the prA' part of Figure 5.7. The result is greater facility and flexibility in the metering arrangements, with some economy. Variable pulse rate metering can be controlled for most calls by the originating junctor as in Figure 5.7 and for other calls by a small proportion of the incoming junction junctors sending back meter pulses over the junctions. In all systems the charge determining and metering arrangements are simplified by the assumption that charges for calls are invariably metered against the calling parties and at rates uniform for all parties and determined by the digits dialled by those parties, but exceptions of necessity occur in practice and others would occur if the processing could cope with them. Freefone and reversed charge calls are obvious exceptions. Pay station charges are different from those for regular subscriber stations. The meter pulses generated for calls from coin-box lines have to operate or be related to a mechanism for collecting deposited coins, which means not only originating junctors special to the coin-box class of service but also the generation of two sets of pulses at different rates if the coin-box rates are different from those of regular subscriber lines. Cross-bar systems with translator common processors and call data transmission to processors as calls are in process of connection, are better able to cope with such complications than are step-by-step systems.

Charge rates are commonly varied according to the time of day and day of the week, to encourage traffic at otherwise slack traffic times. If it is the rates of meter pulses derived from pulse generators common to the exchange which are varied, which is usually the case, only switching to a different pulse generator is involved in a change in the rates.

The meter pulses which are finally emitted are positive battery pulses on the P-wire as in Figure 5.7 to operate the calling subscriber's meter. For two-party lines there are two meters per exchange line, in the line equipment connected through exchange switches to a junctor where the pulses are generated. To avoid having to provide two separate wires through the switches for the selective operation of the meters, as in figure 4.4, two-phase pulses may be used as in Figure 5.14. When a party line calls and a call button is pressed, the line is connected to an originating junctor or.j and the calling party data transmitted via the marker to the originating register which becomes connected. Alternatively, if press-button sending is used, the call button sends an arythmic code signal or a v.f. signal-message to the register which stores the calling party data. When the call is connected, the calling party data is sent as call data to the originating junctor, the data being sent from the register via the translator, to operate a relay X or Y depending on the data. Referring to Figure 5.14, X party meters are connected to the phase 2 pulse supply and Y part meters to the phase 1 supply, with rectifiers to prevent operation of the meters to earth potential on the P-wire. It will be seen that phase 1 pulses applied to the P-wire operate meter X and phase 2 pulses operate meter Y, when a meter pulse operates relay MR. If the phase pulses are short and integrated by the meters, the duration of the MR relay operation is not critical, but if the period is long enough to operate the meter, the MR relay has to be phased to operate coincidently with one phase pulse.

It may be gathered from this and other features of two-party line operation that the complications introduced into exchange operation are not negligible, and this

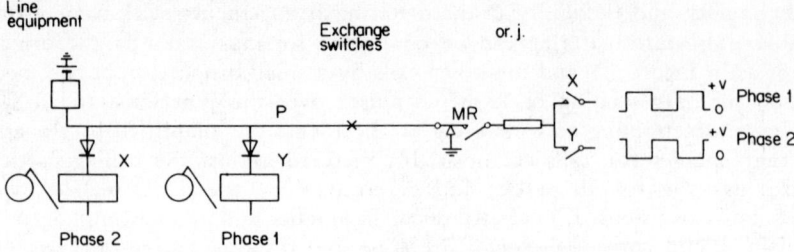

Figure 5.14   Party line metering using phased pulses

together with doubtful economics of line plant saving in large cities and other places, causes party line service not to be offered in many areas, and for its use generally to be discouraged. On the other hand, the difficulties become less with electronic control systems and the economics may once again become favourable. Here, irrespective of their future use, party lines and their problems have been useful as an example of a structure class of service requiring elements in the structure, namely in the line equipments, trunking and originating junctors, for its existence. In electronic systems party line operation can be a control class of service for call charging.

Coin-box operation is also a structure class of service, requiring junctors which can respond to coin deposition signals and emit coin control signals. To reduce the exchange equipment required, coin-boxes have been constructed to control the connections when made including the coin collection, independently of the exchange except to receive meter pulses over the exchange lines as if the pulses were to operate private meters. Structure classes of service provided by such means simplify the exchange equipment and operation, but at a high cost in the case of coin-box class of service for the coin-boxes.

The problems and difficulties of knowing when an established connection is to be released were discussed in section 2.8 to which was added in sub-section 4.2.4 the necessity of ensuring that when a junction between automatic exchanges is released at one end, it cannot be seized for another call until the other end has also released the first connection. Commonly the calling and called line loop signals are transmitted to every exchange involved in a connection. That the called line has answered is of necessity communicated to the originating exchange to start the metering, which is best accomplished by transmitting the called line loop signals back through all the exchanges to the calling exchange. The similar transmission of the calling line as well as the called line loop signals over all the junctions between exchanges involved in a call is also desirable to provide supervisory signals to operators who become involved in dialled connections. It was described with reference to Figure 4.9 that after calls are established and the registers if any are released, no datum transmission other than the subscribers' loop signals is available with electro-mechanical exchanges, and one circuit processor in every exchange holds its part of the connection in its exchange and releases its part when the calling line is cleared. The holding and releasing of connections is thus under the control of the calling party, but this by itself is not sufficient. A subscriber who becomes connected to an operator is very likely to flash to recall the operator and thus to release the connection if the holding is not under her control: a called line may be held accidentally or intentionally but indefinitely by the caller not clearing the calling line. In early exchanges an alarm is given to the maintenance staff to investigate and to release if appropriate a connection held for a long time after the called party has cleared. Maintenance staff are not in continuous attendance in electro-mechanical exchanges and exchanges normally unattended will eventually be common. Automatic release after a time-out period is a solution except that the necessarily long period, several minutes, for the time-out still gives scope for malicious holding. It has been known for example for bookmakers to engage and hold all the lines of a rival office at vital times in the betting business. Means of holding such calls by the called line, so that the offenders could be traced, would be valuable and also to police and fire station lines, to identify the origins of calls when the callers for some reason are not able to give the information. Thus both subscribers and operators need to have some control over calls incoming to them.

Subscribers who are called by other subscribers do not normally operate the cradle switch in any way that resembles quick flashing, and flashing could be used to gain release from an incoming call held at the far end. The optional holding of incoming calls is possible if instead of calling party release, last party release, that is to say that calls release only when both the calling and parties have cleared, is used in all exchanges except in the called exchange where the release decision is made. For release to be controlled in this way, terminating junctors t.j. except those of a special class of service, on receiving the calling line clear, send back the called line clear if it is not already established, and then release, which is equivalent to calling party release. Special class of service exchange lines hold incoming connections until they themselves are cleared, which is last party release to be given to special lines and temporarily to ordinary lines troubled by holding by calling lines which it is required to trace. A difficulty occurs with operator last party release, that few operator services give the normal answer signal because it may cause the caller to be incorrectly charged for the call. An opportunist solution is to arrange that the clearing of a call incoming to an operator and still held by her, causes the answer signal to be given before any of the switches in the connection can be released, metering being made dependent on both a calling loop and an answering signal: if the caller recalls, the answer signal has to be ceased before metering can be started. Some complicated circuitry and timing is clearly involved but there are systems in practice which operate in that way. All the difficulties including that one end of a line may be seized for a new call before the other end is ready to receive it, are the result of the limited signal and data transmission capabilities of existing systems, and from which a new system should not suffer. All the difficulties disappear with two kinds of answering signals, one causing and the other not causing metering to start, and with a release datum transmission independent of the exchange line loop signals.

5.6.2 Line Finding

The exchange line and state of line data influence the processing of a new call originated by an exchange line, the address available for the retrieval of the data being the equipment number. Referring to Table 4.1, the class of service line data ordinary and p.b.x., coin-box, party line and possibly also private meter, select the kind of originating junctor to be used. The states of line data o.b. and s.o. and s.s.o. temporarily connected when required, may also select junctors which respectively send n.u. tone to the caller and give access to observation services; but since a maintenance man is required to change the data, the advantage over changing the structure as in step-by-step systems is much reduced. Full advantage requires the class of service data to be written and erased by a remote control centre.

The disabled subscriber and long distance and international calls barred l.d.b. and i.b. classes of service are transmitted as originating call data, as in Figure 5.7, to the register which becomes connected. A register receiving the disabled subscriber class of service immediately, without waiting for digits to be dialled, connects to a trunk translator to which it gives the class of service as call data, and the trunk translator causes the calling line to be connected to an operator service junctor. The l.d.b. and i.b. data are also given to a trunk translator as call data but along with dialled digits in the normal way, for the connection of n.u. tone if a call which is barred is dialled.

Line and state of line data for incoming junctions if needed at all usually specifies the kind of data transmission over the junction so that a register able to receive data in that form is connected. Class of service or other data relative to the call being connected may be communicated from register to register as part of the data transmission between those equipments, but it is not common to local service to do so.

5.6.3 Call Connection

Call connection processing is initiated and controlled by the register brought into the connection. The registers contain wired logic programs which are selected and executed at each phase of the processing, the processes in conjunction with the translators and markers comprising, for the system of Figure 5.3, the following.

(a) When selected for connection, receipt of call data for the call being connected.
(b) If the call data indicate that no digits will be dialled, by a disabled subscriber or other call, the omission of the next two processes.
(c) When first connected, the return of dial tone from an originating register or of a proceed to send datum from an incoming junction register.
(d) Receipt and storage in the register of directory number digits dialled by a subscriber in response to dial tone or extracted from a distant register in response to proceed to send datum signals.
(e) Translation of the call data and the dialled digits as they are received to determine the type of translator required for the call.

(f) Connection of the register to an appropriate type of translator and transmission to the translator of the relevant digits and call data.
(g) Determination if the digits transmitted to the translator constitute a valid number and if not, connection of n.u. tone to the calling line.
(h) If the number is valid and a local exchange line number, determination if the corresponding line or lines may not be connected because it or they are parked, out of order or temporarily out of service, and if so, connection of n.u. tone to the calling line.
(i) If the number is a valid and connectable local exchange number, determination of the class of service and thus the kind of terminating junctor required, which may be ordinary, party line or p.a.b.x. with d.d.i., and determination if a free path exists from the junctor to which the calling line is connected, via the required kind of terminating junctor to a free called circuit and if not, the connection of busy tone to the calling line.
(j) Selection and connection of a free path determined to exist to a connectable called exchange line, and the communication of data to the terminating junctor and to the originating or incoming junctor, the data being necessary to the further processing of the call by those equipments: also the communication of data to the register or registers still connected to cause them to release.
(k) If the number is valid and not a local exchange line number, selection of the group of service junctors or outgoing junctors to a free one of which the calling line is required to be connected, the determination if a free path exists from the junctor to which the calling line is connected to a free circuit in the selected group and if not, the connection of busy tone to the calling line.
(l) Selection and connection of a free path determined to exist, and the communication of data to the selected junctor for use in subsequent processing and to the register to instruct it to release if the connection is complete or to await a proceed to send datum if connection is not complete.
(m) If connection is not complete, to send data in response to proceed to send data and finally to determine that the register has completed its processing and to release it.

The objectives should be understandable from the previous descriptions of requirements and operations, together with the following comments.

Call connection processing is very dependent on data storage and retrieval. The desirable minimum requirements for exchange line data are listed in Table 4.1 with directory numbers d.n. as addresses. Referring to the table, ordinary and p.b.x. lines have to be differentiated by class of service in step-by-step systems to control the hunting over p.b.x. lines but not for the same reason in cross-bar systems where the jumpering between the translator and marker produces the equivalent of hunting. Step-by-step systems use trunking to distinguish between p.b.x. groups and with and without d.d.i., as in Figure 4.5: cross-bar systems may use class of service data to cause an appropriate kind of terminating junctor to be included in the call connection, in one case to call by ringing and in the other by line loop, and to instruct the register to release when connection is complete or to interchange call data with the p.a.b.x. before releasing, respectively.

Referring to the state of line data, busy or free are provided by the structure, namely the P-wire potential. The parked state of line requires an additional relay per line in the line equipments and is rarely provided in cross-bar systems. Instead the manual connection of a fault circuit analogous to the manual system test and plug-up circuit is usual in pre-electronic control systems. Manual operations by the maintenance staff are also involved in data storage and in equipment provision for temporarily out of service and s.v.i. intercept states of line, as mentioned in section 5.6.2 for o.b., s.o. and s.s.o. originating call states of line and as previously described for step-by-step-systems.

Call connection processing, like all processing, comprises the assembly of data, the processing of the data and the issue of operational and other data as the result of the processing. The registers and the translator-markers share the data processing and issue of operational and other data between them, mostly as determined by economy and with a facility which is adequate in relation to the services which the subscribers expect. Data assembly involves data communicated by the callers by dialling and which again is adequate, and data which has to be stored in the exchange and retrieved as each call is made. Data storage and retrieval can be seen from the descriptions given to be difficult and expensive enough to compel their very economical and sparing use. A basic class of service requiring no data to be stored, group classes of service, cross-bar line switch files limited to lines with a common characteristic, can all be seen to be related to the need for the minimum of data storage, and of processes and processing. Nevertheless these devices are not a great handicap to the design and operation of telephone exchanges, and the simplification and economy of equipment which is their prime objective may still be of value and is still found when electronic equipment makes large data storage capacity available at a relatively low cost and almost unlimited processing possible. The major advance which electronic control equipment makes possible is in respect of services and operations which in electro-mechanical systems are not rendered at all or achieved in inconvenient ways mostly involving operators and maintenance staff and many of which have been mentioned in the foregoing pages. These advances became possible not only because of the increased data storage and processing capacity but also because of greater data retrieval and transmission facility.

5.7 Conclusion

The description and analysis of cross-bar systems are both complicated and tedious but nevertheless important and necessary to the understanding of the problems and the design of later systems and of future systems, which systems differ from cross-bar systems in that they use electronic devices partly or wholly in their construction. All of the problems can be found in all of the systems, and the solutions if not the same can in many cases be seen to have analogous features, notwithstanding the fact that time division time sharing is found only in systems with electronic control.

Because of their limited data storage and transmission and processing power, step-by-step and cross-bar systems although based on what the service and the customers require, are heavily biased by what can be achieved at little expense. For new systems it is not sufficient to continue and improve the development of existing systems but necessary to start from the beginning to understand what is needed and to relate the needs to the current technologies.

Step-by-step and cross-bar systems differ little in the services offered but greatly in the speed of call connection and to some extent in reliability. It will be appreciated that as the range of subscriber dialled calls extended up to a possible twelve exchanges in series and more, the time elapsing between the completion of dialling and the receipt of ring tone became more noticeable and thus important. Connections through cross-bar exchanges can be made in less than 2 seconds, compared with an average of 5 seconds for transit and 10 seconds for terminal step-by-step exchanges. Press-button dialling makes the difference even more noticeable. Press-button dialling cannot be applied directly to step-by-step selection. Press-button dialled digits may be stored in registers but the subsequent step-by-step selection being no faster than with rotary dialled digits, the time elapsing between the completion of dialling and the completion of the connection is increased. Press-button dialling in conjunction with cross-bar system operation is capable of very fast connection service which electronic control still further improves.

Despite the increasing advantages of cross-bar systems over sliding contact step-by-step systems, it was not until their overall costs for installation and maintenance became comparable, that cross-bar systems became so universally accepted — an important fact which has to be bourne in mind for all new systems.

Chapter Six
ELECTRONICALLY CONTROLLED EXCHANGES

6.1 Electronic Devices and Techniques

The first use of relays and that which caused them to be so designated, was amplification. A quantity of available power is used to operate a relay a contact of which transmits or relays a larger quantity of power. The two other functions of which relays are capable, those of data storage and logical processing, came later. It was understood perhaps vaguely as thermionic vacuum and gas-filled diodes and triodes became available, that they could be used for the three functions: the Eccles–Jordan trigger circuit was published in 1921 but apparently with little realization of its potentialities for data storage. As copper oxide, selenium and germanium diodes became available they were used as electronic switches but the large scale use of electronic devices in communication switching systems was prohibited by cost, bulk, power consumption and circuit difficulties, until the appearance of the transistor in 1947. The consequent economic support for the considerable device and circuit technique developments required to bring electronic switching and control systems into larger scale use came not from the communications industry but from the computer and some other industries. For that reason, electronic switching and control techniques are not individual to exchanges as they were for electro-mechanical exchanges but dependent to a considerable extent on devices and equipments not manufactured exclusively for exchanges.

To explain and understand the switching and control techniques now being used in exchanges it is not necessary to be familiar with the constructions and details of the switches and data stores employed. Components and integrated circuit assemblies of components available as packages to performance specifications are chosen and assembled to attain some overall result. Here description, which is necessarily limited to relatively simple operations, will use the gate symbols and circuitry of Figure 6.2, a circuit based on bi-polar transistor technology being shown in Figure 6.1 as an example of integrated circuit construction and the external electrical conditions required for operation. To the gate circuit there are four inputs, A, B, C and D to separate emitters fabricated on one transistor and shunted by diodes which limit the negative voltage excursions of the inputs for the protection of the emitters. With all the inputs at the +v supply voltage of a few volts, none of the input emitters conducts current, and the current from the base via the collector of transistor 1 causes the transistor 2 to saturate, which means that

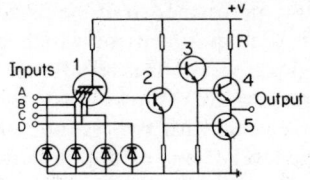

Figure 6.1 Integrated circuit four-input gate

its collector to emitter potential becomes a small fraction of a volt and almost independent of the current flowing. The output transistor 5 is also saturated and the output thus clamped at its saturation potential. Transistors 3 and 4 are non-conducting under these conditions. If any one of the four inputs is near to earth potential, the input transistor 1 saturates and transistors 2 and 5 become cut-off, and neither of them conducting between collector and emitter leaves transistors 3 and 4 to conduct and supply positive current to the output limited only by the registor R. To perform binary logical operations the inputs need to be near to earth or the +v supply potentials. The output provides these conditions provided that the output current in the near earth condition does not take the transistor 5 out of saturation nor in the +v condition is there too much potential drop across the resistor R. With these conditions observed, the outputs of units may be connected in series with the inputs of other units and the required logical operations will occur correctly. The outputs of different units may not be connected in parallel directly but may be indirectly by connection separately to the inputs of another unit which produces at its output the logical conclusion of the outputs of the units applied as inputs. In these ways logic functions in series and in parallel may be realized very similarly to relay contacts in series and in parallel. The output of the four-input gate of Figure 6.1 is at earth only if all the inputs are at +v potential, which is the negated AND called NAND function for logic 1 and 0 represented respectively by +v and earth potentials. An inverter changing logic 0 to 1 or 1 to 0 and following the unit output is needed for the AND function to be performed. The NOR function which requires logic 0 output for A or B or other input at logic 1, is executed for logic 1 and 0 represented respectively by earth and +v potentials, which means that the same units can be used in the same processors to perform NAND and NOR functions provided that the necessary inversions are incorporated.

Units of different internal constructions to perform the same logical functions differ in cost, power consumption, speed of operation and electrical conditions corresponding to the logic 1 and 0 conditions, which differences influence the choice of units for particular applications. Compatability between units of different constructional types occurs frequently by design so that units of more than one type may be used in the same equipments. An AND or OR gate with one output and as many inputs as may be required, up to ten and more occurring in practice, is represented by the symbol of Figure 6.2(a): a small circle before the output as in Figure 6.2(b) denotes negation for NAND and NOR gates. The symbol & or the word OR may be written in or against the gate symbol to denote the AND or OR function performed if it helps toward the understanding of the operation of the unit in the equipment of which it is a part. The particular case of a one-input gate for logical inversions is represented by the symbol of Figure 6.2(c).

Figure 6.2(d) shows two four-input NAND gates connected to produce a trigger circuit with two groups of three inputs. If all three inputs of both gates are at logic level 1, one of the units will have the logic output 1 and the other 0 and the circuit is stable in that condition, and will not change no matter how the inputs to the unit with output 1 may change. If, however, one of the inputs to the unit with output 0 becomes logic 0, those of the other unit being all at logic 1, the output of the first unit becomes 1 and that of the other unit 0 and the circuit is stable in the alternative condition. Figure 6.2(e) is the general symbolic representation of bi-stable circuits such as Figure 6.2(d): the bi-stable part is the two rectangles each with an input normally at logic 0 and an output, one at 0 and the other at 1. An input at logic 1 causes its output to assume the logic 1 state, that of the other output becoming 0, and the outputs remaining stable in those states. The inputs are the outputs of AND gates by which the bi-stable circuit is controlled. Figure 6.2(f) shows a bi-stable circuit used as a one binary digit, or one-bit, store with a three-bit address. An address generator has complementary binary outputs each consisting of two wires either of which may be at logic level 1 or 0, the other being then at 0 or 1 respectively. The outputs provide three-bit word addresses individual to eight stores, a store being addressed when the three address outputs to which it is connected are all at logic 1. The fourth inputs of the two input gates are normally at logic 0, and one gate is used to write digit 1 into the store and the other digit 0. If, when the store is addressed, neither of the write inputs is at logic 1, the bi-stable circuit condition remains unchanged, but either of the inputs being 1 causes the bi-stable circuit to take up the condition corresponding to the input. Also when the store is addressed, its content is read out by an output gate. The eight stores each addressed by a different one of the addresses provide a store of eight one-bit words, into any one of which when addressed the digit 0 or 1 may be written and out of which when addressed the value of the digit stored is available at the output of an OR gate to the inputs of which the outputs of the digit stores are applied. With more address digits larger stores are made, for example sixty-four one-bit words in one integrated circuit and referred to as a scratch pad memory. Stores for n-bit words are made by operating n one-bit word stores in parallel with a common address source, by which means stores aggregating many thousands of bits are fabricated and commonly referred to as bulk stores.

For the logical processing part of electronic processors, semi-conductor devices are used exclusively. Data stores based on remanent magnetism have been extensively used in the past but are being superseded by semi-conductor types. The functions performed by the many kinds of semi-conductor packages which are available are many and various and with them electronic data processors comprising data stores and logical processing may be constructed as required. Where circuits and circuit operation are used in the following pages to illustrate the points being made, the semi-conductor units are limited to those symbolized in Figure 6.1 as being adequate and conducive to simplification of description, but with the

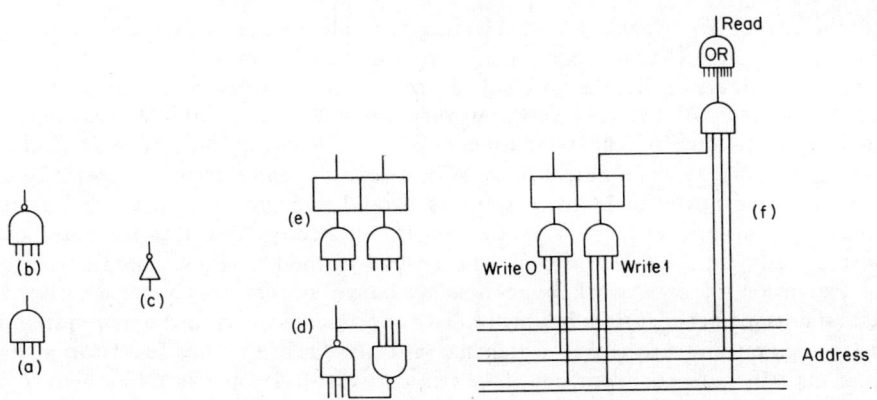

Figure 6.2   Gate symbols and typical circuits

understanding that in practice more complicated units based on one of several constructional technologies might be used.

6.2 Objectives for Electronically Controlled Exchanges

The theoretical equivalence of electro-mechanical and electronic devices for switching and control does not mean that either can be used at will in any circumstance and that the choice can be made on the simple basis of cast and performance. The use of electronics requires major changes of well-established practices and of systems in production and demands clear answers to the questions how and why the changes should be made. Initially the question how was paramount and had to be resolved to some extent at least before justification of the use of electronic devices in exchanges could be attempted.

The first investigations into the use of electronic devices in exchanges quickly showed that none of the criteria set out in section 1.4 for a change of system was likely to be satisfied, in foreseeable time and in respect of the best electro-mechanical systems, by a simple substitution of existing by electronic components. Nor did the first automatic exchanges satisfy the criteria relative to manual service but they did ultimately and there were reasons for changing which were not properly understood at the time. Possibly the same would prove to be true of electronics in exchanges. Perhaps the exchanges would become so much more reliable in operation than those extant, would cost less for maintenance and take up so little space as to justify the change to electronics even if, as seemed likely, the capital costs would become somewhat higher than those for existing exchanges. In fact, relay and cross-bar switches became so reliable after about 1950 that claims for improvement beyond what was possible for systems based on those switches were hard to substantiate, and costs steadily reduced. Existing electro-mechanical exchanges do not satisfy the needs of some new services, notably high speed data and the combined telephone and television of videophone: so some changes are inevitable, but the changes required are more to the structure for transmission than to the control. Also, there is no challenge to the telephone as the dominant service and the interests of the relatively few subscribers to other services must not be detrimental to that of the telephone majority. Electro-mechanically controlled exchanges being weak on data storage and processing, and electronic control being so much more powerful in these respects, would seem to make more and better services at no greater cost a matter of no difficulty, but this has not been conspicuously the case in practice. The cost of introduction and continuation of the expansion of a network by a new exchange system to replace an already well-established other system has proved to be formidable, even if the new system is itself economically attractive, which has not always been true. Justification has come, as will be seen, from new facilities and features some not clearly seen if perceived at all at the beginning.

An alternative and general approach to the question of change and independent of how it is achieved, is to ask the three bodies most concerned with service, namely the subscribers, the administrations and the maintenance engineers, in what ways do they think the service needs to be improved, excluding greater reliability and lower costs which will always be achieved when possible. If this could be done the answers might well cause an exchange system designer to conclude that what follows are worthwhile objectives for a new system.

From the point of view of the subscribers, the standard of speech transmission is not universally good on all calls. Laboratory investigations, and field testing under controlled conditions, show a very wide range of telephonic abilities among subscribers and that the least able find that speech transmission is not adequate to easy conversation on a significant proportion of the calls which they make. Moreover, chance variations between transmissions of different circuits produce a significant probability for some connections of unsatisfactory transmission even for subscribers of average telephonic ability. One of the incidental consequences of these conditions is that loudspeaking telephones which leave the hands free cannot be relied upon to operate satisfactorily and thereby a service for which there is a considerable potential demand is largely suppressed. Known means of improving this situation require exchange switches to provide four-wire message circuit transmission and switching for all interconnections of circuits, two- or four-wire, terminated in the exchanges: that four-wire analogue transmission paths through the exchanges should include amplifiers: that p.c.m. channel to p.c.m. channel connections be made digitally, multiplex channel to multiplex channel, and should not require demodulation to analogue and remodulation to digital transmission. In practice, in the low hierarchic level exchanges where transmission improvement is most needed, neither four-wire switching nor amplifiers are provided except rarely, because of the high cost if metal contact exchange switches are used, and digital p.c.m. channel switching is not possible with metal contact switches. Digital p.c.m. channel switching, which cannot be indefinitely avoided, is technically dependent, and four-wire switching including amplification is economically dependent, on the use of electronics in the structures of exchanges as well as for the control. Chapter 10 includes a discussion of these questions.

The time which elapses after dialling or press-button sending of a directory number and the return of ring or other tone is something which subscribers would certainly wish to be reduced to a few seconds at most. The time is dependent on the times of call connection through exchanges and of information transfer between exchanges. Times of less than one second for each have to be achieved for the overall times to be satisfactory and are possible with electronic control.

Private automatic exchanges having no problems of transmission over distance or of charging for calls which do not extend beyond the private network, and often being designed to suit the needs of individual customers, they are able to incorporate facilities additional to those available on public exchanges. Callers may camp-on-busy, which is to wait for an extension which is engaged to become free when it is immediately connected to the waiting caller without further action on his part: or finding an extension busy, a caller may hang up after signalling that he wishes to be connected to the busy line when it becomes free, called ringing re-call, the line when free being held and rung after the caller has been recalled by ringing. Extensions engaged on calls may be informed by a tone signal-message that an incoming call is waiting for them to become free. Three-way conversations and multiple way conference connections can be established and for which the participants, particularly those in conferences, do not have to use handset instruments: microphones and loudspeakers can be satisfactorily substituted where the p.a.x.s. provide four-wire switching. There would be a demand for some at least of these facilities on public exchanges if they were satisfactory operationally and provided freely or at low cost. Undoubtedly there would also be an increase in the interchange of information by means other than speech, by data, telewriting, pictures and line drawings, if the services were more satisfactory, quicker and cheaper than they are with electro-mechanical exchanges.

Improvements in existing public exchange facilities would certainly include control by the subscribers themselves of some services hitherto involving an operator, notably call transfer and early morning calls. Early morning calls, so called but not limited as to time, require an operator to be instructed to ring a line at a particular time. Pre-arranged and follow-me transfer facilities by which calls incoming to an exchange line are diverted, if previously instructed over the line by the subscriber, to a designated other line, are described in sub-section 3.3.3, as are subscriber controlled enquiry and transfer.

Another approach to new facilities is to explore those which were previously impractical because of lack of suitable means of realization but now become possible because of the availability of new components and techniques. Abbreviated dialling is an example which has come out of the availability of large capacity electronic data stores. It enables a station to designate in some systems as many as one hundred other stations to any one of which it is connected by dialling code digits followed by one or two digits together being fewer digits than those of the directory number of the called station.

From the point of view of administrations, better control of the services by fewer staff more concentrated into operator assistance, accounting and administration centres would take high priority. As services become more and more automated and effective over longer distances, the need for operator assistance decreases; those operator services which still remain become more complicated and difficult and the cost per call progressively rises. As a consequence, the operators themselves need assistance from the machine in the shape of more information and more control facilities. These may affect the design of the structures of the local exchanges and certainly affect the control. Traditional facilities such as manual hold and trunk offering are still necessary. The optional holding and releasing of calls was discussed section 5.6.1. Manual hold originally limited to operators is the facility of holding a call over as many junctions as may be involved in the connection and despite any other control operations. The term trunk offering derives from the original reason for the provision of the facility, that of permitting an operator to connect to and to speak on an exchange line already engaged on a call, to offer a long distance call if the subscriber would accept it, first clearing the one already in existence. This use is no longer valid, but the facility remains to enable operators to connect to engaged exchange lines to monitor them, chiefly to investigate reported difficulties. Preferably the operator trunk offer access to the lines is separate from the normal access wherein may lie some difficulty which has caused the trunk offer facility to be invoked. Additional facilities in the future, and effective for connections possibly over several junctions in tandem, should include at least calling line identification, coin-box control and delayed ringing, ring back and re-ringing. Calling line identification c.l.i. provides a visible display to the operator of the directory number of an exchange line to which she happens to be connected. Preferably she is able to cause the number of the line to be displayed at any time during the progress of the call. Coin-box control means the control of the money deposition and collection. Delayed ringing, ring back and re-ringing were mentioned in section 5.6.1: the first enables an operator to call and hold an exchange line without it being rung until she sends a ring instruction, and the others enable an operator to re-call a subscriber who has hung up on a call which the operator is holding.

The keeping of charge records and the rendering of accounts to subscribers is an expensive item of exchange administration of which it is commonly a separate branch. The ultimate objective is accounting centres able to supervise and collect the charging information of exchange lines in the exchanges within a considerable area of country for each centre, automatically at any time, and to prepare the bills for mailing, without human intervention other than the control of machines. Modern computers and business machines are easily capable of fulfilling the objective given the necessary data which the exchanges have to be designed to provide. One solution is period stores, the equivalent of exchange line meters and located in the exchanges but capable of being read by remote control by the accounting centre, all bulk billed charges being aggregated on those stores and the details of the calls not so charged being recorded on magnetic tape. Also included should be means for the automatic recording or collection of charges for operator controlled or assisted calls, which requirement involves data transmission, between the local exchange and the operator control point, of directory numbers and charge rates or units to be charged.

Administrations would take the opportunity presented by increased capability to store and process exchange line and state of line data to offer a greater variety of classes of service to subscribers and to improve the day to day control of exchange line services where the control is meagre and clumsy. The involvement of maintenance staff in o.b., t.o.s., s.v.i., s.o. and s.s.o. exchange line states of line was described in relation to Table 4.1. As the maintenance effort required in exchanges decreases because of greater reliability of the equipment and mechanization of the operations, so the proportion of exchanges regularly staffed by engineers decreases, and services dependent on engineering effort become still less attractive. What administrations need is means to control services from administrative centres, without delays or manual effort outside the control centres. The basis of exchange line control of this kind is the ability of a control centre to write into the state of line stores of the exchanges under its control the states of line as they change for particular lines, the exchange controls being programmed to render the services indicated. There are also occasions when the administration wishes to change the directory number of a station or other line data not involving a physical change to the structure. With remote control of the data recording, similar means of reading the data is necessary for supervision and verification.

The point of view of the maintenance engineers would be influenced by the fact that few exchanges would have continuous attention by engineering maintenance staff and most would be visited only when faults were known to exist. Exchanges must in consequence be designed to take care automatically of operating irregularities, by parking lines which are permanently looped and releasing called lines held by calling lines and so forth. They must also be designed to limit the effects of equipment faults, automatically to detect those which occur and detect them within periods of time related to their consequences, and to inform the maintenance centre as faults occur. Means to satisfy these requirements have to be built into the exchange equipment or into routine testing equipment. The maintenance staff will be required to supervise the maintenance equipment and records and to clear faults when they occur. The exchange equipment should be designed to assist the location of faults by supplying information to aid the men concerned. In large exchanges it may be economic to provide equipment to locate faults down to small units of equipment. In unattended exchanges which involve an hour or more of travelling time to visit, automatic fault location equipment is less easy to justify.

Having established by some means the objectives for a new system to follow cross-bar and there could be variations to suit different administration, the application to exchange design of the electronic switching and memory techniques developed and in large scale production for the semi-conductor and electronic computer industries can be considered.

6.3 Structure and Call Connection

The first decisions to be made for a new exchange system concern the basic elements of the structure. Relay and cross-bar exchange switches having evolved as best for electro-mechanical systems, none better is available for semi-electronic

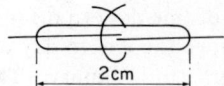
Figure 6.3   Reed contact

systems, and both are in use in practice. Those countries in which cross-bar systems are already established have tended to keep to cross-bar switches miniaturized in some cases to save space and cost, an exception being the U.S.A. where the Bell System is superseding its established cross-bar system with a reed relay system. The reed contact which is the basis of all reed relays, was invented in the Bell System which uses a particular construction called ferreed in which the reed contacts are maintained in both the open and the closed conditions by remanent magnetism without the continuous expenditure of power. Figure 6.3 shows the contruction of a reed contact. Two thin iron reeds are fused axially into the ends of a glass tube, the tips of the reeds being just not in contact. An axial magnetic field produced by current in a coil surrounding the tube magnetizes the reed tips to opposite polarities causing attraction and contact between them. The enclosed space of the tube is filled with a suitable inert gas, and the tips of the reeds are treated usually by gold plating to improve the contacting. If not magnetically latched, up to four reed contacts may be operated within one coil without difficulty due to an effect which increases with increasing quantities of reeds, that of unequal magnetic effects in the different reeds. With magnetic latching, more than two units per relay becomes difficult.

Although more reliable contacting has been claimed as a reason for using reed relays instead of cross-bar switches, practice shows that the difference, if any, is not sufficient by itself to decide the issue between them. Reed relays are more expensive than the equivalent in cross-bar switches. To off-set the difference, use is commonly made of the greater flexibility which individual relays possess in the ways in which they may be connected together to provide the necessary paths through the exchanges. Depending on the availability of cross-points to trunks at each switching stage, which means in general the cross-points per vertical trunk at each stage, so the quantity of switching stages varies, increasing as the availability decreases. By varying the availabilities and therefore the quantity of switching stages, the aggregate quantity of cross-points required in an exchange can be minimized. The minimum can be more nearly approached with reed relays which can be assembled and wired into switches of many different availabilities and sizes, than with cross-bar switches manufactured of necessity in very few standard sizes. An advantage of greater importance is the greater speed of operation of reed relays as a result of which it is possible to limit the path search and connection to one call at a time for the whole exchange even the largest. Mutual interference between calls being connected simultaneously is then not a problem because it does not occur. For this to be true, the time taken to complete a connection through an exchange must be 10 milliseconds or less, which is possible with reed relay contacts and electronic control. Otherwise in principle there is no difference between reed relays and cross-bar switches and the use of either may be assumed.

One system of trunking in use in cross-bar exchanges has been shown in Figure 5.3 and another which trades some increase in the complexity of the control for some decrease in that of the structure, in Figure 5.4. With processing power and flexibility as the main feature and justification for using electronic control, the trading of control complexity for structure simplification is naturally exploited to the maximum in semi-electronic and electronic exchange systems. In semi-electronic systems it is common for the exchange switches to make only one kind of connection which is 'folded' to produce whatever connections the transmission circuits terminated on the exchange require. This is illustrated in Figure 6.4(a) for the trunking and Figures 6.4(b) to (d) for connections made. On one side of the exchange switches are connected exchange lines and line equipments together with service circuits and junctions each with a junctor, possibly including a processor, to suit the signal and data transmissions individual to itself. Also on the same side are connected registers and other common processors and some special purpose circuit processors. On the other side of the exchange switches are loop trunks often called cord circuits because of their analogy with cord circuits in manual exchanges, and the equivalent of single-ended cords terminated on tone and other circuits. All connections involve finding and connecting a path from one side of the switches to the other, and complete connections are made with as many such paths in series as may be required. An originating call from an exchange line, Figure 6.4(b), is connected to a straight-through loop trunk which is then connected to an originating register o.r. for the receipt and storage of the subscriber dialled digits. The call requiring to be routed out of the exchange over a junction, the register connects to an outgoing junction to which the calling line can also be connected using a cord circuit with a suitable originating junctor or.j. The connection between the exchange line and the junction is either made, but with the transmission interrupted in the cord circuit until the register is released, or the connection is 'reserved', which means that the trunks and cord circuit eventually to be connected are busied against seizure for any other call. The register finishes it processing of the call, then releases leaving the connection in the exchange to be continued by the cord circuit between the exchange line and the junction. An incoming junction call similarly engages a register, i.c.r., which is released to leave the final connection established. If it is a terminating call, Figure 6.4(c) shows processors and connections used in some systems and which are analogous to those described for the minimal structure exchange of Figure 2.5. Until the called subscriber answers, the called line is connected to a processor which sends the appropriate kind of ringing to the called line, and the incoming junction has ring tone transmitted over it because of a similar connection to a ring tone circuit. When the called subscriber answers, the ring and ring tone processors are released and connection between the called line and the junction is made using a cord circuit and terminating junctor t.j. simplified to the extent that it does not have to provide ringing, ring tone and ring trip facilities. The quantity of terminating junctors is not reduced because of the necessity of junctors being reserved while ringing is taking place, which feature actually increases the exchange switch traffic and the quantity of switches required. The economy comes from the transfer of the ringing and the ring tone processes from the terminating junctors to processors less in quantity than the terminating junctors. It is important that the scale of the economy realizable by such methods be known and understood for system design. A two thousand exchange line of average traffic loading would require about one hundred and twenty terminating junctors and twenty ring and twenty ring tone processors. Because of the cost of mounting and accommodation on the racks, and the duplication of much transmission equipment in both the terminating and the temporarily connected junctors, the cost of one ring and one ring tone processor is three or more times that of the equipment eliminated from each terminating junctor. Including the increased cost of the exchange switches and the control, the overall cost saving is small, and proportionately less and more respectively for exchanges with less and more traffic than has been taken as an example. The economy of semi-electronic systems is an aggregation of similar cost difference with the result that semi-electronic are competitive with electro-mechanical systems for large exchanges but not for small ones, the full significance of which will be apparent later.

Busy and n.u. tones are connected similarly to the ring tone.

Figure 6.4(d) illustrates the way in which some special processing is introduced into exchange line calls in response to line data or state of line data read as the line is about to be connected. Service observation or special service observation being indicated for an originated call, connection is made to a processor providing the required facilities and the call then proceeds as if it were originated at the processor. An incoming call finding a required exchange line with the interception state of line would be connected to an interception processor which was then connected to the line. The manual connection by maintenance staff of special processing equipment to a line temporarily requiring unusual facilities, as described for wholly electro-mechanical exchanges, is thus avoided, which together with the necessary state of line data storage remotely controlled from an administrative centre makes the whole operation automatic.

A trunking system which is used in practice is chosen to satisfy some feature of the operation or to achieve some economy of structure or some combination of both. In these respects there is no basic difference between the trunking of Figure 6.4 and that of Figure 5.3. All the features of one, Figures 6.4(b), (c) and (d) for example, are realizable in the other. Despite apparent differences, there are similarities between all path search and connection systems resulting from the need for and use of the same basic data to produce basically the same result. The data required are the structure members in terms of transmission circuits and trunks available and of switches between them, and the state of the structure in terms of transmission circuits and trunks already in use: the data are used to determine free paths. Generally more than one free path exists, which makes one-only selection a basic process to produce one free path for connection. The selection should be random but that, although possible, is impractical. Systematic selection to improve the traffic capacity of the trunking is sometimes advocated and used, but the economic gain is trivial and generally not worth the sacrifice of an important principle satisfied by one-only selection which is at least quasi-random, that a

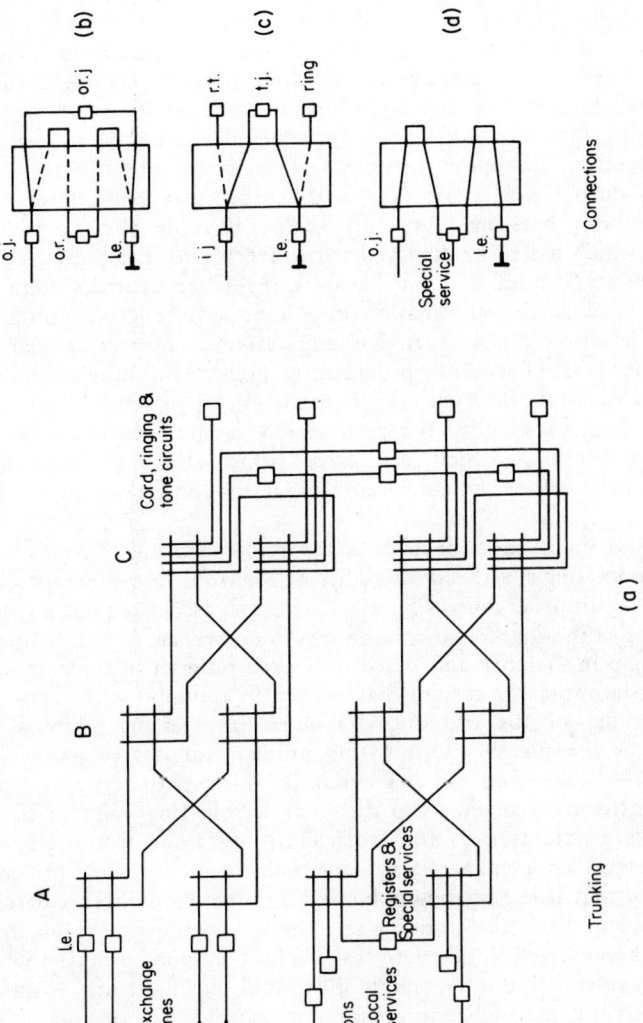

Figure 6.4 Folded trunking and typical connections

second attempt at connection because the first has failed, should take a different path or it will also probably fail.

A path search and connection system using electronic components is not limited by the speed of operation and durability of the components, as is the system of Figure 5.7 to suit relay processing. Nevertheless, economy of time and of processes and processing are still important. If the total search and connection time is such that only one call need be processed at a time while still keeping up with the traffic in the largest of exchanges, lock-out is required to ensure that no two registers can attempt to set up connections at the same time but some major restraints on trunking and operation disappear, namely sectionalization of the trunking and circuit lock-outs during call connection, with consequent simplification of the equipment and the processing. For this to be possible the 10 millisecond connection time which has to be achieved is the average for complete connections including all attempts, if more than one has to be made because the system can test for a free path to only one destination circuit at a time. Restrictions on the trunking or the selection can have an economic effect in respect of quantities of switches and trunks which have to be provided. As pointed out in relation to Figure 5.3, a regular and easily defined pattern of interconnections between the ranks of switches may be used to simplify the path search equipment, but at the cost of restraint on the trunking which can have other effects. Other important characteristics of path search and connection apparatus and processing will become apparent later.

Path search and connection processors for semi-electronic systems naturally exploit the high operating speed and durability of electronic components relative to those of relays. Circuit processors using wired logic and associated directly with the exchange switches as the markers of cross-bar systems are, are generally one of two kind known as map in memory and based on stored program logic is described in chapter 9. Interrogator-marker systems have much in common with relay markers. Systematic patterning of the trunking is required. A starting point is marked and a free one of the possible terminating points, and if free paths between them exist, one will be selected and connected. If no free path exists, a second and possibly a third attempt is made using different terminating points if they exist. Interrogator-marker systems are not sufficiently different from relay marker systems in their effect on exchange design for them to be described in detail. It is sufficient to know that if the starting point and terminating points are determined, and these are obvious from the trunking and the connections to be made, a path may be found and connected. Guide wire systems have unique properties which will be described with reference to an example illustrated by Figure 6.5. No particular pattern or arrangement of trunks and cross-points is needed, in one operation path search and connection is completed and a free path found if one exists, with a minimum of processing and through any number of switching stages in series, and at the high speed of electronic wire logic. Guide wire systems are thus more flexible than others and achieve the maximum of efficiency in the trunking whatever system might be used, but which are invariably of the type of Figure 5.3 because folded trunking has no intrinsic advantage but has some complications in searching in one operation for a path which traverses the exchange switches more than once and in both directions of search. Electronic exchange systems outlined in chapter 10 are better able to take advantage of guide wire path search and connection than are semi-electronic systems in which its use has been very limited.

Figure 6.5 shows path search and connection equipment using semi-conductor gates as in Figure 6.1 to control the cross-points of switches of two ranks of switches in series and as defined in the trunking shown schematically as well as in circuit form. The control for one switch in each rank is shown in circuit detail, each switch with two horizontal and two vertical trunks with cross-points at the intersections of the trunks, except for one intersection to demonstrate that the pattern does not have to be regular. Drawing space limits the detail which can be shown: in practice switches have many more than two trunks in each direction and partial availability, that is intersections without cross-points, looks more sensible. There may be some systematic patterning of the inter-rank trunks but because it is an advantage, not because it is essential, nor need the patterning be exactly precise and regular. The trunking schematic of Figure 6.5 indicates some possible other switches in the ranks and a random arrangement of cross-points. There is no limit to the quantity of switches in the ranks nor to the quantity of ranks in series.

Still referring to Figure 6.5, each cross-point is operated by the output, marked with a cross, of a bi-stable circuit individual to the cross-point, when that output is at logic 1. When operated it connects in series the horizontal and vertical trunks which intersect at the cross-point. Cross-points are assumed to be reed relays with, in the message switching path, three contacts switching two message wires and a P-wire for the electro-mechanical equipments including a meter operated by pulses over the P-wire, as in Figure 5.7. The P-wire of Figure 6.5 is a separate one for the electronic path search and connection circuitry and it will not transmit data other than the busy-free P-wire data. The cross-points could also be electronic for an electronic system. Different but equivalent circuitry may also be used for a system with cross-bar switches: instead of a bi-stable circuit per cross-point, one per column is needed to operate the bridge magnet, together with a mark per horizontal to operate the select magnet temporarily. Each cross-point bi-stable circuit in Figure 6.5 has a NAND gate G1 which when operated operates the bi-stable circuit to close its cross-point individually. No other equipment or operation is individual to a cross-point but is common to a row or to a column, which means that it is associated with a horizontal or a vertical trunk. The release of operated cross-points is by release data applied to rows of cross-point bi-stable circuits. The cross-points are controlled over three wires of the trunks between the ranks and into and out of the first and last ranks. The three wires are an electronic P-wire, and guide wires designated mark forward mkf and mark back mkb. The P-wire is at logic 1 = +v potential using the units of Figure 6.1, when the trunk is engaged, and at logic 0 = earth when free. Both of the other wires are normally at logic 0 which is changed to 1 to mark forward or backward. The horizontal trunks to the switches on the right of the diagram are assumed to be terminated on processors which according to whether they are free or engaged, control the potentials of their P-wires. Each cross-point bi-stable circuit has an output which is an input to a NOR gate G2, a gate G2 having inputs from all the cross-points of a column, all the inputs being at +v potential when the vertical trunk of the column is free and one being at earth when engaged. Hence the output of the gate and which is the P-wire for the trunk, is at earth when the trunk is free and at +v when engaged.

A horizontal trunk is marked forward if its mkf wire is marked and the trunk is free, conditions which change the output of a NAND gate G4 to logic 0. Every vertical trunk has a NOR gate G3 with an input for every cross-point of the trunk, the inputs being connected to the outputs of the G4 gates of the horizontal trunks of the cross-points. Hence when a horizontal trunk is marked forward by its G4 output, it marks forward all the vertical trunks to which it has access through cross-points, irrespective of whether the vertical trunks are free or busy. The free-busy switching is performed by the G4 gate of the horizontal trunk in the next rank of switches to which the vertical trunk is connected and so on through as many ranks of switches as there are in the connection. A mark on the mkf wire of a vertical trunk indicates that there is a free path from that trunk back to the processor from which the forward marking starts, or if the marking starts from more than one processor, there is a free path to at least one of them. Path connection starts with the marking back of one of the end trunks by a mark in its mkb wire. If there is any possibility of more than one being marked, one has to be selected by one-only selector, not shown in Figure 6.5, to ensure that only one mkb wire is marked. A mark on the mkb wire of a vertical trunk is applied as an input to NOR gates G5 of all the horizontal trunks to which the vertical trunk has access via cross-points. If a horizontal trunk is on a free forward marked path, its P-wire will be at logic 0, its mkf wire will be marked and its G5 gate operated, conditions which cause the NAND gate G6 to change its output from 1 to 0. The outputs of all the G6 gates of a switch are applied to a one-only selector o.o.s. which selects one of them which it indicates by a logic 1 output to the mkb wire of the trunk. Thus there is one vertical trunk directly marked and one horizontal trunk indirectly marked by the backward mark, to cause the G1 gate of the cross-point at the intersection of the two trunks to be operated and thereby to operate the cross-point. The cross-point is operated because the gate circuit to which the G1 gate is connected changes its output to 0, but the cross-point bi-stable circuit is not put into a stable state because the P-wire remains at logic 0 and the other gate of the circuit is unchanged with its output at logic 1. Nevertheless, the mkb wire of the horizontal trunk is marked and transmitted to a vertical trunk of a switch in the next rank, a cross-point in that switch is operated and so on back to a processor which responds to the connection thus made by busying its P-wire. As the P-wire changes to logic 1, the partially operated cross-point bi-stable circuit to which it is connected completes its operation to the stable state of the cross-point operated, and by changing its output to a NOR gate G2 to 0, busies the vertical trunk of the cross-point by changing the state of its P-wire to logic 1. This has the effect of completing the operation of the selected cross-point in the next switch and so on for the whole of the connection until it is complete. The one-only selectors are designed to hold their outputs unchanged for sometime after the inputs change to give enough time for the selected path to be established. The connection is held so

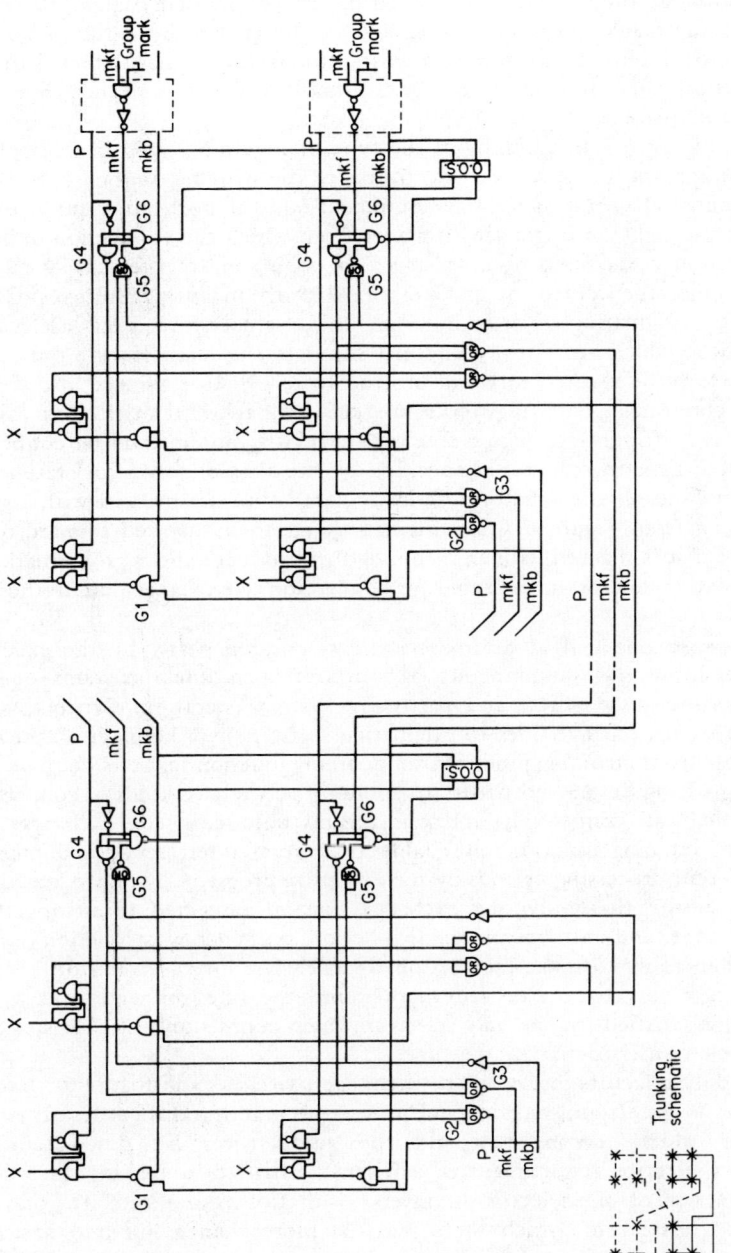

Figure 6.5 Guide wire path search and connection

long as the processor holds its P-wire at 1: change to 0 releases the bi-stable circuit and cross-point of the trunk connected to the processor, that changes the P-wire potential of the trunk to the next switch where the cross-point is released and so for all the cross-points. If no free path exists, the backward mark applied to start the connection does not meet a forward marked trunk, and no cross-point operations take place.

Relating Figure 6.5 to trunking of the type of Figure 5.3, and in particular to subscriber originating calls, the vertical trunks to the switches on the left of Figure 6.5 are terminated on exchange line equipment and the horizontal trunks on the switches on the right on originating junctors from which the connection paths are continued through further switching stages to originating registers. New calls are connected under the control of an operational program during the execution of which all the originating registers but only the originating registers, are marked forward. Those which are already engaged suppress the mark leaving only those which are free to be marked forward into the switches. An exchange line which is calling for connection to a register is marked by backward mark from its line equipment, and if a free path to a free register exists, the line will be connected. The type of originating junctor is specified by the class of service or state of the exchange line and all the junctors of the required type are marked with a group mark. As seen from Figure 6.5, an originating junctor is marked forward of the junctor only if it is marked with an incoming forward mark and is group marked as well. Terminating junctors are selected for connection to exchange lines in the same way.

The necessary one call at a time feature is ensured partly by the executive program which prevents simultaneous execution of operational programs requiring to use the same guide wires, and partly by system operation. Originating call connections from exchange lines to originating registers may be made at the same times as those from incoming junctions to incoming junction registers because, with the trunking of Figure 5.3, no paths are common to the two kinds of connection: but originating call connections and connections terminating on exchange lines, services and outgoing junctions are liable to mutual interference and must be allotted different processing periods by the executive program. One-only selection is required to ensure that only one exchange line is connected to an originating register at a time, and only one incoming junction to an incoming junction register. Registers when ready to make connections to exchange lines, services or junctions have to gain access to a translator-marker and the one connection at a time requirement is satisfied by one-only selection which permits only one register to be connected to a translator-marker at a time.

The one-only selectors individual to exchange switches as in Figure 6.5 have up to twenty or so inputs and outputs and present no practical difficulties other than security for which reason they and the guide wires are duplicated. The quasi-random selection requirement is sufficiently satisfied by making the order of one-only selection of one selector the inverse of that of its duplicate. The one-only selection of registers of which there may be more than a hundred has some problems but is possible using static gates, the equivalent of the relays of Figure

5.6. The one-only selection of exchanges lines only one of which may be marked at any one time for an originating or terminating call, presents some difficulties because of the quantities of exchange lines and for which reason the sequential scanning described in the next chapter is preferable.

6.4 Analogue Cross-bar Semi-electronic Systems

It has been shown that given components with the characteristics of electronic devices, a relay or cross-bar switch exchange structure may be controlled and operated in ways not possible with only electro-mechanical devices and in ways which give advantage to the system. Notably if the switches and the control operate fast enough, one connection at a time is possible, thus avoiding the complications of connections made simultaneously, and there may be some small advantage in respect of structure and control equipment economy. Clearly it is not difficult to produce electronic units which are the equivalents of cross-bar circuit processors, registers and translators and which together with electronic path search and connection and other equipment may be used to produce a semi-electronic system which is the analogue of conventional cross-bar systems generalized in Figure 5.5. This was the first and obvious application of electronic devices to exchange systems and it showed that the advantage to be gained over conventional cross-bar is trivial. There is no marked effect on the defects and weaknesses of electro-mechanical systems in respect of very limited data storage, transmission and processing to which attention was drawn in section 5.7 and which limit the services that may be economically offered and involve so much manual assistance from the operators and maintenance staff. Clearly the durability, high operating speed and large data storage capacity of electronic devices have to be still further exploited to produce a major change such as is usually associated with a new principle of design.

Chapter Seven
SEMI-ELECTRONIC EXCHANGES

7.1 Introduction

In addition to the high speed of operation and durability of electronic devices which together increase manyfold the quantity of processing that may be performed in a given time, manufacturing techniques make available at relatively low costs and in very small volumes large capacity data stores and large processors which together increase many times the quantity of processes which may be included in exchange control systems. The main effects on the control systems include a large increase in data stored and used, the time division time sharing of processors as well as space division time sharing, and the use of stored program logic for part or all of the processing. In this chapter the principles, characteristics and applications of space division and time division time sharing of processing and processors will be analysed and defined in detail. A clear view of wired and stored program logic applications and their effects on exchange system design is less easily seen and is approached by the consideration of systems using only wired logic as in this chapter and of systems not so limited as in chapters 8 and 9. The importance of the increased data processing and storage capabilities of electronic processors will be apparent from these expositions.

Theoretically, wired logic or stored program logic may be used in any situation but the practical differences of cost, speed of operation and design lead to differences in exchange systems satisfying the same performance specifications. The differences will be brought out in the comparisons but so also will be the similarities. Much of the wired logic processing illustrated by circuits in this chapter is achieved in stored program logic systems by programs which are the equivalent of the wired circuits: hence the circuit descriptions are useful to both techniques.

7.2 Time Division Time Sharing — Scanning

As explained in chapter 2, an operator as a common processor time shares her activities among the circuits and calls which she has to control, some by time division notably for the detection of new calls, and some by space division in particular for call connection processing which in automatic exchanges is controlled by registers. With time division time sharing reintroduced into exchange operations after being lost initially in passing from manual to electro-mechanical automatic exchanges, space and time division find application in much the same way as in manual systems so long as the processing is limited to wired logic, as will be seen in this chapter.

Figure 7.1 shows an example of line scanning equipment suitable for exchange line processing for a block of $2^{10} = 1,024$ exchange lines, for new and for terminating calls in an exchange with relay or cross-bar exchange switches as explained in previous chapters, or with electronic switches to be described in chapter 10. An address generator LA has complementary binary outputs which are applied to a common processor and the ten least significant digit outputs to a line and state of line data store LDS and to gates G which together with the transistors Tr2, Tr4 and Tr6 constitute an interface switch and necessary transducers between the common processor and the exchange line equipments. The gates are as described in conjunction with Figure 6.1, using logic levels of earth and +v to produce NAND and NOR function gates, and they are operated in groups each by a group of address outputs. A group of n address outputs produces 2^n partial addresses, the outputs operate one or more groups of 2^n gates, one gate for each address and with always one gate in each group operated. An operated gate has its address inputs all at +v. The output of a gate having only address inputs is at earth when the gate is operated, but for gates with an additional input, the output is dependent on the additional input when the gate is operated and is the inverse of the additional input.

Address outputs excluding the three least significant digits are used to mark one of 128 wires mk each serving eight exchange lines. To do so there are sixteen groups of eight gates G1 with always one gate in one group operated by three LA digits and the inverted output from one of sixteen gates G2 always one of which is operated by the four most significant LA address outputs which are used. The mark on the mk wire selects eight lines for processing and the lines are individually selected by the three least significant digits of the address applied to gates G3, G5 and G7. Thus, depending on the address out of the address generator LA, one line is selected for processing, data pertaining to that line are communicated to the common processor and the common processor can communicate data to the line equipment.

Each exchange line has equipment which includes a K cut-off relay and message register MR operated by current over a P-wire as in previous systems. Until the K relay is operated, battery and earth potentials are applied to the line via resistors R1, the earth connection being made through the emitter of a transistor L with its base at earth potential. Loop current through the emitter produces collector current which causes the transistor to saturate because of the resistor in the collector circuit. The potential of the collector is thus +v for no loop and substantially earth for loop current, and is the equivalent of a line relay L and its contact. The collector is connected to the base of a transistor Tr1 the emitter of which, protected against excessive reverse voltages by a series diode, is supplied with current from an mk wire through a logic inverter, such that current flows in the collector circuit of Tr1 if the line is looped but the K relay is not operated because no connection has yet been made to the line through the exchange switches; and further the line is one of eight selected by a mark on an mk wire. The 1,024 transistors Tr1 are thus divided into eight groups of 128 transistors each, in which only one transistor in each group can be conducting at any one time. The collectors could therefore be commoned in groups of 128 except that the capacitance of so much wiring would slow down the rate of change of potential too much or a large driving power would have to be used. Instead the 128 collectors of a group are commoned in groups of eight, each group driving a transistor Tr2 which provides power for driving long connections, the output from the transistor operating a logic inverter the output of which can be +v, that is logic 1, only for an exchange line looped but not connected. There are thus sixteen transistors Tr2 per 128 line group and they are commoned to one inverter via a diode close to the inverter to reduce the wiring capacitance which each transistor has to drive. There are eight inverters and the output of each is connected to one input of a four-input NAND gate G3 the other three inputs of which are operated by the three least significant address digits every combination of which operates one of the gates. The outputs from the gates G3 are combined by an eight-input NOR gate G4, the output of which is applied to the common processor, with the result that the common processor receives the state of L datum for one exchange line at a time, the line being the one at the address then existing out of the address generator LA. Clearly service for the 1,024 lines could be hazarded by one fault, for which reason all the electronic equipment other than the L transistor and circuitry per line, is duplicated. It is then possible to have faults in both sets of scanning equipments and still to give service provided that for no fault affecting part of one equipment is there a fault affecting the same part of the other equipment.

The states of the P-wires of the lines are similarly made known to the common processor using transistors Tr3, Tr4 and gates G5, G6. When a P-wire is earthed the potential of its Tr3 transistor emitter is slightly below earth, for which reason the potential divider resistors operated by the mk wire potential are needed to take the base potential of the transistor still further below earth to cause the transistor to conduct.

An exchange line is marked as available for connection by +v potential applied to the MK wire of the line by a transistor Tr5. The emitters of eight transistors Tr5 at a time are raised to +v potential by a marked wire mk and eight gates G7 select one of them at a time to receive the operational data o emitted by the common processor. Again because of wiring capacity, transistors are interposed, i.e. the transistors Tr6. A logic 1 datum applied to the mark output o from the common processor causes the line at that moment addressed by the LA output to be marked on its MK lead, and if a connection results a cross-point assumed to be a reed relay is operated and held in series with other cross-point relays in the connection, the P-wire potential changing when the connection is complete, as described with respect to Figure 6.5 for guide wire path search and connection which might be used. The common processor detects when the connection is completed by the change of P-wire datum communicated to it.

The store LDS in Figure 7.1 contains for each exchange line a data word comprising semi-permanent line data and period and temporary state of line data. The word corresponding to the LA address at any instant is read, according to the needs of the system, either as soon as the address is established or on instruction by operational data o from the processor. The word is read out to the common processor which can also write a word into the LDS store at the then existing LA address. The LDS store may be comprised of separate stores appropriate to semi-permanent, period and temporary data.

Reference to Table 4.1 shows that some line and state of line data are required when the directory number is available as the address, and some when it is the equipment number which is known. It is apparent from the description of the equipment of Figure 7.1 that the address supplied by the LA addressor cannot be the directory number. So far as the control is concerned it is an equipment number because it defines through the interface switch an exchange line equipment and its input to the exchange switches and all data retrieval and processing for the line connected to that input is achieved with that address. Generally the equipment and its position in the structure is arranged so that the address also defines the physical location of the equipment individual to the line, a matter of importance to the maintenance staff.

Equipment numbers and line data both refer to individual lines, the first thus being convenient addresses for the second. The same is not true of directory numbers: each number defines a quantity of lines from zero if the number is not in use up to possibly hundreds of p.b.x. lines and unpredictably variable with time. The lines are moreover more or less randomly distributed over the equipment number positions. Hence the extraction of line data for given directory numbers presents problems. As part of the solution to those problems, it is often the practice to write into each word of the LDS store the line data appropriate to both equipment number and directory number addresses, together with the directory number of the line at the equipment number address. By scanning all the lines, those with a given directory number and their data can be identified one at a time. In this way one store suffices for both equipment number and directory number addressed data but the control must be able to tolerate the time that it takes to access directory number addressed data by scanning.

As an example of line data storage, consider the classes of service given in Table 4.1. Some of the classes are mutually exclusive in consequence of which they may be coded with fewer bits than there are classes of service. Table 7.1 shows two groups of mutually exclusive classes of service each using two storage bits for more than two classes of service. On the other hand, because some of the classes of one group may occur in combination with one or more of the classes of another group, separate groups of storage digits are required. The Table 7.1 shows the data written into the four digits of the store which are numbered 8 to 11 out of sixteen numbered 0 to 15, the values of the other digits not being shown. The remaining digits are also allocated to mutually exclusive groups of line and state of line data, the combinations of data in the various groups commonly amounting to hundreds and a notable advance on what is possible in electro-mechanical exchanges. Some of the processing of the data is described in chapter 8 in which Table 8.2 is a typical list of facilities required of modern exchanges, the facilities being distinguished by classes of service requiring line and state of line data storage as described.

Referring again to Figure 7.1, at every instant the common processor is receiving

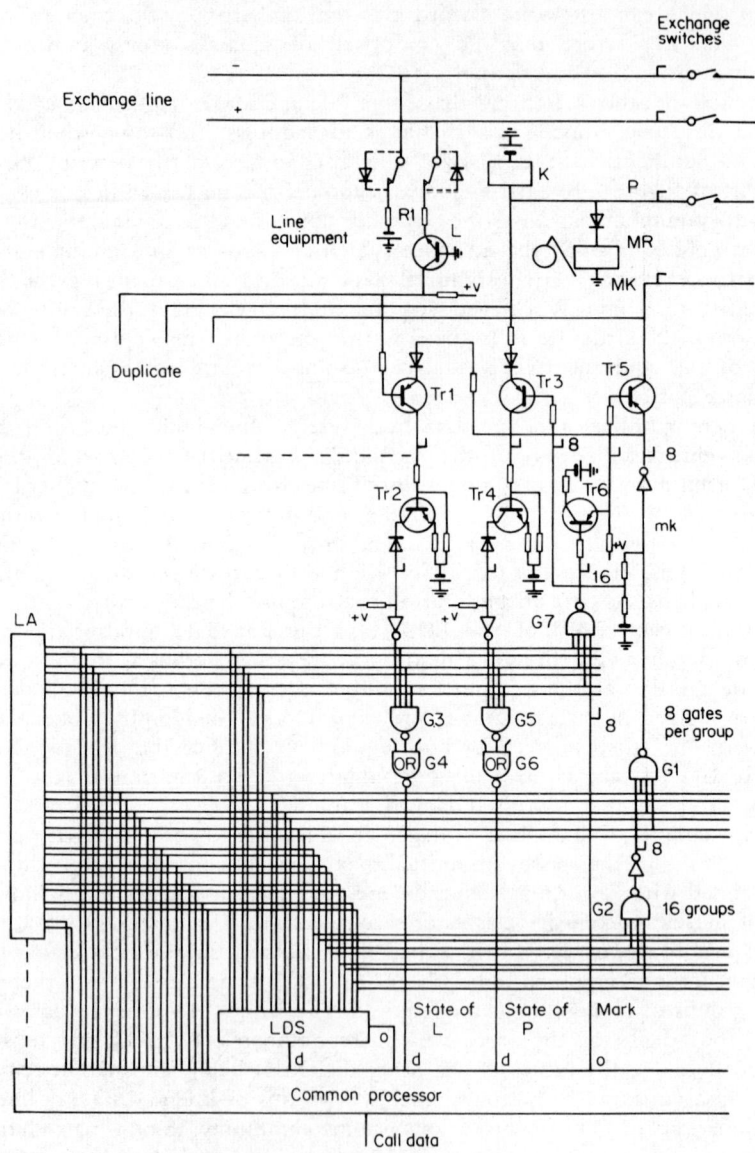

Figure 7.1 Time division scanning

Table 7.1 Part of line data store

Store digits											Class of service
15	14	13	12	11	10	9	8	7	...	0	
						0	0				Ordinary and p.b.x.
						0	1				Coin-box
						1	0				Party line
						1	1				Disabled sub.
			0	0							Distance not limited
			0	1							l.d.b.
			1	0							i.b.

the state of loop datum of an exchange line defined by the LA address together with the state of the P-wire of the line. The processor is also receiving or it can read the line and state of line data pertaining to the line. If the data indicate that the line is looped but not connected through the switches, conditions associated with new call origination, a new call may not be allowed because the state of line is parked, or o.b. or t.o.s., and in that event the processor takes no action to connect the line. If however, the data signify that the line is originating a new call which is allowable, the line is marked by an MK datum and connection follows over a free path to a free register, if such exists, due to the action of control equipment not shown in Figure 7.1. In the cross-bar system of Figure 5.7, an exchange line L relay being operated causes a one-only selector to select a file of lines and another one-only selector to select a line in a file to produce a unique line for the path search and connection processors to connect to a register. The L relay then releases, the whole operation taking its own time subject only to delays due to lock-outs by other connections. Two or more calls originating together are connected one when it is connected followed by another. If a line is to be barred all originating calls, manual action is needed in the exchange to prevent the L relay of the line from being effective. In the system of Figure 7.1, the lines are scanned one at a time thus giving the effect of one-only selection and simplifying the path search to the extent of eliminating the one-only selection processes necessary on the exchange sides of the A switches of Figures 5.7 and 6.5. If a line is found to be calling, whether the call is or is not to be allowed is ascertainable from the line data electronic store LDS and if allowed, the line is marked for connection at once or its identity is stored for connection later as will be described in this section. Alternatively if a line is not to be allowed originating calls, it may be connected to a register which receives and stores the o.b. state of line as call data, to instruct it to permit only emergency calls. If the state of line is s.o. or s.s.o., when the line calls it may be connected to an s.o. or s.s.o. processor as well as to a register, and with s.v.i. state of line, the line on being called using directory number scanning may be connected via an s.v.i. processor. The action needed to impose o.b. and other conditions on a line is manual, namely to cause data to be written into the LDS store, but the action does not have to be taken in the exchange. It may be taken more efficiently operationally at an administrative or maintenance centre having remote control means of writing data into the store

The capacity of electronic stores being so much greater in practice than that of wired stores, the classes of service and other data stored and the consequent processing may be more complicated and extensive than in electro-mechanical systems. Clearly the electronic control is operating to the same effect as the electro-mechanical control but with greater facility. There are, however, some problems to which the discussion in section 2.13 of time division time sharing by manual operators, is apposite.

The frequency with which lines are scanned depends on the delay which can be tolerated in obtaining service when starting a call. Five tests per second is commonly accepted. There may be ten thousand or more lines to be scanned, and because of other exchange operations with which scanning would interfere, possibly not more than half of the total time is available for scanning. Hence the average 'dwell' of the processor on any line may not be more than 10 microseconds, which is insufficiently long by a factor of hundreds or more for relay-type exchange switches to be operated. Three possibilities exist. Firstly, to use a structure which is fast enough to be operated in the time available: this solution is limited to electronic exchange switches. Secondly, if the switches although not operable in 10 μsec are fast enough for one-call-at-a-time call connection through the exchange, which is the case with reed relays, the scanning may be stopped while a connection is made and resumed generally but not necessarily at the same place. Thirdly, if the exchange switches and call connection processing are not fast enough for one-call-at-a-time call connection, which is typical of cross-bar switches, the scan can be used to determine and store the addresses of lines to be connected, the actual connections being later operations possibly conventional cross-bar system processing. In the first case the scanning rate is uniform and chosen to suit the call connection processing. The address generator LA of Figure 7.1 is typically a binary counter driven by clock pulses from a constant frequency oscillator, the cycle time of the counter being the required period between scans. Lines are scanned only once per cycle of the counter, using as line addresses the least significant digits of the counter. With one more significant digit in the counter, as in the figure, the scanning takes up half the total time, thus leaving half for other exchange processing. With two more digits in the counter, the line scanning would be limited to one quarter of the total time and so on. Scanning which is cyclic at a regular rate controlled only by clock pulses is designated synchronous cyclic time division scanning, or just synchronous scanning. Asynchronous scanning although cyclic is not at a uniform rate entirely controlled by clock pulses. The common processor dwells at one address only long enough to process the data at that address, then itself advances the address by one unit, which is indicated in Figure 7.1 by the dashed line between the processor and the generator over which the processor may send one pulse to change the address. Alternatively the LA equipment may comprise bi-stable stores which the processor may set to any number, with the advantage that the processor can start the scanning at any point in the cycle that it may choose. With asynchronous scanning, the minimum dwell time of the common processor being the minimum time to process data for one line, is less and may be much less than the constant dwell time of synchronous scanning, which time is the maximum ever needed to process the data of one line and the maximum dwell time for asynchronous scanning performing the same operations. Clearly asynchronous scanning makes better use of processor time than synchronous because it keeps the processor continuously operating and not waiting for the address to be changed by a clock pulse. Clock pulses are needed to time regular intervals at which to start the scan cycles but the time taken to complete one scan is not constant, nor are individual lines scanned at exactly equal intervals, the times for those at the end of the cycle being the most variable. The variation is due to the quantity of processing required per scan being dependent on the existing states of the lines or other equipments being scanned and is thus subject to the vagaries of traffic.

With cyclic time division processing, the positions in the cycle are also called slots, and the time or times of dwell in the positions the slot time or times.

Synchronous or asynchronous scanning may be used to detect and record the addresses of lines to be connected as required by the third of the possibilities previously mentioned. The addresses are stored in a 'hopper' from which they are taken one at a time for processing in an order which is thus acyclic. For each line as it comes to be processed it must be possible to set the address generator LA to the address of the line, to read the line data and to mark the line. Acyclic time sharing keeps the processor continuously employed on effective processing and thus makes the most efficient use of the processor while so engaged. If it also has to be used for the cyclic scanning to detect and store the addresses of lines to be connected, the effective use of the processor is reduced, for which reason a simpler processor may be used to execute the scanning and thus to leave the main processor free to operate simultaneously with scanning, provided that between the processor programs and scanning there is no mutual interference. The use of the third method occurs generally because neither of the others is possible in the system being designed. If there is a free choice between synchronous and asynchronous scanning, synchronous is to be preferred because it requires less and somewhat simpler processing and less equipment: the address generator may for example be common to many time division time shared processors all working simultaneously in the exchange.

Some other possibilities which exist and can be found in practice are developments of the three main methods described. For example, instead of waiting for call connection processing, which takes milliseconds to complete, the processor may initiate the operations and leave them to be completed by markers and other processors, itself going on to other processing without having had to wait for the connection processing to be completed or to store the addresses of lines to be processed later. Also, although five scans per second is required for good service, occasional delays of a second or more are tolerable in new call detection and possibly in terminating call connection. Line scanning in consequence is not a first priority operation and it may be interrupted in favour of higher level processing with which it would interfere.

The analogy between manual operating and electronic control of exchanges thus begins to emerge clearly.

Closer examination of the systems of Figures 4.1, 5.7 and 7.1 is needed to see exactly the operations and their significances. All of the systems are producing the same results from the same causes and with the same data, namely the connection of a line when it calls or is called and if connection is allowed, to an appropriate space division time shared transmission trunk and circuit processor. The connection of a line when it is called has not been described for the system of Figure 7.1 but it is not difficult to see that it requires and follows the marking of lines as for the system of Figure 5.7. When a line is looped to initiate a call or is marked for a terminating call, connection through the exchange switches is indicated by the P-wire datum which, operating the K relay, disconnects the L state of loop detecting circuit so that the processor receives L = 0 state of line whatever the true state of line loop, to leave the state of line processing to another processor. The K relay of Figure 7.1 is, however, redundant so far as processing is concerned, because the common processor given the state of the P-wire and the state of the loop data is quite capable of processing them every time they are presented. The K relay is included firstly to eliminate the transmission loss which the R1 resistors would otherwise cause, and secondly, when the transmission loss is negligible, to disconnect the battery and earth connections which would otherwise prevent insulation testing of the lines and which is an essential maintenance operation. The order of insulation resistance which is significant is one megohm. If the K relay contacts are replaced by diodes, as shown by dashed lines, the reverse resistances of the diodes being much greater than one megohm, the insulation resistances of the line wires can be measured using potentials which reverse bias the diodes. Only the transmission loss remains as a reason for using the K relay and a transistor circuit for the L relay being so much more sensitive than a relay, the resistors R1 may have resistances high enough for the loss to be negligible, when the K relay may be omitted.

It is not difficult to see that the individual meters MR may also be substituted by a bulk store similar to if not part of the actual LDS store used for other line data. To operate a meter MR and also, if connected, a private meter as shown in Figure 3.5, meter pulses over a P-wire commonly have a duration which has to be greater than 0.1 sec and which is less than 0.2 sec. Therefore if the line scanning at 5 Hz for new calls is also used for the meter pulse scanning, a meter pulse may occur between two successive dwells of the processor and be missed in the accounting. Moreover the meter scanning must be at regular intervals with little variation allowed and no interruptions, and these conditions cannot in general be guaranteed by asynchronous scanning. Thus there are difficulties in using the same scans for all line processing including meter pulse accounting other than with synchronous scanning at a sufficiently high scan cycle frequency. If the meter processing cannot be included with other line processing in the same scan, it may have synchronous scanning equipment individual to itself. For each address in the scan a state of MR datum is available, logic 1 indicating the presence and logic 0 the absence of a meter pulse. The scanning frequency must be high enough for every meter pulse to be detected at least once, which means that it may be detected on two or more successive scans, but only one unit is to be added to the meter total however many times the one pulse is detected. In the bulk store there is at every

Table 7.2 Meter pulse scanning and recording

State of MR	'Last look' state of MR	Operational data	Comment
0	0	None	Normal state
1	0	Write 1 in 'last look'	Meter pulse begins
1	1	None	Pulse continues
0	1	Write 0 in 'last look' add 1 to meter total	Pulse ended
0	0	None	Normal state

address, in addition to the meter pulse total which may require sixteen binary digits, a 'last look' digit. When a line and its store word are addressed, the value of the last look digit is read out to the processor which writes into the digit the present state of the MR datum. Thus the processor detects the present state of MR and reads the state during the previous scan, the last look state. The processing is indicated in Table 7.2, successive lines of which show the data available and the operational data issued on successive scans of a line at one address. In the normal state between meter pulses, the state of MR and the last look digit are both logic 0; no operational data are given and these are the conditions on every scan until a meter pulse begins. The state of MR is then found to be 1 and digit 1 is written into the last look digit. So long as the pulse continues, both the state of MR and the last look digit are logic 1 and no instruction is given, but when the pulse has ceased, the state of MR is found to be 0 with the last look at 1. Finding these conditions, the processor writes 0 in the last look digit and adds one unit to the meter total stored at the address. Successive scans will then find the state of MR and the last look digit back to the normal logic 0 for each. To give protection against noise pulses which might cause false metering, it may be required to use a frequency of scan which guarantees at least two scans per meter pulse duration, and processing which adds one to the meter total only if the state of MR = 1 and last look = 1 data occur on at least two successive scans.

The limited quantities of wired and relay data storage per line of the first exchange systems become almost unlimited as electronic storage in electronic control systems, and time division simplifies the processing and eliminates the need for the K relay as previously mentioned. On the other hand, the time sharing requires additional equipment for data transmission between the individual processors and the common processor, for the interface switches and for transducing from one kind of transmission to another. In the present example, Figure 7.1, the additional equipment comprises the transistors Tr1 to Tr6 in transducers and the gates G1 to G7 and the inverters as interface switches, some of which together with the common processor and the LDS store must be at least duplicated for security of operation. As a result, although electronic control achieves more than that of previous systems, it tends not to produce savings in capital costs. Cost advantage accrues from the savings in operating costs such as those due to fully automated charge accounting.

7.3 Time Division Time Sharing — Processing

The main features of time division time shared processing are included in the scanning example of the previous section which also usefully includes the starting point of the processing required for calls controlled from subscriber stations.

Independent of processing speed, not all processing can be effectively space division time shared, as the conditions for space division time sharing apparent from previous sections and stated in detail in the next section, 7.4, make clear. Provided that the processing speed is high enough, theoretically all circuit processing can be transferred to one or more time division time shared processors and therefore all relays, meters and the like which perform only data storage and logical operations may be eliminated from the structure and their functions transferred to the common control. Those devices, mostly relays, which also perform other functions, mainly switching and data transmission, may still have to be provided in the structure to execute those other functions. The meter of Figure 7.1 is easily eliminated from the structure because it has no other function than data recording, but a parallel connected PR relay of Figure 3.5 if it exists cannot be eliminated because its function is data transmission. The K relay has a switching as well as a processing function and its elimination is by substitution of preferable components, resistors and diodes. The processing functions of the ringing relays RG and F of Figure 3.6 can be transferred to a time division time shared common processor but the transmission of ringing current still requires the contacts of a relay to switch it and only a space division time shared processor, because it is continuously connected while processing, can remove the relay from the circuit processors. The important point is that to satisfy the operational and minimum cost requirements, the total processing of an exchange has to be divided between circuit processors and space and time division common processors according to principles which are set out in section 7.4, and to the characteristics of the components used and which give rise to different exchange systems in practice. In this section the important features of time division time shared common processing are described and illustrated.

The essential features of time division time sharing are the connection of a unique common processor for brief periods in turn to similar individual processors, the assembly by the common processor of all the data available and relevant to an individual processor to which it becomes connected, and the processing of that data to achieve an aggregate effect the same as if the common processor had been permanently connected to the individual processor. The common processing is therefore very similar to individual circuit processing but some differences are compelled by the intermittent nature of the processing. Table 7.2 has already shown that the addition of one unit to a message register total involves more processing than that of the addition itself. A cause of difference is operations involving time. The individual processors of electro-mechanical systems use slugs on relays, slowly charging capacitors and other methods for measuring time which clearly are not usable in an intermittently connected processor the time of dwell of which is less than the time to be measured. If the intermittent connection is regular at a clock controlled rate, the counting of elapsed time division cycles provides a means of time measurement and this procedure can be more accurate than is obtainable by other means.

The term scanning is applied to the connection of a common processor in some particular order to individual processors one at a time for any purpose. Scanning always implies the detection of changes of data in the individual processors and may use synchronous or asynchronous time sharing cycles: but the processing consequent on a change of data may be included in the scanning or executed later by acyclic connection of the processors concerned to the or another common processor.

Again the best approach to the understanding of a new technique is through a practical example in continuation of previously established techniques. The use of time division time sharing for the line equipment processing of cross-bar systems generalized in Figure 5.5 was illustrated in conjunction with Figure 7.1. Time division time sharing may also be applied to the processing of the junctor circuit processors and register common processors of Figure 5.5. The connections from an originating junctor or.j. to an originating register via an interface switch are given in Figure 5.7. A similar arrangement may be used in a semi-electronic system with cross-bar or reed relay interface switches and the register processing may be individual to each register or time division time shared among all the registers. Figure 7.2 is given as an illustration of one of a quantity of space division time shared registers electronic except for necessary data transmission transducers, the logic processing for the registers being executed by a unique common processor time division time shared among the registers. Figure 7.3 is a binary decision program on which part of the processing is based and Figures 7.4 and 7.5 show possible circuits for some of the operations. The example provides the opportunity to introduce shift registers as data stores with some advantages and to convey the characteristics and scale of the equipments needed for those processors, namely the originating and incoming junction registers, which have the largest quantities of processes and the most complicated processes to be found in telephone exchanges.

Referring to Figure 7.2 and relating it to Figure 5.7, on the exchange side the four wires between the junctor and the register comprise one wire, DB, over which current flows in the collector circuit of a transistor to operate the relay DB of the junctor when a bi-stable circuit DB in the register produces current in the emitter of the transistor, a wire P which when earthed in the junctor operates transistor Tr1 in the register, and two wires L over which loop signals are received and detected by the current which flows round the loop from the battery on one wire to earth on the other via the emitter circuit of a transistor Tr2. The current in the loop changes the collector potential of the transistor from +v = logic 1 to earth = logic 0, which needs to be inverted, like that of transistor Tr1, to present the correct logic conditions to interface switch gates. Two of the four wires on the outgoing side transmit loop signals Lo from a metal contact reed relay A, and one wire is a P-wire designated Po which, when supplied with current from a transistor Tr4, holds exchange switches connected forward from the junctor. The fourth wire MK is a marking wire supplied with current from a bi-stable store MK. In the outgoing loop controlled by the A relay is a relay Lo with a contact which indicates when current is flowing in the loop. Operational data communicated by the common processor have to be remembered by bi-stable circuits, of which DB and MK have already been mentioned. The bi-stable circuit Po operates the transistor Tr4 via transistor Tr3, and circuit A operates relay A. Dial tone and n.u. tone are sent back to the calling line through a transformer in series with the incoming loop, the tones being applied from common tone sources and switched by the diode switches controlled by the bi-stable circuits d.t. and n.u. for the dial and n.u. tones respectively. Each bi-stable circuit has a gate by which it is switched to the on condition and a gate to return it to the off condition. All the equipment mentioned so far transduces data from the junctor to the register or in the opposite direction.

Also part of the register are the shift registers SR1, SR2 and SR3 used as data stores. Shift registers each capable of storing thirty-two binary digits are assumed for this example but registers storing up to thousands of bits are available in practice. The shift registers are driven by short duration pulses derived from a master pulse generator m.p. producing several phases of pulses at a frequency of 256 kHz continuously and used to provide 'clock' timing for the whole system. When a clock pulse input to a shift register occurs, the logic state, 0 or 1, then existing at the input to the store is written into the store to appear at the output thirty-one pulses later and persist until the thirty-second pulse occurs. The output being fed back to the input, a digit once stored in the shift register will circulate within the register and appear at the output every thirty-second clock pulse until by external means the digit is changed. To write digits into the store, two two-input NAND gates G1 and G2 are included in the circulating path, the gates being in series and each with one input in the circulating path. The other input of each gate being normally at logic 1, the store output is inverted twice before being applied unchanged to the input for continuous circulation of the same data. The second input to the gate G1 becoming logic 0 from the output from the gate G3, digit 1 is written into the store no matter what the output from the store may be. Similarly digit 0 is written into the store if the second input of gate G2 becomes logic 0 due to gate G4. The circulating store will be recognized as a bi-stable circuit with a shift register in the feed-back loop.

The digits stored in the shift registers SR1, SR2 and SR3 are numbered 0 to 31 and identified each by the cyclic slot in which it appears at the output of its store. The master pulses m.p. are applied to a four binary digit counter which divides the frequency by four and produces four output pulses n4.0 to n4.3 in rotation and each with a duration equal to the period of the master pulses which is 4 microseconds, very nearly. Pulses at the cycle frequency of the counter, namely 64 kHz, are applied to a divide by eight counter with eight phases of output pulses n8.0 to n8.7, each pulse having the duration of one cycle of n4 pulses. Gates thirty-two in number and each with one n4 and one n8 input produce all the combinations of the n8 pulses with the n4 pulses and therefore the pulses p0 to p31 each of which identifies one of the digits stored in the shift registers, as each appears at the output of its store. Figure 7.2 shows data allocated to the digits of the stores. Those of stores SR2 and SR3 concern only the common processor. Some data of the store SR1 are data to be transmitted to the junctor from the

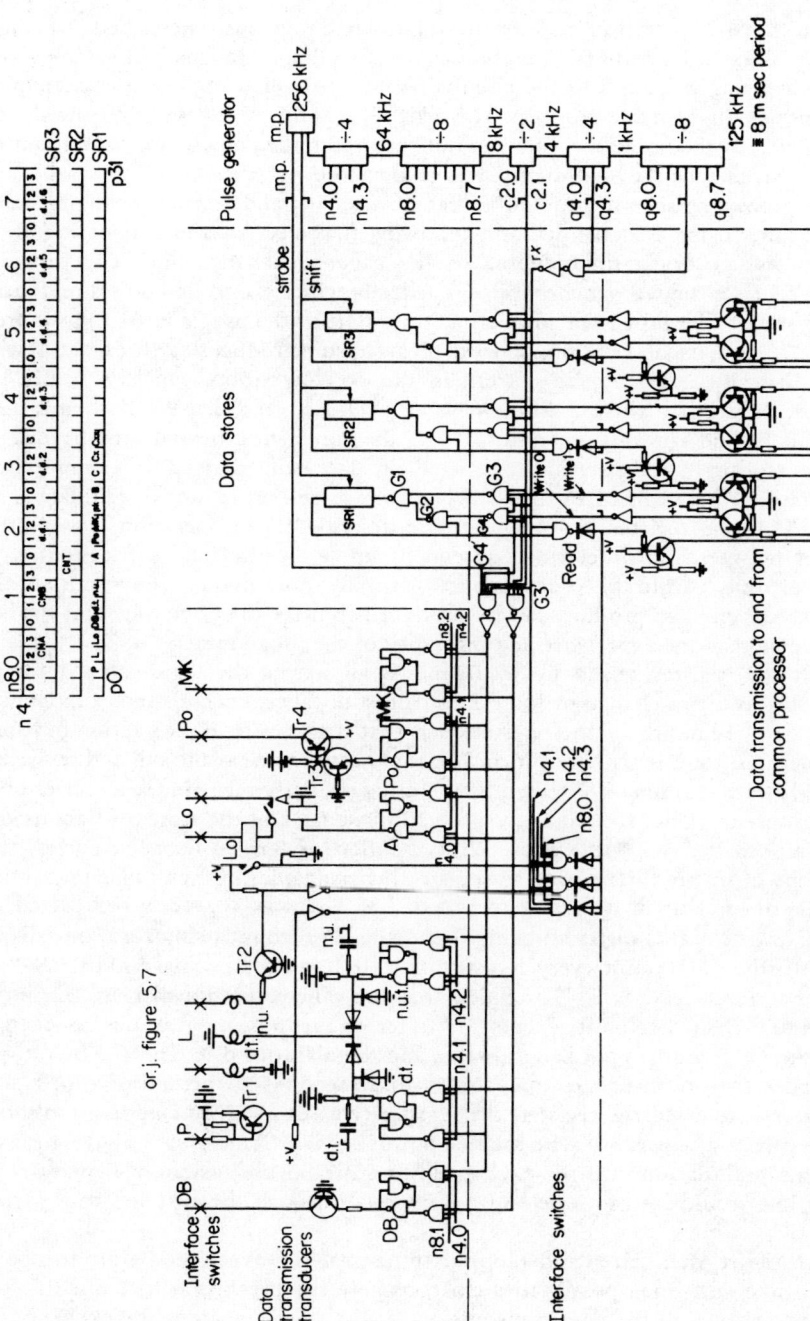

Figure 7.2 Register with processing time division time shared

register and therefore they require the transducers previously described. The digits DB, d.t., n.u., A, Po and MK operate the correspondingly designated bi-stable stores as well as being written into the circulating store SR1. The digits B, C, Cx and Cxx are arbitrary digits of the program. The digits P, L and Lo are data transmitted from the register to the processor; they are not required to be and are not stored, but for circuit convenience they use digit places of the store.

The processing shown in detail is that of receiving and storing the digits dialled by the caller using a rotary dial and following the selection and connection of the register via an originating junctor to the caller's exchange line. The program, Figure 7.3, is executed synchronously cyclically, clock controlled pulses presenting the registers to be processed and defining the start of the program. As follows from Figure 7.2 and its description, seizure of a register produces earth on the P-wire, which logically is $P = 1$, and current in the exchange loop, which is $L = 1$. The program shows that when $P = 0$ the register is released, and that $P = 1$ when $L = 1$ is followed by the arbitrary operation $B = 1$, the connection of dial tone by $d.t. = 1$ and the storing of call data in the register, the data being derived for example from the common processor of Figure 7.1. The storage and processing of call data is not included in the register processing to be described. The operation $B = 1$ differentiates between the processing consequent upon the first and subsequent $L = 1$ states of line. Arbitrary processes are naturally not unique, the results which they are designed to produce can be produced in more than one way, and the best way depends on the experience and ingenuity of the programmer.

With the register seized by P, B and L all having the logic value 1, further processing awaits a change of loop data from the caller, except that a time limit is placed on the holding of the register such that if the caller delays action beyond a set time, n.u. tone is sent to him and he is expected to clear the call and re-make it if he wishes to. Timing is effected by counting regularly recurring scan cycles up to an appropriate value, for which purpose 12 digits CNT of the store SR2 are used to count up to $2^{11} = 2,048$ cycles. With a period of 8 msec between cycles, time durations of up to a little over 16 sec are thus available: in fact, twice that time is possible but not used. Referring to Figure 7.3, when the register is first seized, the n.u., Cx, Cxx and C digits all being 0 and the CNT count being zero, one digit is added to the CNT count every 8 msec, indicated on the flow diagram by CNT + 1, until the count reaches 2^{11} provided that the caller does not dial or clear in the meantime. When the count reaches 2^{11} after 16 seconds, the n.u. digit is changed to 1, the CNT count is put back to zero, and the dial tone disconnected by $d.t. = 0$. The caller then receives n.u. tone for up to 16 seconds at the end of which, if he still has not released the register, the register releases itself by an operation shown on the program as park, to take the line out of service. Some systems have means of writing a park datum into the exchange line store; in the system of Figure 5.7 the calling line would instead have to be connected through the switches to a parking junctor.

With the register seized and a loop datum signal $L = 0$ received before the expiry of the time out, three possibilities exist, namely the break is a brief one due to a fault disturbance in the data transmission circuit which includes the exchange line

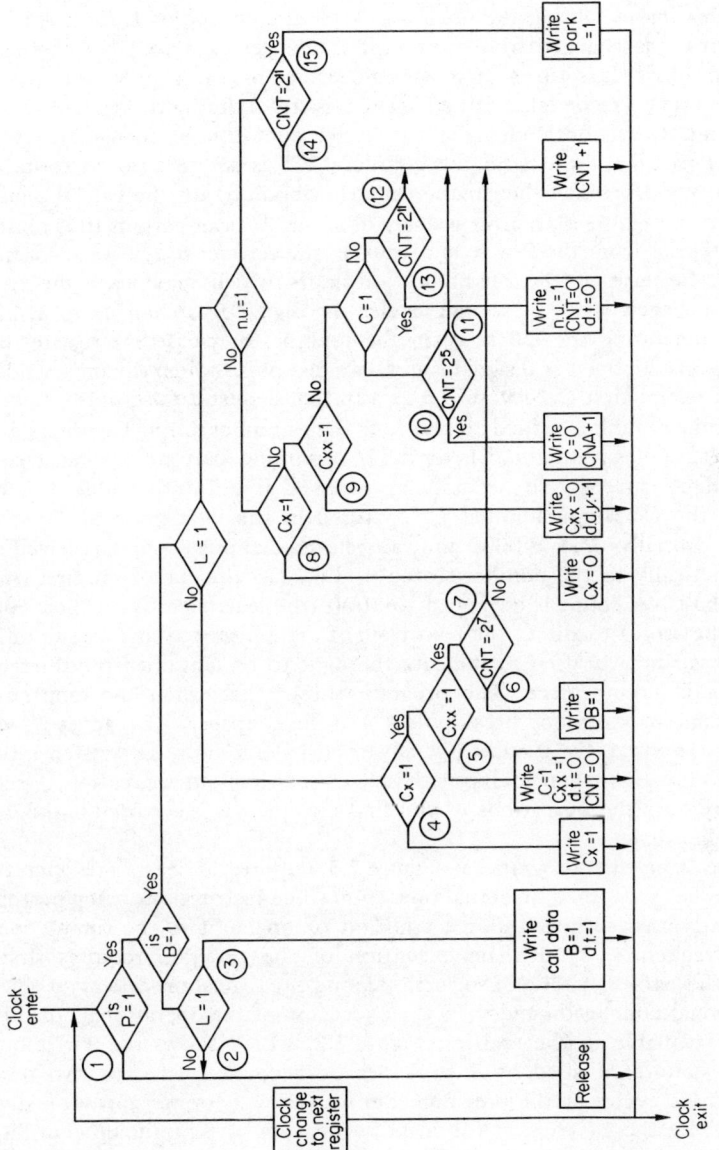

Figure 7.3 Part of flow diagram for register processing

subject to interference from various sources, it is the first pulse of a train of dial pulses each with a minimum duration of about 20 milliseconds and the complete train representing the first dialled decimal digit, or thirdly the caller has abandoned the call by clearing it. During the next scan after the change of L from 1 to 0, an arbitrary digit Cx is changed from 0 to 1: if L changes back to 1 before the scan after that, the digit Cx returns to 0 with no effect by the L pulse thus assumed because of its brevity to be false. Otherwise if L is still 0 during the second scan, the digits Cxx and C are both changed to 1, the digit d.t. is changed to 0 thus disconnecting the dial tone, and the counter CNT is set to zero to re-start the timing. If the loop break is due to the caller having hung up the call, the junctor should release the connection after a delay of about 2^5 scan periods thus removing the earth potential from the P-wire and causing the register to release according to the program. Because of the possibility of faults which may leave the register permanently engaged, after 2^7 timing cycles the digit DB is changed to 1, having the effect of operating the DB relay in the junctor to release the register by an alternative means. If the L = 0 signal is in fact a dial pulse, when the pulse ends and L becomes 1 again, first Cx becomes 0 as a buffer against a false pulse, then Cxx becomes 0 and one unit is added to the four digit number d.d.y. The digits dialled by the caller are designated d.d.1 to d.d.6, assuming that six are received and stored. The digit currently to be or being received is d.d.y, the value of y being indicated by the four digit number CNA stored in the data store SR3 with the dialled digits. Initially CNA is 0000 and causes each dial pulse as it is received to be added arithmetically to the number stored in d.d.1. At the end of the first train of dial pulses, the CNA count is advanced to 1000 (the least significant digit coming first out of the store) to direct the next train of dial pulses to d.d.2 and so on. The number CNB is similarly used to indicate the digit to be sent out from the register when sending is taking place. The program shows that when the loop current becomes continuous after loop break pulses, and the counter CNT reaches 2^5 which it will do in 256 msec, C = 0 and CNA advanced by one unit are written into the stores, the CNT counter continuing with the 16 sec time-out thereafter. A second dialled digit repeats the operations of the first except that it is stored in d.d.2, and so on for further digits.

Every exit from the program of Figure 7.3 requires at least one operational instruction to be performed, in actual operation some instructions being performed more frequently than others and the addition of one unit to the timing counter CNT most frequently of all. The execution of the program requires first the assembly of the data P, L, B and so forth, second the logical processing of the data to an operational conclusion and third the execution of the operational conclusion. The data are available in the register, Figure 7.2, which shows the 8 kHz output from the n8 counter divided by 2 to 4 kHz by a counter c2 with two outputs defining alternate cycles of the previous counter. The c2 counter output is divided by 4 to 1 kHz by the q4 counter followed by division by 8 by the q8 counter to a final overall cycle frequency of 125 Hz, the 8 msec period of which provides the basic timing for the register control system: the frequency dividers are common to all the registers and possibly to much other equipment in the exchange, for example they may be the LA pulse generator of synchronous line scanning equipment, as in Figure 7.1 except that the addresses are to a different base. In the same way as the n4 and n8 pulses produce the pulses p0 to p31, combinations of the q4 and q8 pulses produce thirty-two pulses r0 to r31, each of which causes one of thirty-two registers to be connected to the common processor. An r pulse has the duration of two p pulse periods comprising two register circulating store cycles, and it is used to gate data existing in a register to the common processor, which takes up the first register store cycle, and to gate operational data from the common processor to the register during the second cycle. In Figure 7.2 the pulse rx operates NAND gates which allow the output from the SR1, SR2 and SR3 stores to pass to the common processor, and also gates G3 and G4 which, with the c2.1 pulse as an additional input, allow operational data from the common processor to be written into the stores but only during the second of the register store cycles of the rx pulse. The rx pulse also operates gates which transmit the P, L and Lo data to the common processor, the gates having in addition n4 and n8 inputs which select the p pulse times for the transmission of the digits. Similarly gates G3$'$ and G4$'$ operated by pulse px in parallel with the gates G3 and G4 of the SR1 store, gate operational pulses to gates which, having n4 and n8 pulses on other inputs, distribute the operational pulses to the bi-stable circuits DB, d.t. and so forth. The data pulses transmitted between the common processor and a register must be given time to change and reach steady state conditions before being sensed, for which reason they are strobed in the registers by master pulses m.p. applied as yet another input to the gates G3$'$ and G4$'$. The strobing of the circulating store inputs is incidental to the writing of the inputs into the stores on the occurrence of an m.p. shift pulse. The data pulses in the opposite direction are strobed in the common processor.

Referring to the common processor and to Figure 7.4 in particular, data digits from a register are staticised on the bi-stable circuits P, L and B and so forth corresponding to the data digits in the register. Strobed input pulses are selected by n4 and n8 pulses to operate the bi-stable circuits during a c2.0 cycle, all the circuits being restored to logic 0 at the end of the c2.1 cycle. The data digits have the values 0 and 1, and the circuit logic the values 0 = earth and 1 = +v potential. The relationship between the two systems of notation is indicated by digits written against the store outputs, the digits being the data values of the stores when the circuit outputs against which they are written are at circuit logic 1. It is not difficult to see that the processing of the assembled data according to the program requires AND or NAND gates operated by the data digit outputs, in Figure 7.4 the gates in the centre vertical section of the diagram, the outputs of the gates controlling the transmission of operational data. Each gate effectively makes a number of decisions in series corresponding to its number of inputs. The gate G1 for example, is operated if $P = 1$, $B = 0$ and $L = 1$ which according to the diagram results in the transmission of the operational data $B = 1$ and d.t. = 1. The output of gate G1 after inversion is applied to two NAND gates, one of which produces a write 1 pulse at the p12 digit time and the other a write 1 pulse at the p5 digit time, to write digit 1 for B and for d.t. in the SR1 store. The write 1 and write 0 pulses from the various gates are communicated via diodes constituting OR gates to circuits including the transistors Tr1 and Tr2 for the transmission of the digits to the registers. The P = 0 output is unique in that it signifies release and it writes 0 into all the digits of all the register stores irrespective of the output of any other gate. The CNT + 1, CNA + 1 and d.d.y + 1 output pulses add one unit to the respective binary numbers in the SR stores, and the CNT = 0 pulse re-sets the CNT count to zero. The digits of the numbers staticised during the c2.0 pulse are applied to a parallel adder circuit and rewritten unchanged or with the addition of one unit to the count, or re-set to zero, during the c2.1 pulse. The digits of the CNT counter supply to the gates of Figure 7.4 the required CNT inputs. So long as the sixth digit is 0 the count is less than 2^5, which is the CNT $\neq 2^5$ condition, and when the sixth digit changes to 1, the count is 2^5. The eighth and twelfth digits similarly supply the CNT $\neq 2^7$, CNT = 2^7, CNT $\neq 2^{11}$ and CNT = 2^{11} inputs. On the occurrence of a d.d.y + 1 instruction, the CNA digits have to be used to gate the appropriate d.d.y digits to an adder and to gate the output of the adder to corresponding register store pulses, the adder having increased the count by one unit. Figure 7.2 shows provision for six dialled digits to be received and stored. More will generally be needed and incur the provision of additional stores SR in the register: because of the difficulty and cost of increasing the storage once the equipment has been made, the provision for dialled digits should be liberal to allow for possible increase in the future. The quantity of digits to be received for any one call is determined by translation which is not difficult to change and has to be flexible for normal operation.

The register program and equipment which have been shown and described in detail are only part of a much larger whole which is the same or at least the equivalent of the cross-bar system operations listed in section 5.6.3. The timing cycle of Figure 7.2 limits the quantity of registers time sharing one common processor to thirty-two but cycles for larger quantities are easily possible. It will be realized that there are principles and techniques common to all solutions to a given problem but without there being a unique solution or program or implementation of a given program. What is important is that enough of the practical sides of the problems should be understood for the principles and the techniques and their characteristics to be extracted, and for these purposes the detail given should be sufficient.

As can be seen, particularly from Figures 7.2 and 7.4, the transmission of data between the individual processors and the common processor over the interface switch may involve considerable quantities of individual and common equipment mostly transducers. Data for an individual equipment having been assembled in the common processor, the program for the processing of the data requires only logic gates for its execution to produce steady operational outputs which do not change the processor input on that scan but on the next. Thus difficulties of circuit operation and timing which commonly occur in relay processors due to feedback of the output to the input, do not occur. All the timing comes from counting processing cycles and can be very accurate, also unlike most previous circuitry. Although the technique of processing has to be particularly adapted to time division time shared unique common processing, obviously by reducing the

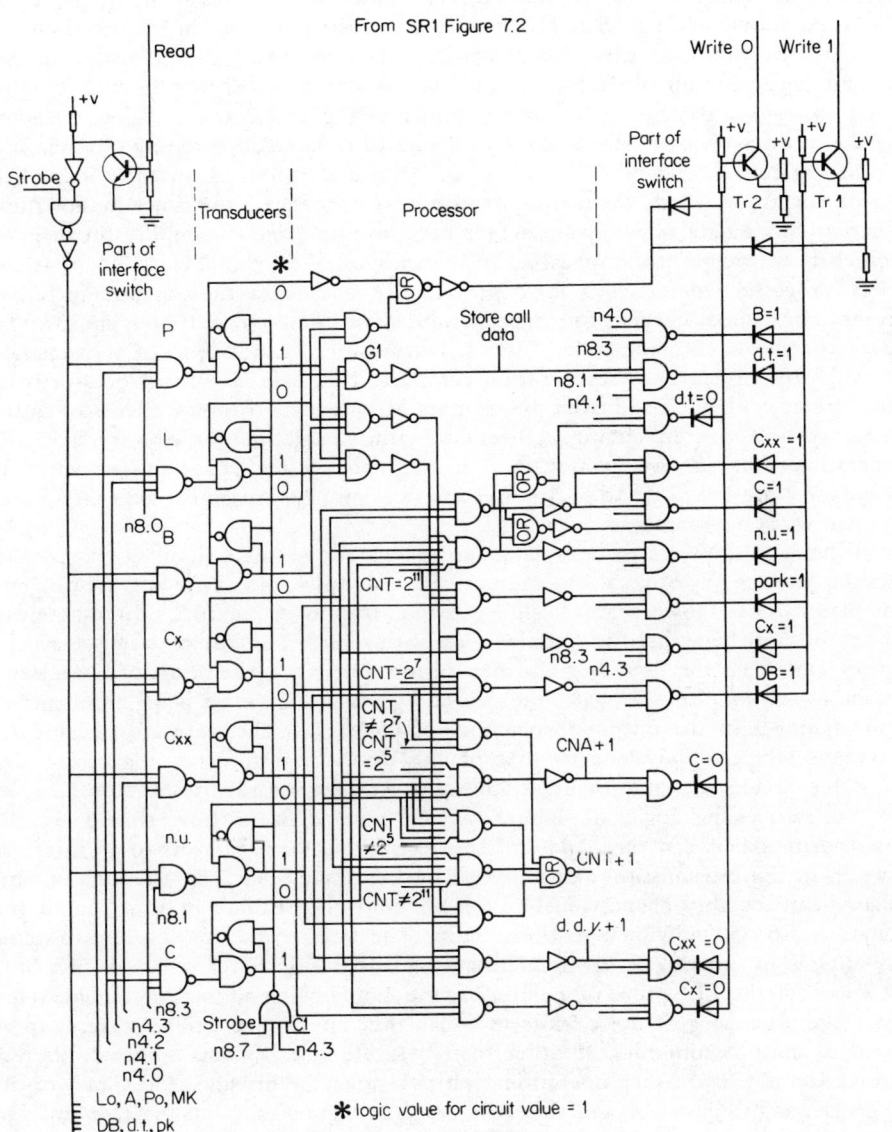

Figure 7.4 Part of register common processor

individual processors in the cycle to one, the processing also satisfies individual circuit processing. All that is required is two-phase clock pulses at the frequency of the time sharing cycle, alternately to apply the existing data to the processor and to execute the result of the processing. This being a useful concept which unifies circuit and common processor design, it is illustrated in Figure 7.5 with parallel data storage in contrast to the serial storage of Figure 7.2 and bringing out some characteristics of logic processing which tend to become obscured by time sharing.

In Figure 7.5, Figures 7.2 and 7.4 have been recombined into the individual processor from which the logic processing was separated for time division time sharing. Some data transmission details have been omitted to simplify the diagram which is rearranged to emphasize, as shown symbolically in Figure 7.6, that the data processor comprises a logic processor P with data transmission including transducers where needed, for input s and d data from, and output s, sm, d and o data to, the circuit controlled, and a d data store M in a loop of transmission through the processor. The operation comprises first data into the processor from the circuit controlled and from the memory M, then data from the processor to the memory and to the circuit controlled, with storage md of circuit data. The operation thus in two phases may be defined by two-phase pulses which in Figure 7.2 are the c2.0 and c2.1 pulses and in Figure 7.5 are similar alternate pulses on two wires to two sets of gates.

Figure 7.5 shows or assumes data storage by static bi-stable circuits, with parallel writing and reading of data into and out of the stores. Logic to execute the program of Figure 7.3 is shown symbolically: gates in series as in Figure 7.4 do not appear, gates in series being used in practice merely to reduce the cost of gating. Basically every exit from the flow diagram to a decision output is the result of n decisions made by an n-input AND gate, the decision gates of Figure 7.5 being numbered to correspond with the output instructions of Figure 7.3. The store M of Figure 7.5 contains storage equivalent to that of the circulating stores of Figure 7.2, the bi-stable circuits shown in Figure 7.5 being part of the equivalent SR1 store.

The processing logic of Figure 7.5 is time division time shared by the interposition between the logic and the register circuit processors of an interface switch in the transmission paths crossed by the dashed line XX. The logic is time shared but memory M individual to circuits controlled cannot be time shared and requires storage individual to the circuits. The storage may be discrete bi-stable circuits as in Figure 7.5, or circulating stores as in Figure 7.2 or words in a bulk store as for the LDS store of Figure 7.1, the words being addressed coincidentally with the processing of the circuits to which they apply. The choice of store to be used is one of economics. If other than bi-stable circuit stores are used, discrete stores for n.u. and other operational outputs must be provided individual to the registers, as in Figure 7.2 and the md storage of Figure 7.6. Thus the economics of time division time sharing can be seen to be very much involved with data transmission between the common and the individual processors, with data storage and with logic processing. The data transmission comprises the interface switch and very often transducers, the costs of which detract from the savings due to the time sharing.

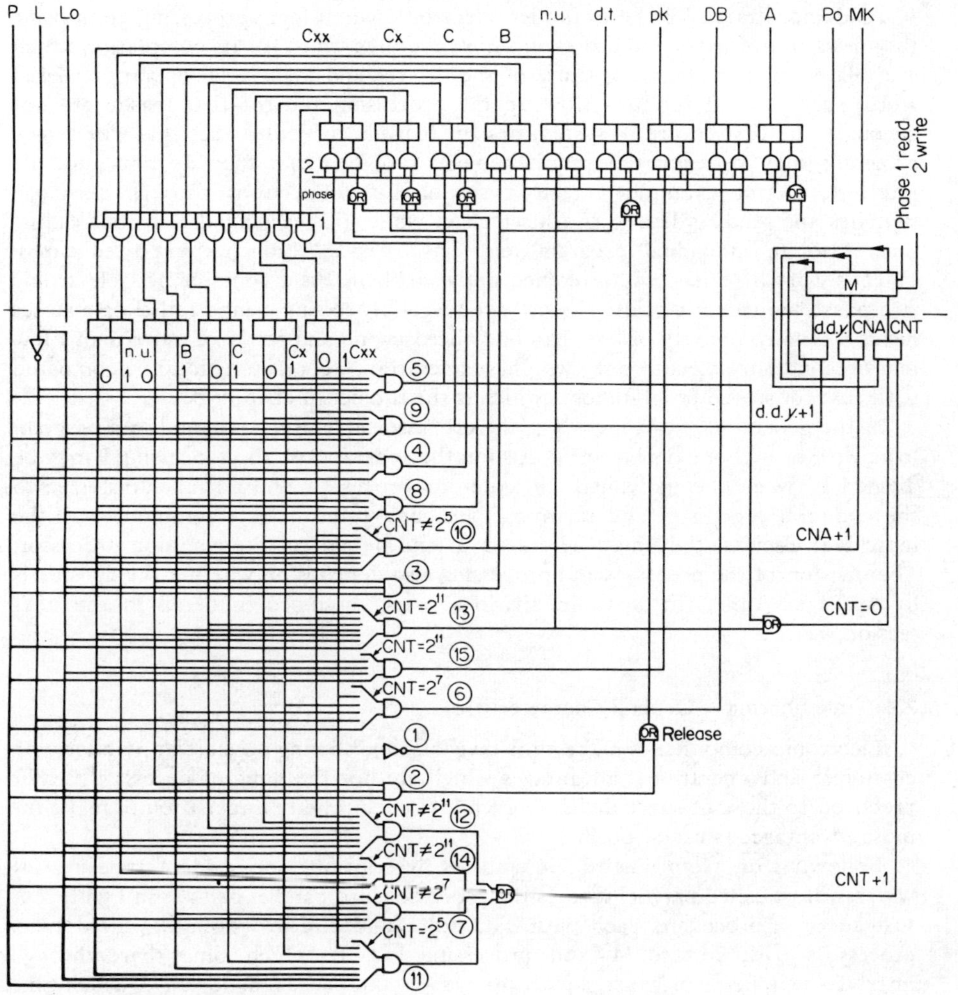

Figure 7.5 Two-phase processor

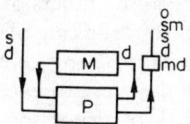

Figure 7.6 Symbolic processor

As demonstrated by the examples, electronic wired logic processing comprising the operation of gates and stores in series, requires times for its completion which are related to the operating times of the devices and their serial operation. Serial stores such as shift registers slow up the processing and for that reason are not generally the first choice in system design. It is not difficult to see that wired logic programs may be executed simultaneously provided that they do not mutually interfere, as for example a register setting up connections through exchange switches and sending data to another register while still receiving data from a calling line. Also an individual program may have parallel branches executed simultaneously if there is no interference between them. As a consequence, electronic two-phase individual or time shared wired logic processing using parallel data stores needs a time commonly of less than one microsecond and rarely of more than a few microseconds, to execute one two-phase operation, a characteristic to be compared with that for stored program logic which is the subject of chapter 8.

In the general case of Figure 7.6, the processor may use wired or stored program logic with or without time sharing. Also in the general case, the processing P may be divided between the individual and common processors and not all transferred to the common processor, and if any of the data stored in M is common to all the individual circuits, they may be stored in and be part of the common processor. The division of the processes and processing also depends on whether the division is by space or time, the possible divisions being discussed in detail in the next section, 7.4.

7.4 Time Sharing — General Characteristics

Electronic components make time division time sharing possible in practice with economic and operational advantages which are not the same or necessarily to be preferred to those of space division. Each must be properly understood to make the most advantageous use of both.

Following on from Figure 7.6 and its description, one kind of time sharing which will be called parallel divides the processor into parallel parts as in Figure 7.7. Instead of x processors each with data storage M and processing P, there are x processors with storage M' and processing logic P' which time share through interface switches y processors with storage M'' and processing logic P'', which time share through interface switches z processors with storage M''' and logic processing P''' and so on. In Figure 7.7 the interface switches are interposed at points shown by lines X—X and Y—Y. The interface switches connect the processors in parallel by data transmission paths which may include transducers at a number of points. The processors x are assumed to be all members of the same rank, which means that they occupy the same position in the trunking plan of the exchange and if they are not all identical, most of the program of any one is common to them all. A wired logic processor of necessity being designed to execute specific programs, is not able to function for more than one rank of processors unless, as sometimes happens in practice, the programs are the same or nearly the same in the different ranks. Moreover, if a wired logic processor is capable of processing two entirely different programs, it is analytically two and physically divisible into two separate processors: the same is not true of a processor with stored program logic as will be explained in chapter 8, but nevertheless the principles of time sharing are basic to processors of one rank and are either not invalidated by or are easily extended to the apparent exceptions.

The movement and circulation of data shown in Figure 7.6 for one undivided processor becomes as in Figure 7.7 for parallel division. The data incoming from the equipment controlled are available to all the processors, and each processor is able to communicate data to the equipment controlled and also to processors in later divisions and to the data stores in its own and earlier divisions, all of which follows from previous descriptions of the operations and equipments of systems in practice and cross-bar in particular, and from Figure 5.5 for example.

The differences between space and time parallel division concern the ways in which the memory M and processing P can be divided between the individual and the common processors and the resulting quantities of time shared processors y, z ... needed for x individual processors.

The objective and overall requirement of time sharing is economy, that the equipment needed to control some quantity of exchange equipment shall be achieved at a cost which is reduced as a result of the time sharing.

The first requirement for the economical use of parallel division being a rank of processors, a processor of the rank gains connection to a space division common processor and holds it for its own exclusive use to execute some part of its program which the rank processor anticipates is about to be required and will be completed on the average in much less time than that required for its whole program. Thus anticipation and relatively short processing time are necessary requirements, the first operationally and the second economically, for the use and advantageous use of space division time sharing. In illustration of these requirements, registers are space division time shared among originating and incoming junctors because the need occurs always when a junctor is first seized and therefore is anticipated, and continues only until the call is connected through the exchange switches and data transmitted to another register if any, operations which take only a small part of the average call connection time. The quantity y of time shared processors depends on the average number of individual processors simultaneously requiring a time shared processor, that is on the time shared traffic. If some part of the processing of the y processors satisfies the requirements for space division time sharing, then a second division may be made to z second common processors. This happens with cross-bar system registers for which call connection processing is anticipated by examination of the directory number digits received and takes a time which is small compared with the average time that a register is held. Therefore translator-markers are space division time shared among the registers. Obviously there is no theoretical limit to the number of divisions which may be made. The limit is reached in practice when no advantage is to be had by further division; and two divisions generally reducing the quantity z to unity, more than two divisions is not usual in present practice. It is not, however, difficult to see that the division could be extended to ranks of processors the same in different exchanges and time sharing

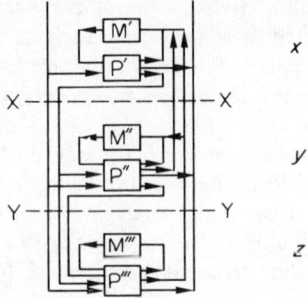

Figure 7.7 Parallel division of processors

processors in a common centre. As exemplified by cross-bar systems, the first division economizes in storage and in processing. The economy of the second division is little in processing but mainly in data storage. A third division is unlikely to be useful to processing but of considerable value to data storage in the form of a data centre for exchanges in an area of administration.

With time division time sharing, a unique common processor is connected cyclically to all the x processors of a rank of individual processors to execute any processing which may be found to be required, or to record the addresses of processors for their acyclic connection later to a common processor to complete the processing required. Because the incidences of processing do not have to be anticipated, all the processing P may be transferred to a unique common processor which being unique seems to make any further parallel division pointless. However, further division to make use of data stored at a common centre is a possibility with some practical advantages and there are some other problems and possibilities concerned with the division of the processes and the data storage best explained later in conjunction with a practical example, that represented by Figure 7.9.

Before the differences between parallel time sharing time and space division are pursued further, it is necessary to understand that before such division can be arranged, the total control of an exchange has been divided, for what will be called series time sharing, between ranks of processors which are selectively connected in series to provide individually for each connection the control which it needs and as little more as cannot be avoided. What it needs is deduced from the classes of service of the terminal circuits connected and the data dialled or otherwise indicated for the connection.

It is necessary to distinguish between quantity of processes which is related to programs, quantity of processing which is related to the product of processes and the number of times that they are executed, and quantity of processing equipment which is the quantity of physical components used to construct a processor, a rank of processors, all the processors in an exchange or whatever is specified for the quantity.

The objective of series time sharing is the minimization of the cost of processing equipment necessary to an exchange for all the calls to be connected, by division of the processes among processors in ranks and, for each call, the selective connection in series by data transmission circuits of processors in the different ranks; the processors thus selected and connected being collectively able to execute all the processes for that call but not for any and every call. The basis of economy is that no calls require all the processes such as ringing and coin-box supervision but only one or some of such processes, and that the processes required for each call are known and can be selectively added to the connection as each processor is connected. In the limit the processors connected for a call provide no more processes than that call needs, but the cost is not necessarily minimized by designing the system for this to be the case. In fact the division of the processes between the processors to the best advantage is not always definite and obvious. By comparison, the objective of parallel time sharing is the minimization of the cost of a rank of processors allocated processes most if not all of which are common to them all and may be needed by any call which is processed without prior knowledge of the need. Thus all cross-bar system calls need register and translator-marker processing to achieve a satisfactory connection without the prior knowledge of whether a satisfactory connection is possible or not. Again there is a limit, reached when as many processes as possible are time shared, which does not necessarily lead to minimum cost and the most advantageous division is not always definite and obvious.

If the whole of an exchange is regarded as one processor and the discrete processors which it contains as partial processors, then for each call partial processors are connected in series and in parallel to provide a processor suited to that call and that kind of call but to as few other kinds as possible in the interest of economy.

In order to see the principles by which the series and parallel division may be best accomplished, it is helpful to examine how it has already come to be done in practice by evolution as much as by design and therefore almost certainly the best for the existing conditions. Referring to Figure 7.8(a), the processing starts with line equipments which are at the entry point of control data transmitted by calling subscribers. The line equipment processors are circuit processors inevitably associated with the message circuits to which the data apply, because the two wires of the exchange line have to be used for both message and data transmission. If the line equipment processors are the only processors in the exchange, each has to be capable of executing all the processes for all the calls over its line and for which it is equipped with data storage M and processing logic P. If universal processors are used the M and the P are the same for every line but most of the M and the P are redundant in every processor because no line needs all the processes. For example, only coin-box lines can use the coin-box processes. Therefore the processors of the line equipment rank are provided in groups each capable of a limited range of processes. The M and P of the processors in one group are the same, but different in different groups, although with some proportion which is common to them all. It is assumed that the one processor associated with each exchange line in Figure 7.8(a) is able to control message path connections through its own exchange but not through other exchanges, which assumption is valid so long as all data transmission is message path associated. A call between two exchange lines on the same exchange inevitably involves the line equipment processors of the two lines which means that some of the processing for a line after it is connected for an incoming call can be performed by its own processor. For the control of connections incoming over junctions, junctions at their incoming ends are equipped with a processor, the incoming junctor i.j., with storage Mj and processing Pj, all such processors being the same as all junctions are expected to carry calls of all classes of service. An exchange line becoming connected to a junction, the line equipment processor interchanges data with the i.j. processor at the distant end of the junction and leaves that processor to make and control the connection in the distant exchange, the exchange line processor still retaining overall control. If the connection passes through several exchanges, it is established and controlled by a processor in each exchange with data interchange between the processors. In the first automatic exchanges the data communication was by signals s, signal-messages sm, data messages m-d and signal messages s-d as previously defined and over paths associated and switched with the message circuits, because it was the simplest, cheapest and only practical means at the time. The alternative of transmission over disassociated paths, although known in manual systems with order wire working between operators, was probably not even considered. Disassociated data transmission uses message switching methods for data transmission over a network of order wires now more often called data links between exchanges, so that the telephone connections are controlled by a kind of telegraphed data service also with the problem of the telegraph service of collection and distribution of the messages at the two ends of the communications. With sufficient processing power available with electronic processors, this is now possible and is demonstrated in chapter 9, and from which it will be clear that the series division of the processes now being discussed is not changed in any important respect by data transmission over paths disassociated from message paths. The associated method being easier to describe and understand, is assumed for Figure 7.8.

With line equipments per line for originating and terminating exchange line calls and incoming junctors for the junctions as in Figure 7.8(a), together with exchange switches as shown, all call connections are possible in theory but with practical difficulties which are insuperable. Hence the series division, Figure 7.8(b), to leave with each exchange line storage M0 and processing P0 for processes individual to the line, and to transfer to series connected processors in groups processes which are common to groups of lines. As the message circuit trunking concentrates the exchange lines to fewer trunks than lines, the new rank of processors is economically associated with the trunks and becomes the originating junctor or.j. rank with storage M1 and processing P1. The class of service required by an exchange line is provided by its line equipment processor or the or.j. junctor to which it becomes connected for an originating call or by both with each contributing part. The division in practice is mostly a problem of economics although technique mainly of data transmission also plays a part. When a calling line is being connected, class of service data are necessary to associate the appropriate kind of originating junctor and if that junctor is suited to more than one class, to select the service to be given by the junctor for that call. Similar problems occur for exchange line terminating calls. Either the line equipments must provide services like ringing individually for their own lines or the or.j and i.j. junctors must provide all services for terminating calls of all classes, both of which are impractical and uneconomic; or series connected and class of service selected processors t.j. are required to be included in terminating calls as for originating calls. Hence Figure 7.8(b) shows terminating junctors t.j. with storage M2 and processing P2 associated with the trunks leading to exchange lines for terminating calls.

Figure 7.8 does not include services other than exchange line connections. Other services require special processors which are added without difficulty. The important point is that no further series division of the processes can be seen to be useful or has occurred in practice except for the use of outgoing junctors o.j. associated with the outgoing ends of junctions and these are often concerned with no more than data transmission over the junctions as indicated in Figure 7.8(b). It is also possible to transfer to processors o.j. and t.j. some processes otherwise necessary in the or.j. and i.j. processors but such small variations in detail do not affect the main question, nor does the series connection of processors by variable trunking as in Figure 6.4 constitute a fundamental change to the solution of the problem of the series division of the processing.

Processors in ranks in series are, except for the individual exchange line processors, space division time shared along with the message circuit trunks and junctions with which they are associated. In electronic exchanges described in chapter 10, the message circuits may be p.a.m.–t.d.m. or p.c.m.–t.d.m. transmission channels. Processors associated with such circuits are time division time shared among the multiplexed circuits but without affecting the fact that they are series space division time shared as represented by Figure 7.8.

Having series divided the processes between ranks of processors, the parallel division may begin. If limited to space division as in pre-electronic days and as in Figure 7.8(c), there is no advantage in the parallel division of the exchange line processors because for none of the processes can the times of occurrence be anticipated. The originating junctors, which include most of the processes required for originating call connections, are advantageously divided to leave the individual processors with storage $M1'$ and processing $P1'$ for supervision processes, these being processes which are liable to occur at any time during a connection. The remaining processes which are concerned with the establishment of connections are transferred to registers. The registers exchange data with the calling stations and other registers to which they may become connected, but transfer the actual connection processing to translator-markers. The translator-markers take memory $M1'''$ and processing $P1'''$ from the registers leaving them with storage $M1''$ and processing $P1''$. Similarly the incoming junctor processes are divided leaving the junctor with storage Mj' and processing $P1'$ and the incoming register with Mj'' and Pj'', the incoming registers sharing the translator-markers with the originating registers. A register must itself anticipate when to connect to a translator-marker, a problem discussed in the chapter on cross-bar systems. The solution requires the register to perform some part of the translation if only to count the number of digits which have been received. Also described for cross-bar systems is the possibility of the register connecting to the translator to find out if it has received enough digits, and if not how many more it requires, which is an elementary form of time division time sharing.

Figure 7.8(c) also indicates the possibility of parallel division of the terminating junctors t.j. but most of the processing being supervision which cannot be advantageously space division time shared, it is not usually economic to time share the remainder which is mostly ringing and ring tone transmission.

At each stage of space division time sharing there is an economy of storage and of processing equipments. Quantitatively, x individual processors sharing y common processors sharing z second common processors require a total of $x(M' + P') + y(M'' + P'') + z(M''' + P''')$ storage and processing compared with $x(M' + M'' + M''' + P' + P'' + P''')$ if undivided, with a saving due to y being less than x and z being less than y. Against the saving of storage and processing equipment has to be set the cost of the interface switches and the data transmission through them. In the case of originating junctors, the ratio between x and y and between y and z is more than ten to one for each: the M' and P' are a small part of the total M and P and the M''' and P''' being the whole of the path search and connection translators and markers, constitute much the greatest part of the M and P. Therefore the economy is unquestionable. For terminating junctors, the ratio of x to y can be less than ten to one, and the M'' and P'' only a small part of the total storage and processing required, and the economics thus doubtful normally. When time sharing is used it is usually by the method of Figure 6.4(c) which adapts the exchange switch trunking to provide the necessary parallel connection of processors instead of using as in Figure 5.3 interface switches not part of the message transmission for the purpose; but as was shown in section 6.3, the economic significance of such difference is not great. In Figure 6.4(b) the originating register is effectively paralleled with the originating junctor by the same method. Compared with Figure 5.3 the originating junctor does not have to be connected until the connection through the exchange to another circuit is made, and therefore there is a reduction in the quantity of originating junctors required and thus a further but minor saving. The total quantities of storage and processing equipments previously deduced are scarcely changed by the alternative methods of parallel division which can thus be seen to be theoretically equivalent and what is deduced for one to apply very nearly if not quite exactly to the other. The economics of the special service junctor of Figure 6.4(d) series connected by the trunking method instead of by manual connection are different from those of the parallel connection of processors for normal processing. The trunking of the system of Figure 5.3 can also be adapted to produce the same result.

Electronic equipment makes processing cheaper and more powerful in respect of both data storage and processing, and makes time division time sharing a practical possibility. A time division time shared processor is effectively connected continuously to all the individual processors which have access to it, which means that there is no limit to the processes which may be transferred to the common processor: assuming that it can operate fast enough, one common processor can serve all the circuit processors in a rank or as will be seen later all the circuit processors in an exchange. Data storage provides some problems. In Figure 7.2 for example, the data stores are individual to the registers although all the processing is performed by the common processor with great economy. The data stores for all the registers may be contained within a bulk store, the common processor reading the data pertaining to a register each time it processes the register in the cycle, and if necessary writing data for the register into the bulk store. There is no economy in storage measured as storage digits required, although there may be a saving in cost if bulk storage cost per bit is less than that of individual register storage. There can also be a great saving of time: a bulk store provides parallel read out and writing in of data, simultaneously for all digits, in a read-write cycle of perhaps 1 microsecond instead of the 250 microseconds of Figure 7.2. The value of such time saving is the use if any which may be made of the time saved.

The registers of Figure 7.2 are space division time shared among the junctors to economize in data memory and processing logic equipment as previously demonstrated. Also achieved is a saving in data transmission transducer equipment which is made obvious by omitting the interface switch between the junctors and the registers when equipment such as tone sending and A relay loop sending apparatus has to be provided in the more numerous junctors themselves. The processing of the registers being transferred to a unique processor time division time shared among the registers provides further economy and demonstration that time parallel division can follow space parallel division of the processing of the same individual processors. The extreme case is, however, time division time sharing applied to all ranks of processors, as in Figure 7.8(d), each rank with a unique processor Pr and store M_r'' common to the individual processors for each of which there is storage M_r', with the exception of the originating and incoming junctors. For those junctors and following the previous notation, there being no discrete registers, the storage M'' associated with the register function has to be provided in the individual processors, making a total of $M' + M''$ for each junctor, and the storage M''' appears in the unique common processor. Thus compared with space division for the first parallel division, there is an increase in the storage required, from $x . M' + yM''$ to $x(M' + M'')$ as well as an increase in data transmission equipment because the economy, mostly of transducers, noted to occur with discrete registers, does not apply. The total storage needs are reduced by a stratagem which produces the effect of space division so far as storage is concerned. Using a bulk store for all the storage, areas of the store each equivalent to the storage M'' needed by a register are allocated not to individual junctors but to junctors when they are first seized for calls. When the common processor detects that an originating or incoming junctor has been seized for a new call, it executes a program of its own to find the address of a free area of its bulk store which it allocates to the junctor for as long as the junctor needs it for that call. The address of the store area is stored against the address of the junctor so that when processing the junctor the processor can read and write temporary data for the junctor out of and into the store. In this way the space division is eliminated: the space division interface switches and registers are saved at the cost of an increase in the time

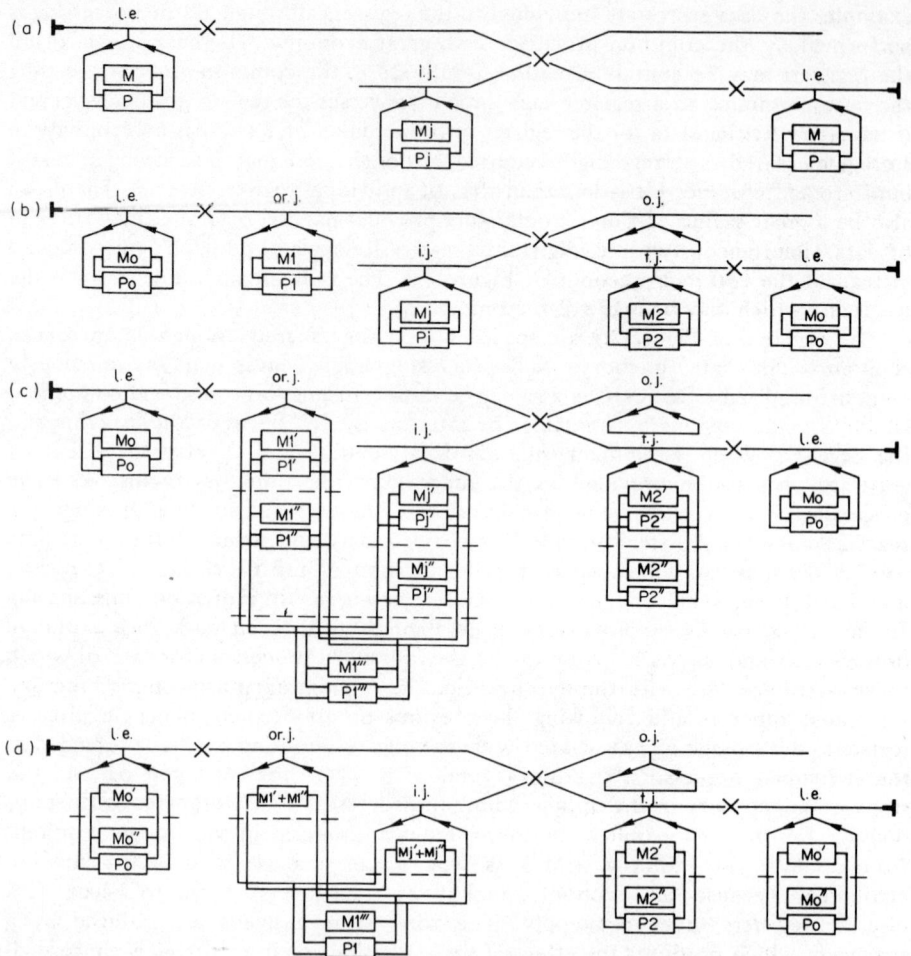

Figure 7.8 Series and parallel division of processors
    (a) basic processors
    (b) series division
    (c) series and parallel space division
    (d) series space and parallel time division
       X = exchange switches in ranks

division interface switching between the common processor and all the junctors instead of the less numerous space division registers, and at the cost of increased processing to manipulate the storage of some inevitable increase in the data transmission equipment, as will be clearer from a practical example to be described. Whether there is an overall saving or increase in cost depends on the costs of the equipments involved. What is theoretically important is that all of the processing can be transferred to a unique common time division time shared processor and none need be located of necessity in the circuit processors: also although there is no point or possibility of successive parallel division as occurs with space division, there is a need for an analogous characteristic and it exists in the form of parallel time sharing of programs in the processor, as will become clear from the practical example illustrating the problems and possible solutions and in particular the originating junctor of Figure 5.7 with time division time shared processing instead of the space division of Figure 7.2.

7.5 Time Division Time Shared Wired Logic

As an example of the operation and characteristics of time division time shared wired logic, Figure 7.9 shows in circuit detail an originating junctor or.j. which receives and sends loop signals, and shows in blocks the time division time shared processors $P1'$, $P1''$ and $P1'''$ which together with their stores $xM1'$, $yM1''$ and $M1'''$ control all the junctors of that kind in the exchange. Line scanning equipment such as that of Figure 7.1, asynchronously cyclically operating path search and connection equipment assumed to be guide wire as in Figure 6.5, causes the junctor to be connected through the exchange switches to an exchange line which is looped to call. Prior to the junctor being connected, the scanner processor, the detailed operation of which is described with reference to Figure 8.4, having found the line looped and unconnected through the exchange switches, has read the line data and state of line data from the line data store, processed it to be sure that the line is allowed to be connected, then applied the MK datum to mark the line. It is now sending call data if any for the line over the call data leads and awaiting the completion of the connection of the line which will be indicated by the line P-wire datum changing to 1. No other connection processing is taking place at the same time, hence the data over the call data leads can apply only to the line being connected. When the selected cross-points have been operated to connect the line to the selected junctor, loop current flows from the transmission bridge including a line transformer, and through the emitter of transistor Tr1. The transistor collector potential changes from +v to earth, which is inverted to apply +v corresponding to state of loop $L = 1$, to one input of each of the gates G1, G4 and G12. When the gate G1 is addressed with logic 1 inputs to its other two inputs, the output changes to earth potential which is communicated via decoupling means, a diode, to a common processor $P1'$ concerned to process supervisory operations which may occur at any time. The processor operates in conjunction with a data bulk store $xM1'$ and a synchronous cyclic address generator AG1 which addresses the junctors one at a time for connection to the processor. Supervision in this case comprises the generation of meter pulses to operate the calling line message register and release of the connection when the called line clears. The processing time and therefore the time of dwell of the processor connection to each junctor during each cycle need be no more than 5 μsec, and the frequency of dwell no more than 10 Hz but a higher frequency is desirable if possible and 50 Hz or a period of 20 msec is assumed, using synchronous cycles of operation. Hence the cycle could accommodate four thousand junctors which is adequate for very large exchanges. The quantity of junctors being x and the capacity of the store xM1$'$ being x data words, each address out of the generator AG1 connects one junctor to the common processor P1$'$ and reads one word out to the processor as the address changes: before the address changes again, the processor can write new data into the word if the need arises. When a junctor is idle, its data word is zero in all digits and its state of loop L = 0, and when the junctor is addressed, the processor applies logic 1 to the release datum wire. The gate G11 is thus operated to ensure that all the binary stores in the junctor are at logic 0. The data word being zero and L = 1 means that the junctor has been taken into use: in response, the processor writes B = 1 into the data word which being no longer zero, the emission of release from the processor is suspended. The processor does not again apply release to that junctor until it has received L = 0 for eight successive scan cycles, counted as cycles are counted in the program of Figure 7.3 and which means that the loop has been broken for 160 to 180 msec. As B = 1 is written into the data word, the call data then existing on the call data leads connected to the processor P1$'$ are also stored in the data word. The B datum is also, when it changes to 1, transmitted to the junctor to operate gate G8 and a bi-stable circuit B an output of which causes current in the coil of a relay B which operates. A contact on the relay earths the P-wire back to the connected exchange line to release the scanner and to hold the connection.

The processor P1$''$ performs the processing corresponding to that of space division time shared registers and is designated the register processor. An address generator AG2 has a synchronous cyclic output of addresses, the period of the cycle being chosen to suit the register program and the quantity of slots to suit the quantity of registers required according to the traffic. The two address generators AG1 and AG2 operate independently of one another. The reception of rotary dial pulses imposes a processing cycle time of not more than about 15 and 8 msec is chosen to suit the digit sending. It is not economic to use separate cycles and processors for digit receiving and sending, hence a cycle period common to both is necessary. Using a time of 8 msec, digit reception is satisfactory and digits to be sent are timed by counting eight cycles for the break pulses and four for the make pulses. The maximum processing time per slot would not exceed 20 μsec which would allow for a synchronous cycle of four hundred slots or the equivalent of four hundred registers which is compatible with the maximum of four thousand junctors of the supervisory cycle. The quantity of slots actually used is designated y. A bulk store yM1$''$ + m has a capacity for y words each storing for one slot the register program data M1$''$ and the address of a junctor to which the slot is temporarily allocated. The addresses require storage additional to that of the register program as indicated by the designation +m. The output from the generator AG2 operates the bulk store. When the address changes, the word at the new address is read out, the $M1''$ part to the processor and the m part to a junctor address store, and the processor can write new data into the word before the address changes again.

A third processor $P1'''$ with its store $M1'''$ functions in the same way as the translator-markers of cross-bar systems, it will be called the translator-marker processor, and it is acyclically time shared among the register cycle slots. When the processor is free and a slot in the register cycle contains sufficient data for a connection to be made, the translator-marker uses the data to make the connection and leaves the register processor to continue its own cycle. Thus the programs of all three processors proceed simultaneously but indpendently. Effectively the processors operate in parallel in time to produce the same result as space division time shared processors operated in parallel in space.

Going back to the point where the $P1'$ supervision processor writes $B = 1$ and call data into its store in response to an exchange line being connected to a junctor, the processor at the same time tests the condition of a temporary store TM1 and finding it free, the processor writes into the store the then existing address issuing from the address generator AG1 and any of the call data not required exclusively for supervision. If the TM1 store is not free, the processor repeats the test during its next cycle and so on until it finds the store free and can write into it the required data. The chance that the writing is delayed by more than one cycle time, assumed to be 20 msec, is small and by more than two cycle times it is negligible in normal operation. During traffic peaks which cause all register slots to be in use, the delay is indefinite as it is when all registers are engaged in cross-bar systems, and dialling must await the incidence of dial tone. The register processor detects that there are data written in the TM1 store, waits until a free slot in its own cycle comes round in the cycle and then writes the call data into the $M1''$ part of the slot word data store and the junctor address into the +m section, and erases the data in the TM1 store. Thereafter and until the register processor erases from its store the data for the slot, the slot is allocated to the junctor at the stored address, for processing by the processor $P1''$.

An address being read out of the +m section of the processor store to the junctor address store, outputs from the address store have the effect of connecting the processor $P1''$ to the junctor designated by the address. When first so connected, the state of the loop L will be logic 1 to which the processor responds by noting that register processing has begun for the then existing slot in its own cycle and by emitting an operational datum to operate in the junctor the bi-stable store d.t. which sends dial tone to the caller. The operation of d.t. is communicated to the $P1'$ processor via gate G7 to inform that processor that the call processing is proceeding. Dial pulses which are subsequently received, the first causing the $P1''$ processor to send an operational datum to disconnect the dial tone, are stored using the principles of the example of Figures 7.2 to 7.4. If the call is prematurely released, the junctor is released by the supervision processor, and the $P1''$ processor independently erases the call data from its own store. The cross-bar system registers translate and count the digits received, as was described, to determine when sufficient digits have been received to enable call connection processing to be commenced, which involves some problems and difficulties which were also described. These problems and difficulties and some of the register processing can be avoided with electronic registers if the translators are also electronic and thus have the operating speed and durability of such equipments. A trunk translator such as that of Figure 5.13 gives an output only if enough digits have been received for an outgoing junction or service call to be established, and a simple addition does the same for local calls. Constructed of electronic components, the translator part of the call connection processing may be executed by the translator-marker processor during every slot of the register cycle, except when the processor is engaged in call connection processing for one of them, and thereby relieves the register processor of a considerable amount of data storage and processing. Referring to Figure 7.9, some of the data read out of a word in the $M1''$ store are communicated to the P''' processor which if it is not engaged on call connection processing for one of the junctors, processes the data and may write back into the store word data resulting from the processing. The suspension of this processing for all of the slots when one of them is holding the processor is a matter of milliseconds which is insignificant to register operation. The data written back into the store word include not only that call connection processing may commence but also data needed after that processing is completed. In particular if the call is to another exchange, after the junction is connected the translator may indicate when the last digit to be expected of the caller has been received and the slot may be released as soon thereafter as all available data have been processed; or that the minimum of digits to be expected of the caller for the call which he is making have been received and thereafter the register processor must use time-out to know when to clear the slot. The translator may also supply digits to be added to or substituted for digits received from the caller, for exchange lines in transfer and alternate route junctions for example; and for an outgoing junction connection, which of the received digits to omit in retransmission to the next exchange. No register processes are required to decide when translation may take place: translation and call connection processing may take place after no digits if the call data indicate that the call is a disabled subscriber class of service, the translator then providing the digits to route the call to an operator, or they may take place after the first, the second or any received digit. If no connection corresponding to the number dialled is possible because no such number is in use, the translator writes digit 1 into a digit place in the slot store to cause the processor $P1''$ to operate a bi-stable circuit in the junctor to send n.u. tone to the caller, the processor then expecting and waiting for the calling line to be cleared. Using the translator-marker processor in this way achieves some operational advantages with fewer and simpler processes and an economy in storage by relieving the register processor and its store of the need to perform, as was described for space division time shared translators, the translation which determines when sufficient digits have been received into the register for a connection to be possible through the exchange switches.

When with the processing just described, the translator-marker processor determines, from the data which it has just received for a register slot, that call connection processing can commence for that slot, it already has all the data that it needs for the processing which it immediately commences by emitting an MK datum to all the junctors and freezing the data which it has. The MK datum is effective in the junctor to be connected because that junctor is at that instant addressed by the junctor address store. As a result, a bi-stable store MK in the junctor is operated to mark on the MK2 lead the start of the connection to be made. The $P1'''$ processor also writes into the $M1''$ word that call connection processing has commenced and this transmitted to $P1''$ causes that processor to operate the bi-stable stores A and Po in the junctor to prepare for and hold the connection to be made. The call connection processing may therefore proceed independently of the register processing which continues in its synchronous cycle which is at no time interrupted. The MK2 datum is also applied to gates G3 and G6 in the junctor, the first being addressed by the AG1 address generator and the second by the junctor address store. Outputs from these gates when operated indicate to the processors $P1'$ and $P1''$ which of their slots is at that instant involved in call connection processing. As no other connection can be in process of being made, call data then appearing on the call data leads must refer to that connection and in this way the $P1'''$ processor is able to communicate data for the call to the supervision and register processors. Charge rate data are received and stored in the supervision processor in that way. The possible termination points for the path search leading to a connection are marked using data emitted by the translator marker, if the terminating points are service or outgoing junctors. For calls terminating in the exchange on exchange lines, the line scanning equipment is used. The lines being in one-thousand-line number groups with the last three digits of their directory numbers semi-permanently written into the line data store, LDS in Figure 7.1, the group to be scanned is determined from the initial directory number digits by the processor P''' which communicates over the call data wires to the common processor of the group the last three digits of the called station number. The common processor compares these digits with those of the lines as they are scanned, and when it finds agreement and the line then being scanned free and allowed to be connected, the processor stops the scan, marks the line and the class of service of the terminating junctor to be included in the connection and waits for the connection to be made. If no connection results because there is no free path, the scanner may go on to the end of the scan in case there is another line in the group. In this way all the lines with the specified directory number are tested in the order in which they appear in the line scanning cycle.

If for any call no connection can be made because there are no free destination circuits or intermediate trunks, the path search and connection processing will connect a service junctor giving busy tone to the caller, and if no connection to a busy tone junctor can be made because they are all engaged or other reasons, the fact is indicated to the register processor as call data for that processor to cause n.u. tone to be applied to the junctor, call data being communicated from the $P1'''$ processor to the $P1'$ and $P1''$ processors when the MK marking through the gates G3 and G6 indicates the junctor to which the data apply. In one way or another a positive result of the call connection processing is assured, and the MK marking removed by processing, that is gating, within the junctor itself. If because of faulty operation there is no positive result, the $P1'''$ processor sends a release datum to remove the MK2 marking and, via the register store, a datum to the $P1''$ processor to cancel the Po forward hold and either make another attempt to connect or send n.u. tone to the caller. Usually the result is a connection through the switches which is indicated to the junctor and hence to the $P1''$ processor by the operation of the I relay when and because current flows in the outgoing loop, the A relay contact in the loop having been closed when connection processing was initiated. If connection is to an exchange line on the exchange, the processor releases itself from the call by operating the DB bi-stable store which releases the register via gate G4 and transfers the control of the A relay from the register to the line loop via gates G12 and G13. If the call is to another exchange, the processor $P1''$ sends data to a register in that exchange before operating the DB store to release itself from the connection.

It is possible for the originating junctors or.j. and the incoming junctors i.j. to be operated in the same time division cycle and by one processor common to both. If separate cycles and processors are used, they would both have to share the same $M1'''P1'''$ processor.

When a connection has been made, the supervision processor $P1'$ continues to control the junctor. When the called line answer signal is received, the D relay operates and communicates the fact to the processor via gate G2. The processor then uses the stored charge rate to charge for the call by the operation of the MR relay in the junctor. The relay is operated via gate G10 during one processor cycle and released through gate G9 eight cycles later to give the exchange line meter or its equivalent, and the private meter if there is one, time to operate properly. The processor finally releases the junctor by a release datum.

The example of Figure 7.9 illustrates that the processing for a circuit processor can be divided between time division time shared processors operating in parallel in time with different cycle times and times of dwell, and synchronously, asynchronously or acyclically. The programs are linked by data transmission within the processors, between the circuit processors and their common processors via the interface switches between them, and between the common processors over call data bus-wires. The programs are otherwise independent so that by time sharing they may be performing the processing for many circuit processors simultaneously. The example also demonstrates that it may be economic to leave still in the circuit processor some of the processes which could have been transferred to a time shared common processor.

The example of Figure 7.9 also shows that comparing only the processing the quantity of storage needed is about the same for time division as for space division, namely $xM1' + yM1'' + zM1'''$ with some extra +m for random addresses in the time division case but also with some saving due to the elimination of translation in the register. It does not follow that the cost of storage is the same: that depends on the type of store and the total quantity of storage needed, and on duplication or other effects of security requirements. Independent stores as in Figure 7.2 have the advantage of not requiring to be duplicated but bulk storage even if duplicated may be cheaper. The quantity of processing equipment is much less for time than for

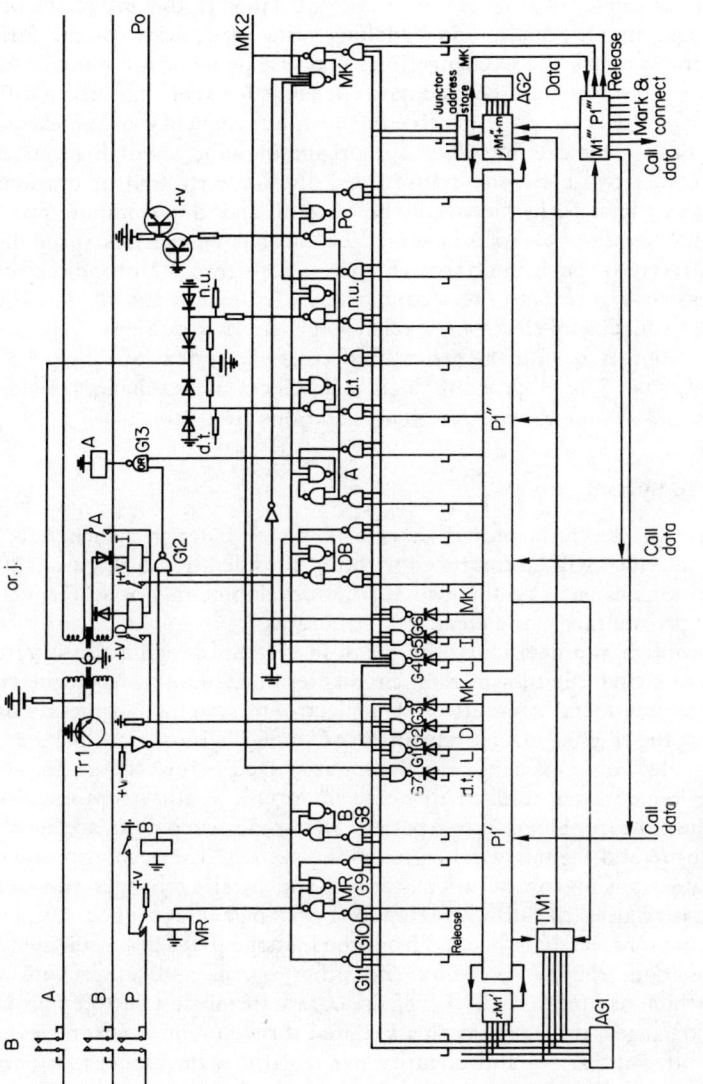

Figure 7.9 Originating junctor with time division time shared processing

space division, namely the reduction of x processors P1$'$ and y processors P1$''$ to one of each kind. Usually there is one M1$'''$ P$'''$ processor for space or time division. Thus on the basis of processing alone, time division costs less than space division, but what the example of Figure 7.9 does not show is the influence of data transmission equipment. Simple loop signalling being inadequate for the future or even for present systems, the exchange lines may use press-button number sending by voice frequency signal-messages and data exchange between registers in different exchanges is commonly at present effected by multifrequency data-messages. To receive these signal- and data-messages, the originating junctors of Figures 7.9 and 7.10 would require two expensive transducers, one for each kind of transmission, and many more data paths between the junctor and the common processor. Economy would require the transducers to be located in registers space division time shared directly through interface switches as in Figure 7.2 or indirectly using the trunking as in Figure 6.4: the second alternative makes the all time division wired logic controlled semi-electronic exchange generalized in Figure 7.10 possible and able to be compared with the generalized cross-bar system of Figure 5.5 using only space division. Stored program logic and electronic exchanges make such changes to systems as requires separate generalizations given later.

7.6 Generalized System

A summary of the principles of exchange systems design which have been deduced and described will be combined with the consideration of Figure 7.10. The principles are not changed by the two further developments not so far included, namely stored program logic and electronic structures.

The first problem and decision to be taken in systems design is the structure of the exchange to switch the transmission circuits terminated on it. Analogue circuits are switched by the metal contacts of semi-electronic systems but metal contacts cannot switch the digital p.c.m. channels of time division multiplexed transmissions. Semi-electronic exchange development started before the p.c.m. channel switching requirement was realized to be so important and has continued on the assumption that the problems are separate. Electronic exchanges are needed to integrate analogue and digital switching into one system. The second problem and decision is data transmission between stations and local exchanges and between exchanges. That relating to calls over telephone exchange lines cannot, for practical and economic reasons, be disassociated from the message paths but is adequate. The data comprise loop, ringing, coin-box and other signals s together with signal-messages sm which are tones and v.f. signals. Data transmission between processors in different exchanges, if message path associated through semi-electronic exchange switches, is limited in variety and quantity particularly with respect to supervision data after calls have been established: unlimited transmission requires either data link communication directly between processors with stored program logic as described in chapter 9 or electronic exchange switches as described in chapter 10. Here data transmission for junction connections will be assumed to be associated with the message paths over which the calls which the data control are transmitted, the data being d, m—d and s—d exchanged between processors in the exchanges and signals s and signal-messages sm to and from exchange lines.

The next decision is the series division of the processing among ranks of processors to be connected in series by the exchange switches as the message trunks are switched. The decision is arbitrary in the sense that there are no rules or theories which define the way in which the division must be accomplished but the economics of the problem leave little room for choice in the general division into line equipments l.e. with processors individual to the exchange lines, outgoing and incoming junctors o.j. and i.j. with processors individual to the junctions, service junctors s.j. with processors individual to service circuits, and originating and terminating junctors or.j. and .t.j. with processors individual to trunks which are available to exchange lines through concentrating switches. The circuit processors are such that the storage M and processing P required for any connection may be satisfied by the collective storage and processing of processors selectively connected in series. The division produces processors in ranks, as in Figure 7.10, each rank providing processes common to all the calls through the rank, but further division of the rank into groups is needed although not shown in the figure, to provide variations in processing to suit different, mostly structure, classes of service. The division into groups requires decisions based on costs. Originating, terminating and service junctors may each be in one group processing all classes of service within their group, a choice which is usually prohibited by cost, or in many groups each processing calls of one class of service or, the common choice, fewer groups some of which process more than one class of service. Junction junctors are divided into few, if any, class of service groups other than for outgoing or incoming traffic. The main distinction between junctors for junctions between exchanges of the same design is in the transducers required for the signal and data transmission over the junctions, such other processing as is included in them being the same for all outgoing and all incoming junctors. The generalizations given, Figure 7.10 and others, do not include interworking with other system exchanges, which occurs when a new system is introduced into a network to replace an obsolete system. The junctors on junctions to such exchanges frequently contain processing to suit the operations of the two types of exchanges to one another.

The processing being divided among circuit processors to be selectively connected in series, division of the circuit processors into processors to be operated in parallel has to be considered. The processes are first divided into those which may be required to be executed at any time and therefore at unpredictable times during a connection and those which occur within predictable periods within (and in general of durations short relative to) the durations of the connections. The first are supervision processes and require data transmission and storage Mr' and processing Pr' and the second are call connection processes which require data transmission and storage Mr'' and processing Pr'' out of which storage Mr''' and Pr''' may be divided for processes requiring still shorter time and so on. Space division time sharing permits x circuit processors of one rank to share y common processors which may share z second common processors, economy if any resulting from the reduction in the processor costs due to the reducing quantities of processors x, y

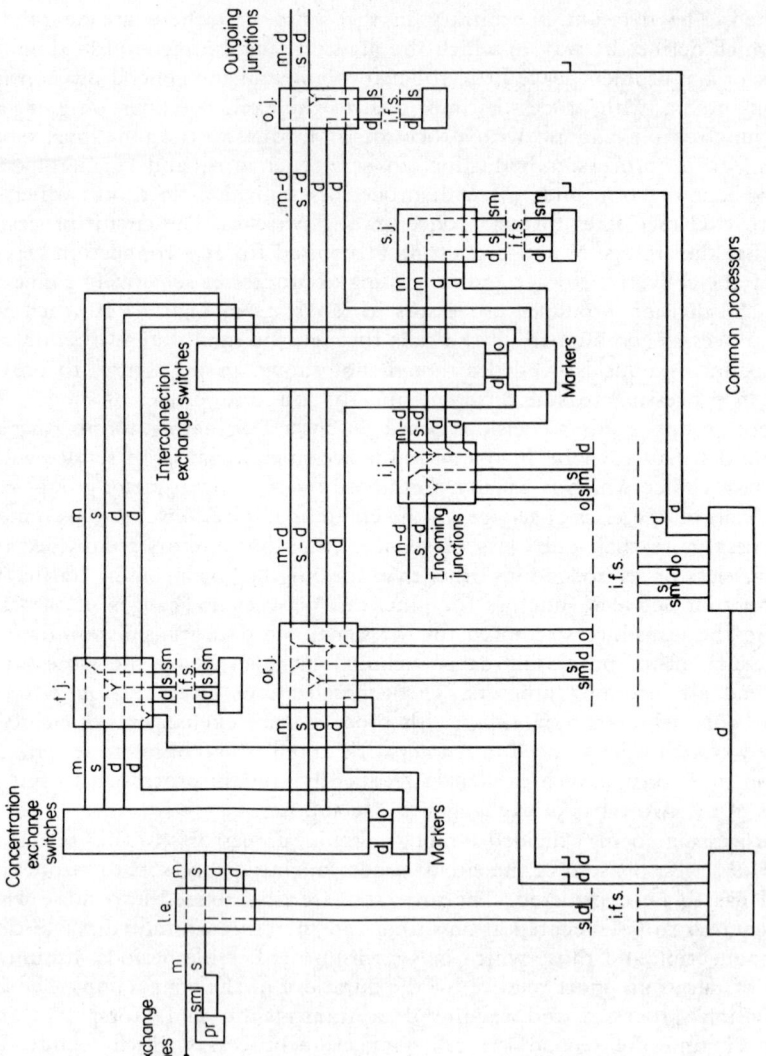

Figure 7.10 Generalized semi-electronic exchange with wired logic

and z exceeding the cost increases due to the interface switches and data transmission equipment necessary to connect the processors in parallel as required. Practice, as exemplified by cross-bar systems limited to space division time sharing, shows that economic advantage occurs only in respect of originating and incoming junctor processing which is time shared with registers and translator-markers together with the possibility that by using the trunking to effect the parallel connection, terminating junctors may economically time share ringing and ring tone processors.

Using time division, x circuit processors each with data storage $Mr' + Mr''$ may time share $z = 1$ common processor, if it is fast enough operating, with storage Mr''' and processing $Pr' + Pr'' + Pr'''$ which in total is more storage and less processing equipment than the corresponding space division time sharing. However, the total storage can be reduced to that of space division, namely $x.Mr' + y.Mr'' + z.Mr'''$, at the cost of storage $+mr$ for each of the y Mr'' to associate it with any of the x Mr' when required together with a small quantity of processing to do so: some small changes in the division of the processing and storage (for translations) between the common processors and the unique common processor, are possible to the advantage of time division to offset the increase $+mr$ and its processing. Thus time division with its greater economy of processing may be cheaper than space division time sharing if the differences in the interface and data transmission costs do not spoil the advantage.

A still further division of the processing and more particularly of the storage, to a processor common to many exchanges, is possible and may have practical application and advantage.

A difficulty not apparent from the systems generalizations is that time division time shared processors for semi-electronic systems are not able to send or to receive data in the form of signal-messages over the message paths. More exactly time division time shared processors can send and receive data only in the form of pulses, and pulses are not easily and cheaply transduced to or from analogue signal-messages. In Figure 7.2 for example, the dial and n.u. tones have to be applied in the originating junctor via bi-stable memory circuits although the tones theoretically originate in the common processor. This difficulty, particularly as regards voice frequency press-button signal-messages and multi-frequency data-messages as already mentioned, causes the need for some form of space division time sharing. In electronic exchanges in which the message transmission paths are themselves time division multiplexed, the situation is different, as will be shown in chapter 10.

As a broad generalization, time division time sharing is not economic for a rank of circuit processors which include electro-mechanical devices to be controlled, unless the quantity of processing or of storage per processor which can be time shared is large and can be time shared among a sufficient quantity of circuit processors, or unless the total quantity of storage is large and more cheaply provided in a bulk store operated by a unique common processor than individually in the circuit processors. Even if these conditions are fulfilled, the time sharing may not be economic if the data transmission costs are high as for press-button dialled data. For circuit processors using electronic components exclusively, the generalization may need some modification as will be seen in section 10.4.

Referring to Figure 7.10, the line equipments satisfy the conditions for the economic use of time division time sharing of a common processor, despite the small quantity of logic processing involved, because the quantity x of circuit processors is large and the quantity of data storage per circuit is also large and demonstrated to be more cheaply provided in a bulk store associated with a unique common processor than in the individual processors. There is also the additional and important feature, that the meter record stored in a bulk store may be read and all the accounting processing up to the printing of the bills may be accomplished by machines and thus satisfy one of the conditions for future systems. Originating and incoming junctors are less numerous than the line equipments but have associated with them much greater quantities of data storage and processing to justify time division time sharing of functions which in cross-bar systems are performed by registers and register-translators: nevertheless there are difficulties already mentioned, with multi-frequency press-button data transmission. Terminating junctors t.j. have only small requirements for data storage and processing and in fact if the ringing is space division time shared by the trunking method, the terminating junctor has little more to do than supply the exchange line with d.c. power and to relay the line loop signals. Time division time sharing is thus of no advantage in general to terminating junctors unless classes of service involving much processing or storage are introduced in the future, calls incoming to coin-boxes being paid for through the coin-boxes for example, when the situation might be different for the relevant junctors although not having much effect overall. Special junctors, such as for interception, are too small in quantity to benefit much from time sharing of the data processing, and the same is true of many service junctors s.j. Outgoing junctors o.j. need be concerned with no more than signal and data transmission.

Thus clear advantage in using wired logic time division time sharing of processin in a semi-electronic exchange system is limited to line originating and incoming junctors.

7.7 Security, Costs and Practical Application

The relative quantities of equipments deduced in the previous section take no account of the security of operation which is required of exchanges in service. The measures needed to ensure adequate service despite faults in the equipments have a profound effect on the quantities of equipments necessary and therefore on costs. Because time division time sharing tends to reduce all processing to a unique common processor operating with a bulk store for all the data storage, it is particularly vulnerable to faults having serious effects. Sufficient security therefore inevitably involves equipment redundancy.

Taking the registers of Figure 7.2 as an example, they may all be put out of service simultaneously by one fault in the pulse generator, or in the time division interface switch or in the unique common processor. The risk of such an occurrence has to be limited to the order of one hour per century. The probability of a fault occurring in the equipments enumerated depends on the construction and particularly on the scale of the integration of components in unit packages but may be as high as one fault per month. As a result, the vulnerable equipment must be at least duplicated. If only duplicated, there is difficulty in detecting that a fault exists and its location, which problems can be solved more or less satisfactorily. Using majority voting or decision for three equipments working in micro-synchronism, which means having three equipments all simultaneously processing the same data, comparing their outputs and accepting, if a discrepancy occurs, the two which agree as the data to be used, faults are instantaneously detected and service is maintained even despite more than one fault existing but in different places in the parallel operating equipments. If all the data storage for each register is contained within a bulk store, not being individual as in Figure 7.2, that too must be replicated with the data processing equipment. It is not difficult to see that time division registers are not substantially cheaper than individual control registers unless the quantity of registers required is large, and that this is a characteristic general for equipments with time division time shared logic. As a consequence, semi-electronic exchanges with time division time shared processors and wired logic are more expensive than cross-bar exchanges of small size and cheaper only for large exchanges.

A new system for universal large scale application such as step-by-step systems achieved in their day and cross-bar systems attained later, requires to possess enough advantage over a system which it supersedes to make the trouble and cost of making the change worth while. The possible advantages are lower costs and more attractive subscriber and operating service, but the first not being achieved and the second not being sufficient, no significant move toward semi-electronic systems with wired logic took place in practice. The attractions of stored program logic were an additional inhibiting factor. Nevertheless as will be shown in chapter 10, electronic exchanges in which the transmission paths are time division time shared, create new technical conditions in which for some processes wired logic is to be preferred to stored program and is an additional reason for maintaining interest in wired logic processing as in this and the next chapter.

Chapter Eight
WIRED AND STORED PROGRAM LOGIC

8.1 General

Electronic devices came to be used in digital computers in the 1950s to increase the speed of operation and the magnitude of the numerical problems which could be solved. To match the increased speeds and data storage needs, the magnetic tape store was devised for input–output data and various kinds of semi-permanent and temporary stores for data within the machines. The wired store which defined the sequence of operations to be performed remained and, except for single purpose machines, had to be wired or plugged individually for each problem. Von Neumann in the U.S.A. perceived that a sequence of operations of which a machine is capable of executing by wired logic, could be defined and controlled by a series of numbers easily written into and retrieved from an electronic store, and which become a stored program. The machine thus created is capable of solving logic problems in general, the digital computer being a special case, of which the Bell Laboratories quickly took advantage by applying stored program logic to telephone exchange control processing. Used in this way, the processors are single purpose machines required in considerable quantities which are not the conditions which stored program logic was invented to satisfy. The value of that kind of logic in exchange design is in new methods of system operation and of manufacture to which it has led and these are the subjects of this chapter and chapter 9.

Stored program and wired logic are interchangeable in the sense that theoretically any given processor can be realized by either, but because of their very different practical characteristics, an exchange control system based on one would not be the same as one based on the other. Each used to produce the same exchange system is not in consequence a valid method of comparison of the relative advantages of the two kinds of logic.

Data transmission, memory M and processing P as defined in section 7.4 still apply with stored program logic processing which may be used in all the processors in an exchange, or in some processors with others using wired logic processing, and it is possible for both wired and stored program logic to be used in the same processor.

8.2 Practical Examples

The application of stored program logic to the control of exchanges and its relation to wired logic will be approached through practical examples. Stored programs frequently comprise in practice a thousand program words or more for processes which can be described in very few spoken words, which means that here only very simple examples can be described in detail. Nevertheless, they are enough to indicate the most important characteristics involved. The examples to be given are based on the common processor of Figure 7.1 and both wired and stored program logic processors and processing will be described the better to bring out the differences between them.

In the scanning systems to be described, as in Figure 7.1 the common processor receives through the interface switch the L and P data for the line being processed, L = 1 indicating that the line is looped but not connected through the switches and P = 1 indicating that the line is connected through the switches. If the line is to be connected a datum is sent to mark the MK wire of the line. The line and state of line data available in the bulk store LDS are the same as or similar to those of Table 4.1. The store is addressed by the line e.n.c. numbers which suffice for originating call scanning which is to be described. A sixteen digit word is assumed for each line of which digits 0–7 are temporary storage for state of line and digits 8–15 are semi-permanent line data mostly class of service as shown in Table 4.1. The park, t.o.s., s.o. and s.s.o. states of line are put into operation by digits 1 written into the digits 1 to 4 respectively and the o.b. state of line, which prohibits both s.o. and s.s.o., by digit 1 written into both digits 3 and 4 of the line word. Digits 8 to 11 are allocated to classes of service as in Table 7.1.

It was previously remarked that if all the operations involved in line data processing followed by line connection through the switches can be completed within 10 μsec, which is possible with electronic exchange switches, clock timed synchronous cyclic time division time shared scanning is possible. With reed relay exchange switches the time for connection exceeds 10 μsec but being within 10 msec, asynchronous cyclic scanning including line connection during the scan cycles is possible. With cross-bar switches taking much more than 10 msec to operate, the scanning may be synchronous or asynchronous cyclic but the equipment numbers and possibly other data of lines to be connected have to be stored for acyclic processing during some other programs. These three possibilities are to be examined.

Figure 8.1 shows wired logic synchronous time division time shared scanning equipment the operations of which are timed by the outputs of a pulse generator LA from which the line addresses are derived. The generator is shown to produce fifteen complementary binary outputs which would typically be the digit outputs of a binary counter driven continuously by pulses at a constant frequency produced by a generator which controls the rate of scanning to be suitable to the equipment employed. The generator also produces read and write strobe pulses together with enable pulses which permit of operational outputs only when all processing circuit operations have become stable. The generator pulses thus produce time phased operations comparable with the two-phase operation of Figure 7.5. The ten least significant digits of the counter LA provide addresses for 1,024 lines and operate the interface switch i.f.s. to connect the lines one at a time to the common equipment, at the same time reading to the output of the store LDS the word in the store for each line as it is connected. Using the output from a five-input AND gate G1 operated by a combination of the five most significant digits to make the addresses effective at the interface switch and the LDS store, the scanning of the block of lines occurs once during each counter cycle. It is convenient in practice to use blocks of 1,024 lines and to spread their scanning for new originating calls throughout the counter cycle instead of concentrating it into one part of the cycle, and in which respect the scanning is analogous to that of Figure 2.14. As the exchange grows, lines are added in 1,024 blocks and scanned during a part of the cycle determined by the input connections to the five-input gate for the block. As shown in Figure 8.1, up to 16,000 lines could be scanned in one half of every cycle to leave half for other call operations. For this to be possible the maximum processing for one line must not take more than 6 μsec if five scans per second is specified.

Data are entered into the store LDS by being written into the register LDW at any time and transferred into the bulk store on the occurrence of an address pulse. The common processor performs both operations during the processing time of a line concerned, to control the pk datum for the line. The administration enters new other data into the store by controlling equipment which first writes the data into the store LDW and then, knowing the address of the line and being supplied with the LA generator pulses, it generates a write pulse at the appropriate instant. The originating call data t.o.s., s.o., s.s.o. and o.b. are entered and erased in this way. The time division being synchronous, meter pulses may be entered and stored in a store MRS similar to LDS and as described in section 7.2. The administration is able to read the meter record of any line by equipment which generates a read pulse when the line is addressed.

At each effective address in the block, the common processor, which is common to all the lines in all the blocks, receives data from the addressed line equipment, in Figure 8.1 the data being L, P and pk, and from the line data store LDS, the data having been read out of the store when the address was first established. The processor is able to send data to the addressed line via the interface switch, the figure showing MK data sent in this way, and to other parts of the exchange control in three ways: firstly over destination common wires to mark all originating registers or n.u. tone junctors or s.o. or s.s.o. equipments, as possible destinations for a connection to the marked exchange line; secondly over group marking wires to limit the originating junctor included in the connection to one providing the structure class of service required by the calling line; and thirdly over bus-wires to junctors and registers liable to be included in the connection made, the data transmitted being stored in the equipments as they are connected and being control class of service and other call data needed at later stages of the connection.

It is assumed that a line is parked by a pk datum transmitted through the exchange switches to the line equipment and thence to the common equipment. No such transmission path is indicated in Figure 7.1 because one is difficult to derive with metal contact switches other than by a separate and additional path through the switches, which is too expensive. Electronic exchange switches for which the scanning being described is suitable, are not all restricted in this way. A line is parked because although inactive it is looped and would interfere with other calls unless prevented from doing so by being parked. The line is required to be taken

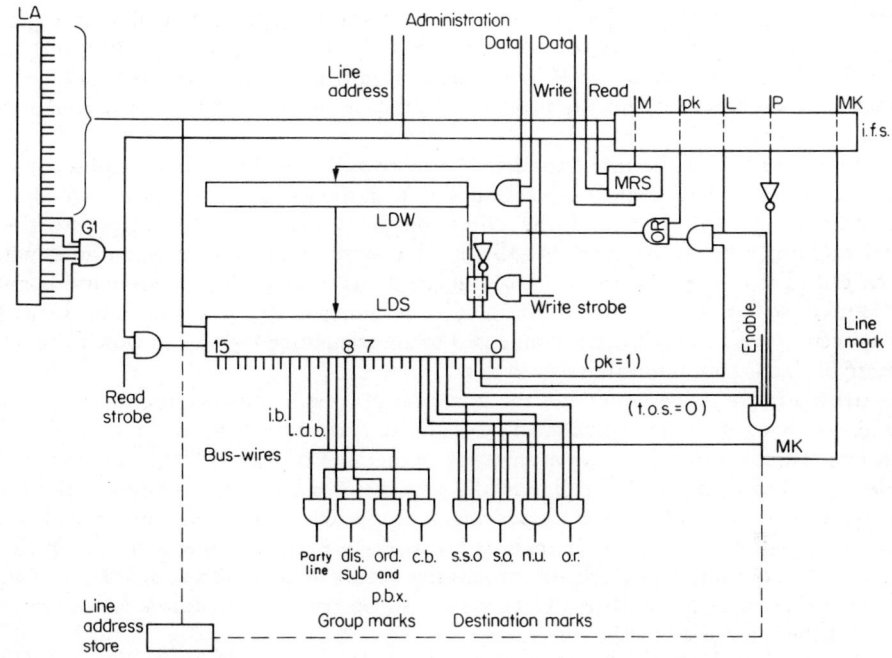

Figure 8.1 Synchronous cyclic time division — processor with wired logic

out of park when it becomes unlooped. Preferably the loop should be broken for some measured time before the line is unparked because some short-circuit line faults are intermittent and allow the loop to be broken for short periods. The timing requires a small addition to the programs to be given.

Figure 8.2 is a flow diagram for the operation which starts with the addressing and reading of a word from the LDS store. Then if the line at that address is already connected, which is P = 1, the processor waits for the next line and data to be presented. If P = 0, L = 1 because the line is looped, pk = 0 and t.o.s. = 0, the conditions for a normal originating call exist and the line is marked with an MK datum. If s.o. and s.s.o. are both 0, the line is to be connected to a free originating register o.r. via an originating junctor providing the structure class of service indicated by line data read out of the store. Relating the operation to the generalized system of Figure 7.10, to the scanning of Figure 7.1 and the guide wire path searching and connection of Figure 6.5, the MK mark is applied to the line guide wire and all the originating registers are marked as possible destinations for the connection, as seen from Figure 8.1. It is also seen that originating junctors of one of four kinds are marked with a group mark and call data may be transmitted over bus-wires to all the registers and junctors. If a free path via a group marked junctor to a free register exists, a connection will be made from the calling line, through a junctor to a register which on seizure earths the P-wire of the line. Earthing the P-wire changes the value of P in the processor from 0 to 1, the line and other markings are removed and the processor waits for the address to change to that of the next line.

In Figure 8.2, if the line data digit 3 is 1 and digit 4 is 0, s.o. = 1 and s.s.o. = 0 and all the s.o., and if digit 3 is 0 and 4 is 1 all the s.s.o. junctors, are marked as a destination group. If one of the group becomes connected, it stores the then existing line address together with the line and state of line data and continues with the call as if it were the calling line. If digits 3 and 4 are both 1, the line is o.b. and required to be connected to an n.u. tone junctor or, if some calls such as emergency are to be allowed, connected to an originating register to which the o.b. state of line is communicated as call data.

With all processing complete, the common processor remains idle until the line address changes. If no connection is made, it is because none is attempted or, if attempted, is unsuccessful. No attempt at connection is made if P = 1, the line being already connected, or if the line is idle which is P = 0 and L = 0 because the line loop is open, or if the line is looped but temporarily out of service and thus P = 0, L = 1, pk = 0 and t.o.s. = 1. If there is no working exchange line at the equipment number of the line addressed, no processing occurs because all the digits out of the store are zeros as are both P and L, which is also true if the address is not that of one of the lines of a block.

The writing of the pk digit into the store is independent of other operations and it occurs when the pk digit transmitted through the switches is 1, which is 'switch pk = 1' on the flow diagram. When pk = 1 has been written for a line, it remains until L = 0, when pk = 0 is written. Referring to Figure 8.1, the LDW store is shown split into one digit, digit 1, for pk controlled by the processor and digits 2–15 controlled by the administration. A write strobe common to both is generated just

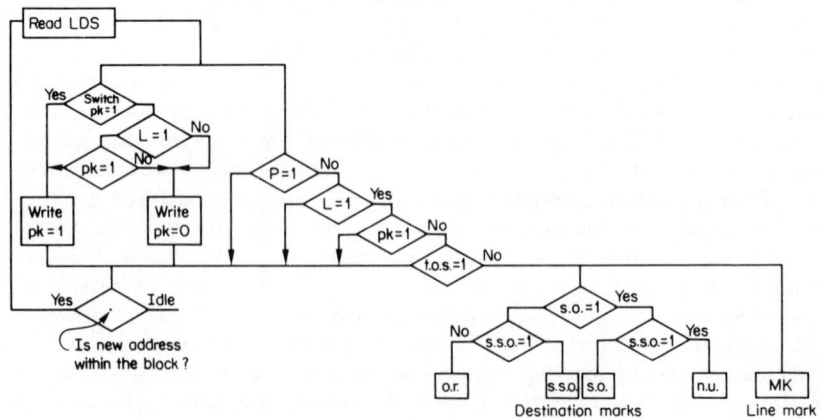

Figure 8.2   Flow diagram for synchronous cyclic scanning with wired logic

before each change of LDS address, the read strobe occurring just after the change. For each line in a block being scanned, a complementary binary digit is applied and written into the pk digit in the store for the line: what, if anything, is written into the remaining digits is transferred to the LDS store on the occurrence of a write pulse from the administration equipment. The complementary binary input to the pk digit is derived from the output of an OR gate which unoperated writes pk = 0. The gate is operated by either the switch pk = 1 datum or the output of a gate operated by pk = 1 and L = 1. A line is parked by a processor connected through the exchange switches to the exchange line, the processor maintaining the pk = 1 datum for at least the period of the scanning cycle. The same processor then releases the connection so that P becomes 0 but although L = 1 and the reason for parking the line, no new call is originated because pk = 1 and is maintained until L becomes 0.

The circuit diagram Figure 8.1 also indicates the alternative to the making of connections during the scanning, namely the storage of the identities of lines to be connected, the actual connections being made later by processing which takes too long to be included in the scanning cycle. The quantity of addresses which the store must be capable of holding depends mostly on the average number of new calls per scan and partly on how quickly other processors can deal with them. When the storage is full, another line requiring to be connected has to wait to be recorded until the storage is found not to be full on a subsequent scan. If more than one address can be stored, state of line data must be written into the LDS store at the addresses stored to prevent them being stored again on later scans. The stored addresses are used one at a time for call connection processing during which if only the addresses have been stored, the LDS store must be read again to extract the line and state of line data for the line being connected. This means that the address generator LA must be able to be set at any address at random as well as cyclically. If the line data are stored as well as the line address, the line connection processing and equipment can be seperate from and independent of that for scanning.

Synchronous cyclic time division time sharing, generally limited to wired logic processing because processing speed is important, is possible only if the product of the number of slots in the cycle and the maximum processing time per slot is less than the maximum duration allowable for all the slots, per cycle. If the condition is not fulfilled, synchronous cyclic processing may be possible by reducing the number of slots or by transferring some of the processing to other processors to reduce the maximum processing time per slot and possibly to increase the allowable time per cycle, all of which are illustrated by the system of Figure 7.9. In that system, by transferring the register processes to a processor $P1''$, the $P1'$ processor is left with the supervision processes which take less time and need to be performed less often than would otherwise be the case, and the processor is then able to be synchronously cyclically time shared among all the originating junctors. In order that the processor $P1''$ may be synchronously cyclically time shared within the cycle time allowable for register processing, the quantity of slots is minimized by provision for only those junctors which at any one time are in need of register processing and the maximum processing time is reduced by transferring call connection processing to processor P''' and its memory M'''. To transfer the register processes to a register processor, the addresses of the junctors requiring register processing are stored in a slot of the register cycle, being the +m of the $y.M'' + m$ store, together with the call data and this makes the register and supervision cycles independent of one another. The processor P''' is acyclically time shared among the registers, its memory M''' holding data for a connection to be made to render call connection processing independent of other processing. Storage for only one connection at a time is sufficient if the connection time is low enough to permit of only one connection at a time through the exchange. If this condition is not satisfied, provision has to be made for the storage of data for more than one connection.

The flow diagram given in Figure 8.3 for asynchronous scanning using a processor with wired logic, differs from that of Figure 8.2 in that within periods of constant duration allowed for scanning, the dwell time of the processor on each line is determined by the processing time for that line. The scan periods occur at constant frequency and are defined by clock generated pulses which with other pulses for other processes control the operation of the exchange. The equipment needed and shown in Figure 8.4 includes a line address generator LA which comprises an eleven-digit binary counter set to zero at the beginning of a scan period, the count subsequently being advanced by pulses from the common processor, each pulse adding one to the count and shifting the scan to the next line in the cycle. When the eleventh digit changes to 1 the count has reached the value 2^{10} and all the 1,024 lines in the block have been processed. If, however, the cycle encounters many new calls to be connected and delaying the scan, the cycle may not be completed during the 'clock = 1' pulse which controls the time allowed for scanning. Asynchronous differs from synchronous scanning processing in two other ways. Firstly, a positive indication for the finish of processing must be obtained for each line in order to advance the scan. Normal operation satisfies this condition in the absence of faults. If a line is not to be connected through the switches, the common processor has all the data it needs to complete the processing. If a line is to be connected, the processor applies the line and destination and group marks required and waits for the P datum to change to 1 to indicate that the connection is complete: it then moves the scan to the next line. If no connection can be made because there is no free path, although not shown in the example the fact should be indicated to the common processor by the path search and connection processors. Nevertheless, because of the possibility of no connection due to faults, the common processor limits the time it can dwell waiting for normal processing to be completed, by time-out which moves the scan to the next line. Writing into the pk digit of the LDS store is parallel processing independent of the main processing and not concerned with scan timing.

The second way in which asynchronous scanning differs from synchronous arises out of the fact that the intervals between consecutive dwells of the processor on any particular line are not uniform. In fact if the clock pulse terminates a scan, some lines are not processed at all on that scan. Therefore operations such as writing pk = 1 into the LDS store as in Figure 8.1 by datum transmitted via the interface switch, become liable to failure. The probability of failure is decreased by increasing the duration of the datum transmission over the line itself, if that is possible and the effect of an occasional failure is tolerable, as is the case with the parking of a line. Such is not the case for metered charge recording. No avoidable risk of failure is permitted, nor can the duration of each datum pulse transmission be prolonged very much without restricting the maximum rate of metering to less than is required in practice. For these reasons, if scanning is used for metering, it must be independent of the scanning for line connection. The equipment which writes into the LDS store for the administration is not affected by irregular scanning rate because its normal method of operation is to wait until the address of the line concerned appears out of the address generator LA. This method depends on the facts that the address of the line is known and the rate at which data are to be stored is so low that one or more cycles can be devoted to each datum recording. By a similar method, if when an exchange line is being connected to a junctor or a register, its address then being indicated by the counter LA output is stored in those equipments as they become connected, the equipments may at any time subsequently store data for the line in the line data store LDS. The method makes use of the scanning processing and that the address of the line equipment currently being processed is that of the counter LA output then existing. An equipment with data to store for a line for which it has the address, waits until the address appears out of the scan address generator, then sends the data on bus wires to the line data store which stores the data at that address. In Figure 8.3, 'bus pk = 1' means a pk datum is being received in that way. The method being independent of data transmission through the switches thus avoids the difficulty to which attention was previously drawn, that of providing sufficient data transmission capacity through exchange switches to satisfy the needs of modern systems.

The storage of line addresses in originating, terminating and other junctors and in registers as the lines are connected to those equipments presents no difficulties other than the cost of the storage, but storage in junction, service or other equipments not connected as the exchange lines are first connected requires so much processing and equipment as to be impractical with wired logic. Such limitation is not, however, important in practice.

Line addresses stored in junctors and registers may be used to address random access data stores for the storage of data independently of line scanning, the chief use being charge recording described in sub-section 9.5.7.

Much of the equipment represented in Figure 8.4 for asynchronous cyclic scanning will be recognized as similar to that of Figure 8.1 for synchronous scanning. The differences include that the scanning cycle is started by a start pulse coincident with the beginning of a scan period 'clock' pulse. The start pulse resets the LA counter to zero and an AND gate G1 output then activates the scanning until either the count reaches 2^{10} or the clock pulse ceases. An AND gate G2 when operated advances the count by one unit. Because the scan processor dwell time is dependent on the processing time, unlike synchronous scanning every exit from the program has to perform some effective operation. In Figure 8.4 an OR gate G3 has five inputs one of which operates the gate when processing for a line is completed.

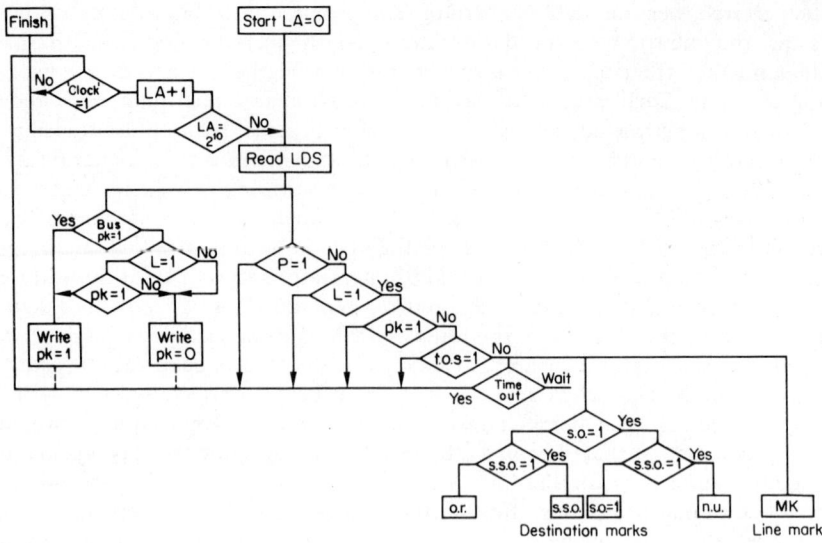

Figure 8.3 Flow diagram for asynchronous cyclic scanning with wired logic

The gate output operates a pulse generator which provides controlled phase operation of the change from one line to the next. It produces first a write strobe pulse to write into the LDS store the existing pk data and, if also instructed by the administration equipment, other data in the LDW store. This is followed by the cessation of an enable output to prevent any effective operations while the address is changed by a step pulse to operate gate G2. Then a read strobe reads new data out of the LDS store and after an interval sufficient for the data to be processed, the enable output is re-established to allow the processing to be effective.

In Figure 8.4 the storage of addresses of lines, instead of connecting the lines during the scan cycles, is a possible alternative and is indicated and operates in the same way as in Figure 8.1. In this particular case there would be little point in doing so because the time taken for line connection is the reason for asynchronism, but other conditions could justify asynchronous cyclic operation with the transfer of some processes to other processors and requiring the addresses of slots to be stored.

Figure 8.5 shows symbolically a stored program logic processor asynchronously time division time shared among line equipments, for originating call scanning and connection of calling lines during the scan cycle, thus being operationally equivalent to the equipment of Figure 8.4. The common processor, which is based on the principle of a very simple general purpose computer, comprises wired logic equipment P including a register, A, capable of storing words in this case of sixteen digits numbered 0 to 15. The program and data store PDS comprises sixteen digit words addressable by the output of a program address register PDA. A line data

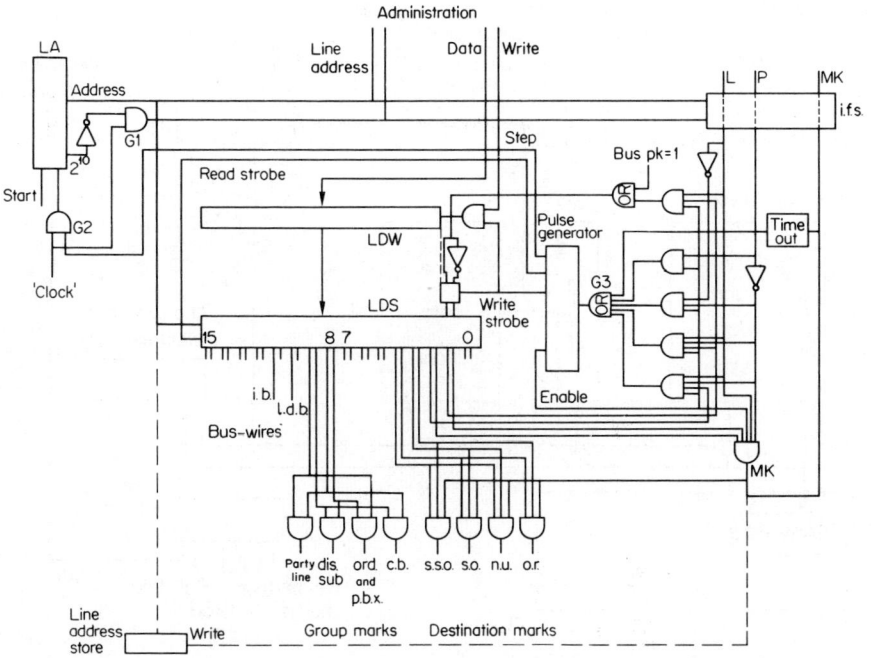

Figure 8.4 Asynchronous cyclic time division — processor with wired logic

store LDS also of sixteen digit words is the same as the LDS stores of Figures 8.1 and 8.4 and is addressed in the same way by the output of a line address register LAR. The output of the line address register also, as in Figure 7.1, addresses through an interface switch i.f.s. exchange line equipments from which state of line data L and P are communicated through the interface to the processor and to which MK data are applied to mark lines for connection. To mark a line, the line and state of line data for the line are stored in a mark line data store MLS which not only stores the data but also outputs destination and group marks and line data the same as the corresponding equipment in Figures 8.1 and 8.4, and the MK datum to mark the line. Data movement into and out of the processor P is made via the A register, the contents of which can be written into the PDA and LAR registers, into the LDS store at the line address then existing and into the mark line data store MLS. Data can be transferred into the A register from the program and line data stores, the line address register and from exchange line processors via the interface switch. The program and data may be loaded from the input/output unit I/O.

A program word read out of the program and data store PDS comprises four bits, 0 to 3, of instruction and 12 bits which may be an address x in the program store or a quantity n, depending on the instruction. A data word is a number in

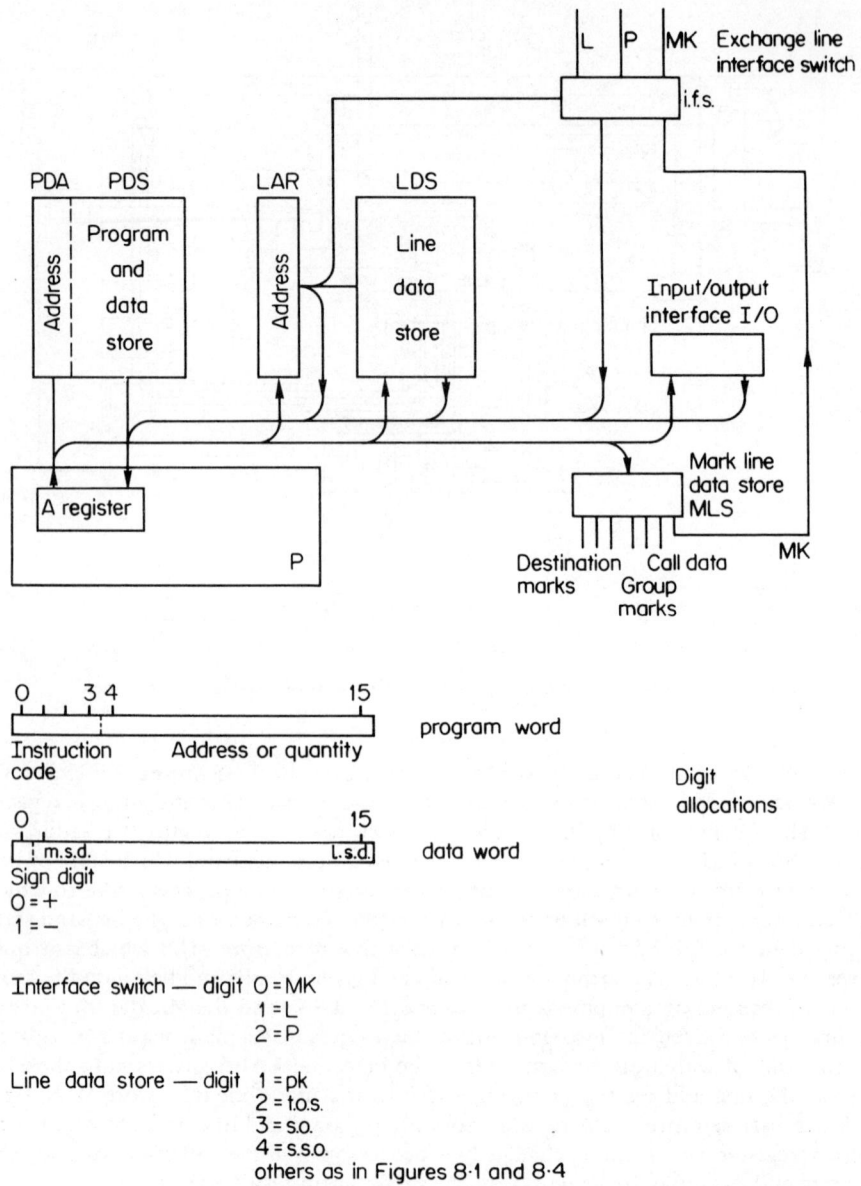

Figure 8.5 Asynchronous cyclic time division — processor with stored program logic

Table 8.1 Instruction Répertoire*

Mnemonic	Address or magnitude	Instruction
CAR	—	Clear A register
LPD	x	Load register A from location x in PDS
LAR	—	Load register A with contents of LAR
LLD	—	Load register A from LDS — location determined by LAR
SPD	x	Store contents of A register in location x of PDS
SAR	—	Store contents of A register in LAR
SLD	—	Store contents of A register in LDS — location determined by LAR
IPS	—	Input data from i.f.s. or I/O interface into A register
OPS	—	Output data from A register to MLS or I/O interface
ADD	x	Add contents of PDS location x to contents of A register
SUB	x	Subtract contents of PDS location x from contents of A register
JMP	x	Unconditional jump to PDS location x for next instruction
JPL	x	Jump to PDS location x if contents of A register +ve
JMI	x	Jump to PDS location x if contents of A register −ve
SFL	n	Shift contents of A register left n places
SFR	n	Shift contents of A register right n places

Notes:—
*When data are read from a store or register, the contents of that store or register remain unchanged.
When data are written into a store or register, the contents of that store or register are overwritten.
Data words in PDS:
    address 61 2^{14}
             62   1
             63   2^{10}

digits 1 to 15, digit 15 being the least significant, and a digit in bit 0 which according to its value 0 or 1 indicates whether the number is positive + or negative —. On receipt of a program word, the processor carries out one of sixteen instructions indicated by the instruction digits, the available instructions being listed in Table 8.1. An unconditional instruction being completed, the processor adds one unit to the number in the program address store without further instruction. Conditional upon the result of other instructions, the processor either adds 1 to the existing program address or jumps the address to that given with the instruction. Conditional jump instructions correspond to the binary decisions of the flow diagram, which is given in Figure 8.6.

A suitable program is given as Program 8.1. It is put into operation by the writing into the PDA address store, under the control of the executive program, of the address given as 00 although the actual addresses are arbitrary and made to suit the addresses for other processing. The continuation of the program is not difficult to follow in conjunction with the flow diagram Figure 8.6 on which the sequence of instructions has been included. As shown, when a line is marked for connection the P-wire is checked for an earth condition by continuous processing round a loop until an earth occurs to indicate that the line connection is complete, and scanning

Program 8.1

Instruction No:	Mnemonic	Address or magnitude	Comment
00	CAR	—	
01	SAR	—	Set line address to zero
02	IPS	—	
03	SFL	2	
04	JMI	27	Jump if P-wire earthed
05	IPS	—	
06	SFL	1	
07	JPL	21	Jump if line not looped
08	LLD		
09	SFL	1	
10	JMI	27	Jump if line parked
11	SFL	1	
12	JMI	27	Jump if line t.o.s.
13	LLD		
14	OPS		Mark line
15	IPS		
16	SFL	2	
17	JPL	15	Continue testing and marking until P-wire earthed
18	CAR		
19	OPS		Cease marking line
20	JMP	27	
21	LLD		
22	SFL	1	
23	JPL	27	Jump if line not parked
24	SFR	1	
25	SUB	61	
26	SLD		Clear park condition from line data store
27	LAR		
28	ADD	62	
29	SAR		Increase line address by one
30	SUB	63	
31	JMI	02	Continue scan if line address < 1024
32	JMP	Executive program	

is continued until all 1,024 lines have been processed when control is returned to the executive program. The scan can also be terminated earlier by the executive program, if the time allowable would otherwise be exceeded. Time out to limit the duration of line marking in case of no connection being made is not shown. It may be included in the MLS store which on expiry of the time would artificially produce the $P = 1$ condition needed, or with some addition to the program and to the equipment to control the interval between successive traverses of the processing loop while a line in marked, it may be timed by counting the number of traverses made. In practice it is more likely that instead of line search and connection during

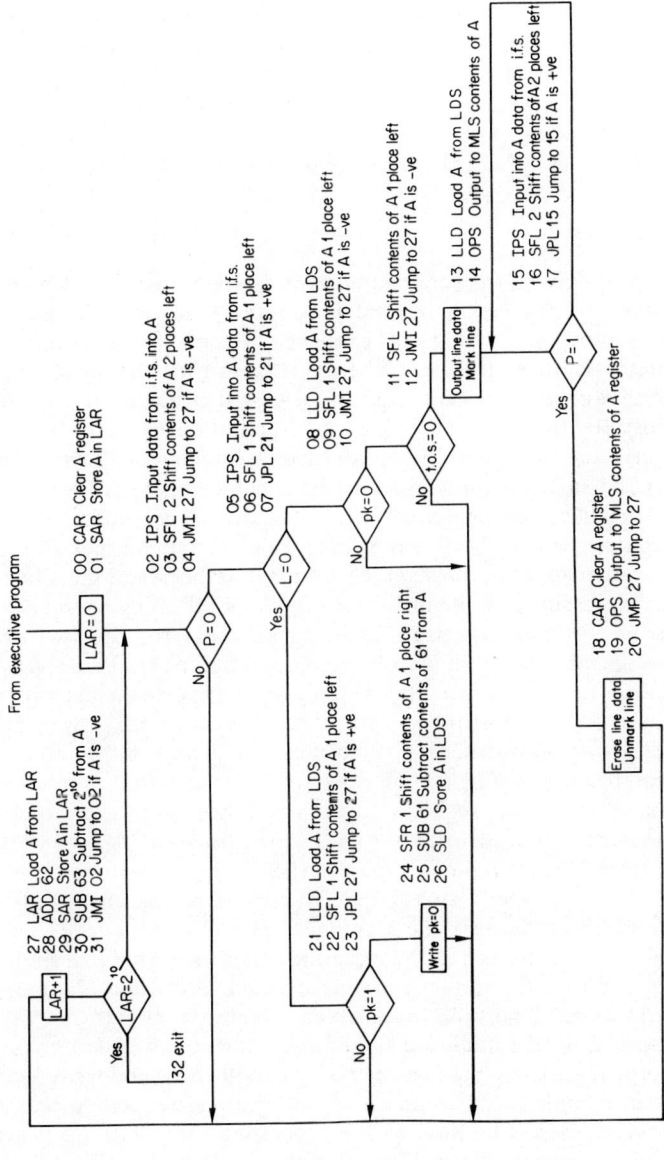

Figure 8.6 Flow diagram for asynchronous cyclic scanning with stored program logic

scanning, the addresses of lines requiring that processing are stored and connected later using a different program.

Stored program processor design and programming are specialized arts of which only a glimpse is given, by the simple example to which space limits the present description, of processors some of which in practice have repertoires exceeding one hundred instructions and programs of more than 10^5 words. It is important to an understanding of exchange systems that the characteristics and capabilities and the limitations of stored program processing should be known, but the details of the processors and of the programs are not essential to that understanding.

8.3 Characteristics and Costs

The examples given illustrate some important characteristics and differences between processors and processing dependent on the kind of logic used in the processors: that, as previously remarked, exchange systems based on one kind of logic operate differently in at least some respects from those based on the other kind; and that, although theoretically equivalent stored program and wired logic processors are possible for any processing, in practice one kind is not always substitutable directly for the other. A quantitative assessment of the two kinds of logic is thus rendered difficult and choice when it comes to be made is generally influenced by qualitative assessments of the most prominent differences. Using either kind, or a combination of both, any coherent overall system specification can be satisfied: and system specifications, which always have been and must be related to what is practically possible, can be influenced by the kind of logic used.

A major cause of systems organizations being different for the two kinds of logic, is processing speed. The serial step by step operation of stored program logic means that in general it takes longer than wired logic to execute a given process, a difference enhanced by the possibility with wired but not with stored program logic, of processes which do not mutually interfere being executed simultaneously within one processor. As a result, stored program logic may take the same time in milliseconds as wired logic requires in microseconds, and some techniques which are possible with wired logic are not with stored program logic in a real time limited control system. The example of Figure 8.1 shows equipment and processing with wired logic which for speed and simplicity cannot be matched with stored program logic: in addition, because the scanning is synchronous cyclic, it permits the pk data through the switches and the metering to be associated in a simple way with the scanning, and thus the system design is affected. Because of the processing time, shift registers as in Figure 7.2 and which because they are individual to exchange registers need not be duplicated, cannot be operated with stored program logic: the data stores have to be the fast operating parallel read and write kind, and if they are part of a bulk store common to all the registers, they have to be duplicated for security. Again because of the processing time, with asynchronous cyclic operation of a common stored program logic processor the quantity of circuits which can be processed within an allowable period per cycle is generally less than the quantity possible with wired logic and often insufficient and compels transfer of some of the processes to other programs, which adds to the processes to be executed and to the data to be stored. Other effects of processing time on system design will be evident in later chapters.

The costs and relative costs of processors designed to execute the same processes depend not only on the processes and on the components used, if different, but also on the methods of manufacture and the quantities manufactured. Originally transistors and other components in discrete or small scale integrated units were used for construction together with ferrite core or other magnetic stores for data, all in large scale production for the computer and other industries. Processor costs were then not mainly dependent on quantity manufactured but on the quantity of components used. Figure 8.7 shows qualitatively the relationship between quantity of processes performed and the quantity of components needed: the quantity of components increases proportionally to that of processes, but with wired logic starting from a small initial value and stored program logic from a much higher value. That this is so can be seen from the description given of the stored program logic processor of Figure 8.5. The minimum cost is that of the wired logic processor even the simplest of which needs many components. The slope of a stored program logic curve is that due to increasing quantities of storage needed for the programs. The quantities of program words and storage are related to the available instructions: the greater the repertoire and therefore of wired logic, the smaller the program store. This is represented in Figure 8.7 by a series of possible stored program processor designs of increasing minimum quantity of components due to increasing repertoire but with correspondingly decreasing programme requirements. Assuming processors designed to suit the uses to which they are to be put, the quantities of components required are given by the continuous curve in Figure 8.7 and which it is assumed will cross the curve for wired logic at some point. Security which demands replication of some if not all of the components and is not the same for both kinds of logic, complicates the comparison. Moreover, the curves are highly idealized. In practice they would be stepped and not smooth, the term components only loosely describes switches and stores used and the proportions of each used are very different in the two cases, and their relative quantities depend on the processes as well as their quantities; but the curves bring out the important points that stored program processing under the conditions postulated benefits economically by concentration into as few processors as possible, whereas wired logic processing does not suffer much economically by being divided among a number of processors.

The large scale integration of semi-conductor electronic components into assembled circuits and into data stores to supersede the magnetic kind, has profoundly affected the costs and the relative costs of processes and processing by wired and stored program logic. By making available as complete units, each of one small package, logic circuits of great complexity and data stores of considerable capacity, complete processors may be constructed with very few units and possibly only one for the logic and one for the data store. Other units may provide for data input and output. With increasing quantities manufactured, the costs of the units approach very low asymptotic values. The quantity almost to achieve the lowest

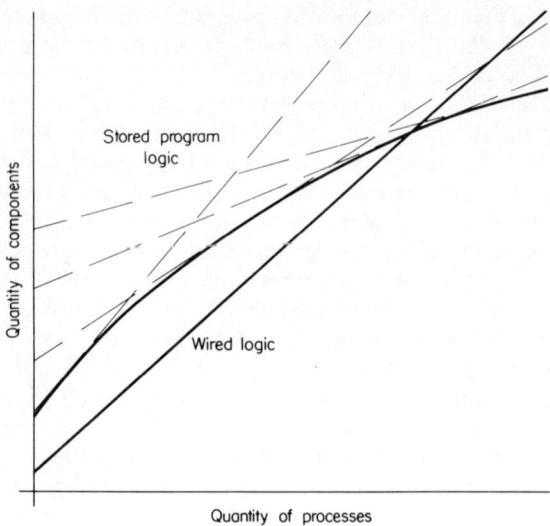

Figure 8.7   Quantities of components required for processors

cost may have to exceed 10^6 units manufactured and sold during a few years, which generally means widespread use over many industrial fields: for 10^5 units the cost per unit may be less than twice the asymptotic value and for 10^4 units the cost per processor may still be well below that for a processor using smaller scale integration units on their lowest costs because of widespread use in many industries. These quantities or whatever they may be at the time of decision, have to be related to the possible demands of an exchange system manufactured over a period of twenty-five years or more to equip possibly more than 10^4 exchanges, and to whether the units are made only for exchanges or for more universal use.

Stored program processors usable in telephone exchanges but not manufactured specifically for that purpose but for general purpose use in many fields comprise major computers, mini-computers and micro-processors. The major and mini-computers have full facilities for variable programming, together with when used in exchanges, security and diagnostics for fault detection and location. The difference, if any, is in the sizes of the repertoire of instructions and the semi-permanent and temporary stores. Micro-processors once programmed can be re-programmed only with difficulty or the substitution of a new program store. They are commonly used in preference to wired logic processors where the difference in processing speeds permits, and the costs are favourable. As such they are analytically equivalent to wired logic processors and hence future references to stored program processors will imply the major or mini type. The descriptions and discussions concerning wired and stored program logic exchange systems in the ensuing chapters, have to be seen against the development over a period of more than twenty years of processors constructed with components manufactured individually or with small scale integration, the costs of the processors being related to the quantity of components curves of Figure 8.7, through mini- to micro-processors using increasing scales of component integration and of costs related more to the scales of production than to those of quantities of components.

Wired logic processors are designed to execute specific tasks and can execute no other except by change or addition to the equipment. Stored program processors

Table 8.2 Exchange line data and facilities

Type of line	Semi-permanent class of service				
	Call charging	Calls barred	Dialling	Authorized facilities	Preference
Spare	Measured rate	None	Rotary dial	Abbreviated dialling	Category 1 2 3
Ordinary & p.b.x. p.a.b.x. with d.d.i. Coin-box	Flat rate 1 2 Ordinary Business	l.d.b. i.b.	Press button v.f. arythmic code Disabled sub.	Transfer pre-arranged follow-me i.c. sub. control	
Pay station Party	Freefone Midnight line Private meter			Enquiry Call waiting Conference Hold i.c. call	
Service					
Videophone High speed data Facsimile					

Temporary state of line controlled by		Common facilities
system	subscriber	
Busy	Transfer in operation	Immediate ringing
Parked	pre-arranged follow-me	Delayed ringing Ring back
t.o.s.	Party line	Re-ring
	X party call	Early morning call
o.b.	Y party call	Reversed charge
s.o.		a.d.c.
s.s.o.		Manual hold
s.v.i.		Trunk offer
Meter record		

have to be designed or chosen to suit their particular uses efficiently and economically but thereafter the processes which they execute depend on what is written in programs, an attribute which gives rise to a number of characteristics collectively known as flexibility. Also, processes can at any time be added up to the limit of the program storage without difficulty of design or of decreasing operational reliability. Because of practical difficulties of circuit design and fault liability, the limit to the processes which one wired logic processor can be designed to execute is less than that of a stored program logic processor. The total processes must therefore be divided among wired logic processors differently to achieve the same result as with stored program logic: which is always possible, but because of the difference in processing facility, exchange systems using only wired logic tend, as in the past, to restrict the facilities and services offered, whereas systems based on stored program logic exploit their processing power by extending existing and offering new facilities and services. Notwithstanding all of this, once an exchange system has been designed and installed, there are some processes which are relatively simple and will not change because they are tied to equipments or operations basic to the system. For these, wired logic is not at a disadvantage and may well be desirable because of speed of operation or of simplicity and low cost, and may be used in conjunction with stored program logic where that is preferable.

Flexibility and processing issues will become clearer with description of stored program logic systems in chapter 9 and electronic systems in chapter 10. The scale of the problem is illustrated by Table 8.2 which is a typical but not complete list of classes of service, states of line and facilities requirements of modern systems, to be compared with Table 4.1 for the first automatic systems. The data are given in columns such that one or more in each column may apply to any one line, except for some obvious contradictions such as X and Y party data for non-party lines, and from which the large number of combinations of the data for all the lines on an exchange can be appreciated. The entries in the table are self-evident from previous descriptions except for the following. The authorized facilities are those for which an extra charge is made and which must be denied to lines for which the charge has not been paid. Lines for which special rates have been arranged for the use of the system during the night when the traffic is at its lowest, are distinguished as midnight lines. Preference categories are invoked when due to service breakdown or other emergency, an exchange is unable to carry all the traffic offered and preference has to be given to police, fire brigade, doctor and other essential services.

Chapter Nine
SEMI-ELECTRONIC EXCHANGES WITH STORED PROGRAM LOGIC

9.1 General

The semi-electronic exchanges of chapter 7 using only wired logic are in this chapter developed to exchanges using stored program logic to the maximum extent, as a means of appreciating the problems and solutions, and the similarities and differences involved with the two kinds of logic.

The starting point for the application of stored program logic is the generalized system of Figure 7.10 in which all the logic except that of the path search and connection processors including the markers, is located in time division time shared common processors which, theoretically at least, may use wired or stored program logic. It was remarked with reference to Figure 7.10, that the cost of data sm transmission equipment such as v.f. to d.c. signal transducers, may compel some space division time sharing. If the trunking method of Figure 6.4 is used to achieve space division of such data transmission equipments, the control processors for those equipments may be time division time shared and it is not difficult to interpret Figure 7.10 on that basis. Arythic coded data being loop and therefore s signalling has no such difficulties. Because of the capacity of a stored program processor to execute many different programs provided that they can be organized for one at a time execution by an executive program, a unique common processor is possible. Figure 9.1 is Figure 7.10 redrawn with a unique common processor except, for practical reasons which will become clear later, for an auxiliary common processor specifically for exchange line scanning and connection.

Referring to Figure 9.1, the line equipments and their interface switch may be as in Figure 7.1 together with an auxiliary processor which for originating calls may include the wired logic equipment of Figure 8.4 or the stored equipment of Figure 8.5. The equipments of Figure 7.9 would satisfy the or.j. and i.j. junctors, the processing of the processors $P1'$, $P1''$ and $P1'''$ being provided by programs in the unique common processor with stored program logic, the programs being executed one at a time and not in parallel as with wired logic. That for the register processing, $P1''$, including as it does directory number reception and transmission for which avoidable errors are inadmissible, has first priority and requires a cycle period of not more than about 15 msec, as stated in section 7.5 with reference to Figure 7.9. The program for the $P1'$ processing may have a cycle period of 0.1 sec and that for the P1''' processing be acyclic and fitted in with low priority because delays of a second or so for that processing would be hardly noticed by the callers. The processing for the t.j., o.j. and s.j. junctors may similarly be transferred to time shared common processors with stored program logic. Transfer to common processors with wired logic was explained with reference to Figure 7.10 to be of doubtful economic advantage. The economics would be still more doubtful with many processors as in Figure 7.10 but all with stored program logic but possibly not with one common processor with stored program logic. The cost of equipment for data transmission between the common processor and the individual units and of bi-stable memory circuits in the units would still limit the economic advantage to something small, but the facility and flexibility of stored program logic would be of operational advantage to the service junctors because of their variety and generally greater logic content.

One processor with stored program logic and a suitable executive program being theoretically possible and economically desirable over several processors to achieve the same result as was indicated with reference to Figure 8.7, it is also technically possible but with difficulty if the same wired logic methods of path search and connection processing and data transmission as are used with wired logic processors, are continued. As was demonstrated in chapter 7, wired logic processors operate very quickly and with different programs simultaneously, which means that generally they are not pressed for processing time and may frequently be simplified by consuming processor time ineffectively: for example by waiting for executive operations to be completed, by performing many simple operations instead of fewer more complicated ones and so on. By contrast, stored program processor processing is slow operating and with programs necessarily in series in time, and to keep up with real time events must be continuously effective. Hence programs are frequently complicated by operations to save processor time which would otherwise be ineffective. Specifically in the present instance, a stored program logic processor having marked the start and the possible terminations of a connection through electro-mechanical switches, cannot wait as a wired logic processor can, for markers or other means of path search and connection to operate even if it takes only a few milliseconds, not only because of the loss of effective processing time but also because of the higher priority programs with repetition times of 20 msec or less which might interrupt the connection processing. Therefore the MK and group marks and any other data generated by the processor for call connection operations of all kinds, must be continued by bi-stable memory equipments adding to the cost of the structure, in order to free the processor for effective use elsewhere. Figure 9.1 shows the additional memory equipments Mk in the marking leads. The processing is also increased by the necessity for the processor to return to confirm that connections have in fact resulted from such operations, which means that the processor must itself keep a record of the operations which it has started in order to be able to come back to them. If a connection is made, the next stage of processing has to be commenced. If no connection is made, either another attempt must be undertaken or busy or other tone sent to the caller. Thus the enlargements of the memory and programming required begin to be formidable. There are further

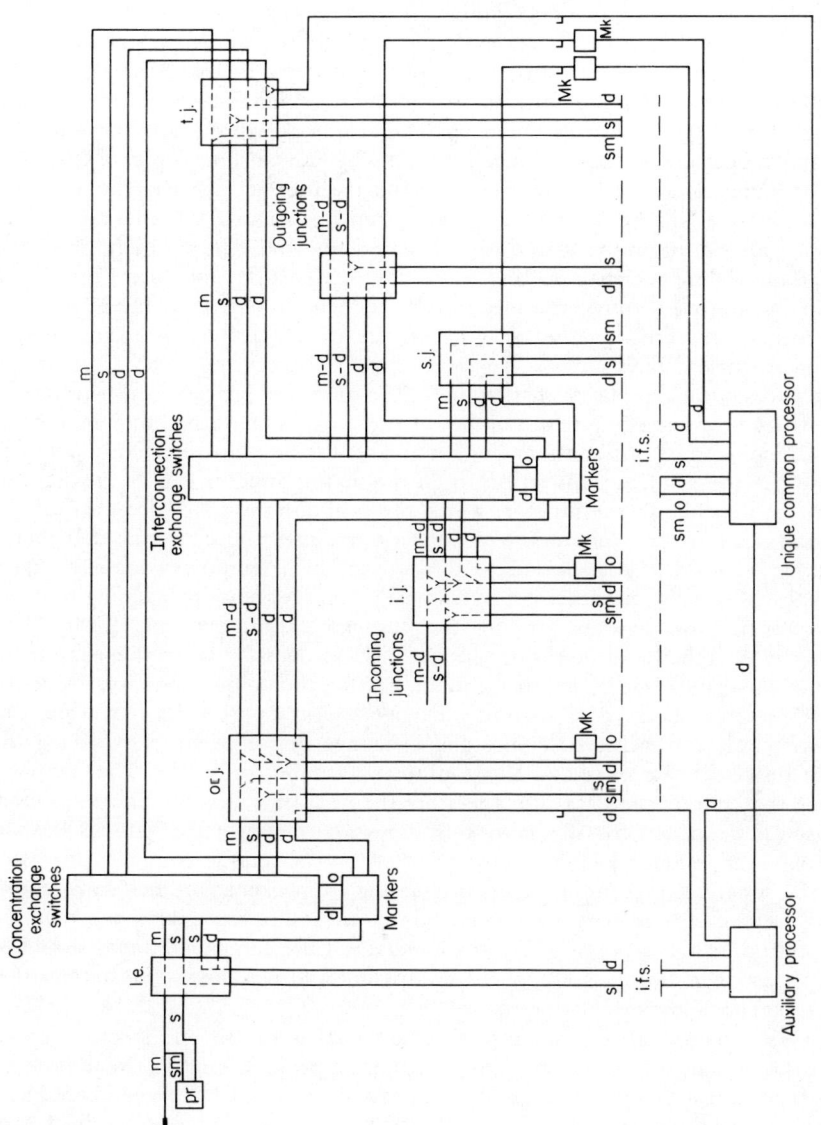

Figure 9.1 Generalized semi-electronic exchange with stored program logic

difficulties with exchange line originating and terminating connections. Scanning in principle if not in detail as described with reference to Figure 7.1, is essential for detecting new originating calls: to mark a line to be connected and to maintain the mark while it is being connected, either the address generator LA, the interface switch and the common operational mark must be held the whole time or every line must have its own MK memory to maintain the MK marking while connection is taking place. The second is contrary to the idea of common processing and too expensive, hence the first is assumed and indicated in Figure 9.1 by an interface switch and processor for the exchange line scanning and marking separate from and auxiliary to the unique common processor and its interface switch. The auxiliary processor operates in the same way as the systems of Figures 8.4 and 8.5 to find lines which are calling and to wait until they are connected. The auxiliary processor having found a line which is calling for a new connection, marks the line by a datum d transmitted through the interface switch, and it marks a group of junctors or.j. or other according to the class of service of the calling line. The unique common processor may not at the same time be executing a program which involves connections to the exchange lines. The auxiliary processor also transmits call data including the equipment number of the calling line to the unique common processor, and it waits until either a connection results or a time-out period expires, then resumes the scan. A connection to an originating junctor causes the state of the loop L to change in the junctor, which the common processor detects during a supervision scan of the junctors. That processor stores the call data then being received, in a word $M1'$ allocated to the junctor in a temporary memory as in Figure 7.9, and causes the junctor to assume the engaged state by sending to it the engaged datum, the junctor sending the engaged datum over the P-wire path through the switches and releasing the auxiliary processor. With the equipment number of the calling line stored in the junctor store, the unique common processor can write metered charge data for the exchange line in a part of its own store, and it can write pk and other data into the exchange line data store by giving the auxiliary processor the data and the equipment number of a line. The auxiliary processor than enters the data into the store the next time that the line is scanned, much as was described for the same purpose with reference to Figure 8.4. Calls originating over incoming junctors are detected by the unique common processor during the supervisory scanning. Calls originating from exchange lines with an s.o. or s.s.o. class of service may be connected to an equipment giving the required service and in the $M1'$ store of which the line data of the connected line is stored. A new call may then be initiated by the equipment and use the stored data as if it were the exchange line originating the call.

When an originating or incoming junctor comes to be connected forward through the exchange switches, the unique common processor marks the junctor by a mark datum through the interface switch, the mark having to be remembered by a store Mk until the connection is complete. It also marks a group of outgoing o.j. or service s.j. junctors by a group mark which must be remembered; or if it is an exchange line terminating call, the processor group marks suitable terminating junctors t.j. and gives to the auxiliary processor the directory number of the called station, the exchange lines then being scanned and one connected if possible as described in section 7.2 for the wired logic system of Figure 7.1. Whatever connection is initiated, the common processor has to return to the call to check its progress and to continue with its processing.

The system described with reference to Figure 9.1 is a practical application of stored program logic processing without making any significant advance on the wired logic original. The most important advances which semi-electronic make over electro-mechanical systems are operational in respect of exchange line storage and processing and economical in the use of time division time sharing, and in both respects wired logic processors are adequate. Indeed the auxiliary processor of Figure 9.1 may easily use wired logic and the cost of the unique common stored program logic processor be greater than that of distributed wired logic processing, not only because of the relative costs of the processors but also because of the high cost of data transmission between the circuit equipments and the common processor. For these reasons and in order to gain advantage, the methods of number translation, data transmission and call connection processing are generally different from those of exchanges with wired logic processors, and adapted to the greater storage and processing capabilities of stored program logic, as will be described in the following sections.

9.2 Number Translation

9.2.1 Trunk

Trunk translation which identifies, from a received decimal directory number, a group of junctions or service circuits to a free one of which the line providing the number is to be connected, was illustrated with reference to Figure 5.13. The identification after one, two or more digits requires a special solution such as that of Figure 5.13 to be worked out individually for each exchange. The end result for cross-bar and semi-electronic systems so far described, is the operation of a route relay which marks all the circuits in the relevant group and communicates call data such as the rate for the call to equipment units requiring the data. Electronically the solution may be implemented by gates and circuitry analogous to the relays and circuitry of cross-bar exchanges or up to the equivalent of a route relay to be operated, by a table in memory as illustrated by Table 9.1 and the description about to be given. The table in memory method is possible with wired logic but more suited to stored program logic which gave rise to its conception.

Table 9.1 is based on the particular case of Figure 5.13 and is the first part of a much larger whole which provides a translation for the first three digits of a received number whatever the digits may be. The special solution of Figure 5.13 reduces the possible 1,000 translations to 721 many of which are in groups all the same and with one output translation per group, the translation implying one route relay and set of call data. The translations and output translations are reduced to still smaller quantities in smaller exchanges with fewer junction routes to other exchanges and fewer local services. The table occupies a section in the exchange semi-permanent store, the words of the section providing the translations and other data. Each word is identified by an address of sixteen digits in three groups as indicated in Table 9.1. The first group, of six digits, defines the section of the data store allocated to translation words. The digits are written by the executive program into the address store of the data store at the start of the translation program and remain unchanged during that program. A group of four digits of the address are the binary equivalent of a received decimal number digit from 1 to 10, the received number 0 being treated as 10. Of the remaining six binary numbers provided by the four address digits, 0000 is used, as will be described, and five are spare, which means that the words at those addresses are wasted. The six address digits between the section and the decimal number digits define a block of sixteen words addressed individually within the block by the four decimal number digits. The processor writes the first block address, 000000 in the present example, into the address store of the data store at the start of the translation program during which the processor changes the address by a change of decimal number digits only or by changing the block as well as the decimal number digits.

One digit of the translation words is a parity digit of such value, 0 or 1, that, added to the sum of all the remaining digits, makes the total odd. Another digit is an indicator which shows according to its value 0 or 1 that the decimal digits received are not or are translatable, respectively.

The stored words corresponding to the table addresses are of four kinds, indicated in Table 9.1 by the remarks in the remarks column. Those words having as decimal number address digits the binary digits 0000, signify that too few decimal digits have been received to constitute a translatable number and another digit is awaited. Those words corresponding to translatable numbers have, reading the translation words in Table 9.1 from right to left, six digits to define the route relay to be operated and eight digits for call data if any. A third kind of word provides a new block number and 0000 as the decimal number digit and which the processor writes into the address store of the data store in substitution for the existing address digits. The words written at the spare addresses are a fourth kind and being all zeros, if read into the processor are indicative of a fault.

With the block digits at 000000 at the start of the program, to translate the decimal digits received for a particular call and held in store by the processor, the processor writes into the decimal number digit of the address the binary digits which it holds in store for the first received decimal digit. If no digit has yet been received, the binary digits in store are 0000, which added to the block number produces a word out of the data store which is all zeros except for the parity bit which is 1. The processor interprets such a word as the end of the translation program for that call, at least one more received digit being awaited before the call processing can proceed. In the example of Table 9.1, all calls starting with the digit 0 are long distance and have to be routed to an exchange of higher level for completion and supervision. If the first digit received is 0, the word read out of the translator includes the indicator digit 1 and digits to specify that the route relay to be operated is number 010011. If the first decimal number digit is 1, the word read out of the store includes a new block number 000001 which the processor substitutes for the existing block number in the address and then adds the binary equivalent of the second received decimal digit to form a new address for reading the translation store. If no second digit has been received, the translation indicates as much by being all zeros with 1 as a parity bit. A second received digit produces a final translation including the number of a route relay to be operated and that completes the program. Similarly if the first digit is 9, a new block number not shown in the table is specified and a second digit received produces a final translation. If the first digit is 2, it produces a new block number 000010 to which the addition of any received second decimal number digit produces yet another block number. A third received digit added to a second block number is translatable. In this way translations for all translatable numbers are produced.

Still using the wired logic path search and connection processors of Figure 9.1, the route relay digits are decoded to operate an Mk remembered mark for a group of junctions or service junctors. The call data digits defining charge rates and other information pertaining to the call being made, are decoded to one or more separate data and distributed each to the junctor for which it is to be effective, by storage in the words allocated to the junctors in the unique common processor store.

It will be realized from the example given that the devising of a trunk translation table is mainly a problem of a system of addressing the translation store to minimize its size. Four binary digits to store each decimal received digit, which is binary-decimal storage, wastes (as described) five out of every sixteen words but at a cost which is not important relative to the cost of the whole exchange. Exchange line translation tables having many more times as many entries, the cost of wastage becomes significant and requires programming ingenuity to reduce it. There are in addition many other problems in constructing an exchange line translation table, as will be seen in sub-section 9.2.2 to follow.

To increase the advantage of stored program logic, the transfer of processing from equipment in the structure to the common processor may be continued. A simple first step is the extension of the trunk translation table so that instead of specifying a route relay to be operated, the route relay when operated marking the circuits of the route, the table provides the equipment numbers of the circuits. This can be seen from Table 9.1 not to be difficult. The call data of the decimal number translations remain unchanged but the route relay numbers become new block numbers. A translatable number being received, the call data applying to all the lines in the group are read. The six other digits of the translation word constitute a new block number which followed by the digits 0000 forms an address at which is written the equipment number of a circuit in the group. Adding one unit to the address produces an address at which the equipment number of a second circuit is located; another unit added to the address produces another equipment number and so on to the end of the list for that group if it comprises fifteen circuits or less. If more, the last address in the series provides not an equipment number but a new block number with which to continue the series. The most useful and general method, and the one assumed in later descriptions, uses a table of equipment numbers. Each word in the table provides an equipment number e.n. and digits implying either another address in the same table or the end of search. Data

Table 9.1   Trunk translation

Address: section	block	decimal number digit	Translation: parity bit indicator bit translation	Decimal number translated	Remarks
010101	000000	0000	1 0 0000000000 0000		Another digit awaited
		0001	0 0 0000000001 0000		New block number
		0010	0 0 0000000010 0000		New block number
		0011	0 0 0000001101 0000		New block number
		0100			New block number
		0101			New block number
		0110			New block number
		0111			New block number
		1000			New block number
		1001			New block number
		1010	1 1 00000000 010011	0	Route but no call data
		1011	0 0 0000000000 0000		Spare address
		1100	0 0 0000000000 0000		Spare address
		1101	0 0 0000000000 0000		Spare address
		1110	0 0 0000000000 0000		Spare address
		1111	0 0 0000000000 0000		Spare address
	000001	0000	1 0 0000000000 0000		Another digit awaited
		0001	0 1 00000000 110101	11	
		0010	1 1 00000000 001101	12	Route relays for
		0011	1 1 00000000 000111	13	services, no charges
		0100	0 1 00000000 011011	14	or call data
		0101	0 1 00000000 110110	15	
		0110	1 1 00000000 110111	16	
		0111	1 1 00000000 000001	17	
		1000	1 1 00000000 000010	18	
		1001	0 1 00000000 000011	19	
		1010	1 1 00000000 000100	10	
		1011	0 0 0000000000 0000		Spare address
		1100	0 0 0000000000 0000		Spare address
		1101	0 0 0000000000 0000		Spare address
		1110	0 0 0000000000 0000		Spare address
		1111	0 0 0000000000 0000		Spare address
	000010	0000	1 0 0000000000 0000		Another digit awaited
		0001	1 0 0000000011 0000		New block number
		0010	0 0 0000000100 0000		New block number
		0011	1 0 0000000101 0000		New block number
		0100	1 0 0000000110 0000		New block number
		0101	0 0 0000000111 0000		New block number
		0110	0 0 0000001000 0000		New block number
		0111	1 0 0000001001 0000		New block number
		1000	1 0 0000001010 0000		New block number
		1001	0 0 0000001011 0000		New block number
		1010	1 0 0000001100 0000		New block number
		1011	0 0 0000000000 0000		Spare address
		1100	0 0 0000000000 0000		Spare address
		1101	0 0 0000000000 0000		Spare address

Table 9.1 (continued)

Address: section	block	decimal number digit	Translation: parity bit indicator bit translation	Decimal number translated	Remarks
		1110	0 0 0000000000 0000		Spare address
		1111	0 0 0000000000 0000		Spare address
	000011	0000	1 0 0000000000 0000		Another digit awaited
		0001	1 1 01010010 101000	211	Route and call data
		0010	1 1 01010100 101000	212	Route and call data

processing produces an address in the table — at the address will be found an equipment number and another address at which there is another equipment number and an address, and so on until the end of search is encountered. The equipment numbers of a group of circuits of any size can thus be written into the table, into words which can be located at random in the table, the members of a group being linked by the arbitrary addresses accompanying the equipment numbers: changes to the equipment numbers or to the size of a group are made merely by writing new data into the table. One store common to equipments of all kinds suffices without the need for division of the store words into blocks which invariably causes some wastage of words, but at the cost of words made longer by having to include an address as well as an equipment number. When the method is used for trunk translation, the translation table provides call data for the group of circuits one of which is to be connected, together with not a route relay identity, but an address in the table of equipment numbers to define one of the circuits, others being obtained if needed from the table by further processing.

Circuits to be connected being themselves connected to trunks from the exchange switches, the equipment numbers ascertained from the e.n. table, are those of the terminals of the exchange switches to which the trunks are connected. Having obtained an equipment number, it may be used to address an interface switch, as in Figure 7.1, through which the state of the circuit, free or engaged, can be determined from the state of the P-wire of the trunk and the trunk marked if free. If engaged, another equipment number may be obtained if there is one, and tested as before and so on until a free circuit is found or none remains to be tested. In this way a possible terminating point is found and marked by the common processor through the interface switches instead of by a route relay. Memory, either individual to the marked circuit or via the interface, is needed to hold the marking while path search and connection is taking place, for which the processor cannot wait but must return later to determine the result of the marking. If no connection is found to have been made because a free path does not exist, another possible terminating point may be determined and be tested and if necessary others until a free path is encountered; or all of the possibilities having been exhausted or a limit placed on the number of attempts to find a path having been reached, busy tone is sent to the caller.

Compared with route relay marking and guide wire search and connection which together test all the possible paths to all the possible destination circuits simultaneously, both the time taken and the processing are greatly increased by the use of a table in memory to define the possible terminating points and to find one which is connectable, but the method has some advantages. Route relays and their electronic equivalent have each to mark via decoupled terminals the junctors of a group which may number one hundred or more. Each junctor has to have an individual wire to a terminal and the wiring changed if the trunking is changed. The memory table method has operational and constructional advantages in that all the interface, addressing and rack wiring is completed when the exchange is installed, much of it having been made in the factory, and the needs of individual exchanges are met by the data written into the memory tables and easily changed in the same way. Also, when all destinations are tested simultaneously, the one finally connected is not known until the connection is complete and additional circuitry or processing is necessary for its determination if the data are needed, as they are in some systems, for later processing. With the table method, the terminating circuit can only be the one marked for connection.

9.2.2 Exchange Line

Exchange line translation differs from trunk translation in that all the digits of the directory number have been received when the translation table is consulted, because the quantity of digits needed is known for the exchange. For every number within the range implied by that quantity a translation is required to determine if it is a working exchange directory number and if so to define class of service or other line data and the equipment number of the line or lines on the exchange with that number. The requirements are clearly satisfied by a table with directory numbers as addresses and words each giving line data and an address in the table of equipment numbers. Groups of lines, most being p.b.x. groups, are in this way accommodated with full flexibility of size of group and location of the lines on the exchange switches. This method is assumed in later descriptions but others are in use for economy of storage or to suit other parts of the exchange system. Many stations, those of private residences particularly, will never need more than one line or any special class of service. For these a group class of service suffices and a block of directory numbers may be allocated and the lines connected to the exchange switches with equipment numbers which can be easily inferred from the directory numbers; in effect the directory and equipment numbers are equated as in the earliest systems. Other blocks may be allocated to groups of lines of different maximum sizes, the translation table giving the terminating call data for the group and the equipment number of one line, the other lines of the group being in the same vertical file as in cross-bar systems or otherwise implied by the trunking which is constrained in consequence. There is in addition the method previously described of giving the directory number to an auxiliary processor concerned with line scanning and leaving that processor to complete the processing, using directory numbers read out of the line data store as the lines are scanned, the translation table being in the line data store. This method is dependent on the use of an auxiliary processor which is a constraint on the system design which in some systems it is an advantage to avoid, and it is also dependent on the system having time to wait for the scanning cycle to be completed.

An equipment number of a line with a specified directory number having been obtained and the terminating call data, it must be determined if the line is available for connection. If not and there is another line in the group, the equipment number of that line is obtained and the availability of the line for connection determined and so on until either a connectable line is found and processing to connect it is commenced, or if all the lines in the group are busy or unconnectable for some reason, busy or n.u. tone has to be connected. As call connection processing of originated calls also starts with the equipment number of a line, the call connection processing of terminating calls has much in common with that for originating calls and the same equipment may be used for both — for example the equipment of Figure 7.1 used as the auxiliary processor of Figure 9.1. Referring to Figure 7.1, if the address generator LA is set to the equipment number of a called line, the line data including those for terminating calls may be read out of the LDS store, and the state of line determined via the interface switch. If the line is not connectable, another equipment number, if there is one, is used to address another line and so on. If no connectable line is found, the unique common processor marks a group of busy or n.u. tone junctors to which the calling line becomes connected. A connectable line being found, its address is held in the scanner to continue to apply the MK mark datum, a group of terminating junctors is group marked and the path search and connection processors associated with the switches are left to make a connection if one can be made, the markings also being held in store as before to leave the common processor free to operate other programs.

By means of translation tables and the associated processing, some wired logic equipment previously required as part of the structure is transferred to the stored program unique common processor. The consequence is an increase in the program to be stored and in the processing and for which the time taken may be more important than the cost. To reduce the searching time to find a free line in a line group of lines of any kind, often the lines are divided into sub-groups with search limited to a sub-group chosen by rotation or in a more or less random fashion.

9.3 Map in Memory Path Search

It is evident that much of the difficulty of path search which has been described is due to the relatively slow speed of operation of electro-mechanical equipment in the processing, and that these difficulties may be expected not to occur in electronic systems. Specifically if the switching is fast enough, marks do not have to be stored so that the unique common processor can operate elsewhere, and the processor program is simplified in that it does not have to include a memory of instructions given so that it can go back after an interval to see if they have been carried out. At least some of these advantages are available to semi-electronic systems by using, where that is possible, electronic instead of electro-mechanical components, for example the electronic version of guide wire path search and connection described with reference to Figure 6.5. If the unique common processor operates with stored program logic, usually and for economic reasons as much as possible of the processing is performed by that processor in preference to distributed wired logic in the structure. Map in memory path search is a common processor program theoretically able to replace any of the wired logic path search methods but with some practical limitations.

Referring to Figure 9.2 which is the C, D, E switch trunking of Figure 5.3 and typical of cross-bar and semi-electronic exchange trunking, path search starts with the marking of one trunk input to a C switch. That switch has output trunks one and only one to every D switch, indicated in the diagram by full lines for the trunks concerned. Let the processor access through the interface the P-wires of the output trunks and write into a store, shown above the switch, the condition of each, free = 0 and engaged = 1. The P-wires may be tested in parallel and the results written into a store all in one operation, and because the switch outputs are fully available to all the inputs, the marked input may be connected to any output trunk which is free. Starting from one marked junctor connected to a trunk output from an E switch, and looking toward the D switch, the conditions are the same and into a second store the free or engaged state of all the input trunks to the E switch may be written. The two stores each having one digit for each of the D switches, comparison of the digits in corresponding pairs shows which of the D switches contains a crosspoint on a free path between the marked start and termination trunks: in the diagram it is the centre D switch. Choosing one of the D switches if there is more than one free path, the co-ordinates of the cross-point in that switch and of the cross-points on the chosen free path through the C and E switches, follow very simply from the stored data. The processor supplies the co-ordinate data via the interface switch as operational data to the relevant switches to close the selected cross-points. The similarity between the path search described and the processing in Figures 5.7, 5.8 and 5.9 which ends with the selection of one of the free paths by the one-only selector o.o.s.3, is not difficult to follow. Nor is it difficult to see that search through any number of switching stages could be programmed and, perhaps less easily, that any pattern of cross-point connections and trunking could be used to produce the same result as marker or guide wire search, but that the processes, processing and the storage requirements increase rapidly with departure from full availability switches, systematic trunking patterns and search and connection operations through a maximum of three switching stages in one operation.

In the systems from cross-bar to Figure 9.1, the path search involves exchange switch wired logic processors supplied with data over marking leads which comprise a fourth wire of the trunks between switches. With the path search accomplished by common processor processing, Figure 9.2, the fourth wires through the switches are eliminated along with the last of the wired logic control equipment, namely the exchange switch marker or guide wire path search and connection processors: they are substituted by unique common processor processing together with the

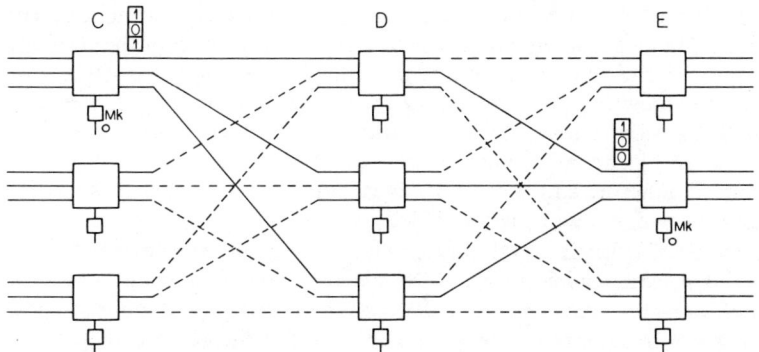

Figure 9.2   Path search by map in memory

transmission, through the interface, of P-wire data in one direction and operational data in the other. In a semi-electronic system, the operational data transmission has to include stores, corresponding in function to the line marking stores Mk of Figure 9.1, to maintain the data until the switches are operated: the transmission must also include transducers to convert the data to power to operate the switches. In Figure 9.2 the stores are indicated by units Mk in the operational data o leads: in practice the stores may be individual to switches or common through interface switches to several switches. As it is the switch operational data which has to be maintained while the switches operate and not the line marking, the need for an auxiliary processor to maintain exchange line marking disappears.

The need for P-wire free-busy data can also be eliminated. Referring again to the C, D, E trunking of Figure 9.2, let every C and every E switch have permanently allocated to it a store into which is written the state, free or engaged, of the trunks connecting the switch to the D switches, and to which the common processor can refer without having to obtain the data by testing the conditions of the P-wires of the trunks. Thus each P-wire is substituted by a binary digit in a store in the common processor which keeps the data in all the stores continuously up to date as it makes and releases connections. When a connection is being made the processor has readily available the data of the connection, including the new states of the trunks to enter into the stores. The conclusion and release of a connection in prior systems results from the removal of earth potential from the P-wire which being continuous through all the units concerned in the connection, causes all the trunks engaged to be released and to test free. With no P-wire data, the common processor has to keep a record of all connections which it makes so that when it comes to release them, it has the data necessary to update the path search data held in store. Thus the processor has in store a map not only of the exchange trunking but also of the states of the trunks and a record of the connections in existence. Specifically the $M1'$ store for an or.j. originating junctor is augmented to include the equipment number of the exchange line which becomes connected to it during originating call scanning and of the junctor which becomes connected during call connection processing. The one-at-a-time terminating points available from the translation table and tested by the map in memory provide the common processor with all the details of a connection being made, for the processor to write the relevant data into the junctor stores. The equipment number of a junctor connected through the switches to an i.j. incoming junctor has similarly to be recorded in the Mj' store of the incoming junctor, and that of an exchange line connected to a t.j. terminating junctor in the $M2'$ store of the junctor.

Map in memory path search and connection eliminates the need for P busy-free and data d marking wires through the switches and transfers processing from the structure, namely path search and connection, to the common control. The processing can be almost if not quite as efficient as the guide wire method typified by Figure 6.5 and using the same trunking, but there is a difference which can be important in electronic systems if not in semi-electronic systems. The guide wire method works without additional complications whether the exchange switches provide full availability or not and with any patterns of connections between any number of ranks of exchange switches in series. This is because the guide wires and their interconnecting gates G1 to G6 in Figure 6.5, construct a map of the trunking and the cross-points whatever arrangement they may make and it enables the trunking efficiency to be improved in some systems. Although theoretically map in memory could achieve the same results, the processing involved may be impractical. This becomes important when the quantities of cross-points and trunks are important and the restrictions of one method lead to greater quantities being needed compared with those of the unrestricted other method. In cross-bar systems the trunking is limited naturally to full availability switches of standard sizes. In reed relay systems there is greater latitude, and in some electronic systems having complete freedom in the cross-point arrangements and trunking, the guide wire method is the only practical way of taking advantage of the latitude.

In prior systems the P-wire serves not only as a free-busy data source but also to hold connections established through the exchange and to provide a transmission path for other data commonly metering. When the P-wire is eliminated, alternative connection holding and data transmission means are needed. That the data transmission facilities of the P-wire are not adequate to the needs of the future has already been indicated. Therefore there is a data transmission problem independently of whether P-wires are used or not and a processing solution not requiring a wire through the switches has been described. For this reason, the ultimate development of exchange control generalized in the system of Figure 9.3 requires trunks comprising only two wires for message m, signal-message sm and signal s transmission. The connection holding and releasing problem remains and requires the switches to be latching mechanically, magnetically or electronically, that is to be operated and released by currents applied for brief periods to change the conditions of the switches which then remain indefinitely in whichever condition they are put, without the expenditure of power if mechanically or magnetically latched. Exchange switches are operated by operational data applied to co-ordinate wires corresponding to the cross-points to be operated: in the system of Figure 9.3 the common processor supplies the operational data via the interface switch and the exchange switches contain storage means md to remember the data and transducers to provide the power. In one system in practice, both the power and the memory are provided by thyristors triggered to conduction by data pulses from the processor and extinguished after a timed interval by quench circuits. The K relays of exchange line equipments have to be latching and similarly controlled, as shown by the md mark individual to exchange lines in Figure 9.3. The release of latched switches depends on their individual characteristics. In the case of co-ordinate operated switches, usually release is dependent on the application of power to one co-ordinate, or to both in a particular order or on the polarity of one of the currents applied. In yet another variation, the cross-points of a connection are not released when the connection is finished. Instead, the operation of any cross-point for the purpose of making a connection through a switch, automatically releases any other cross-point on the same vertical or horizontal co-ordinate.

Electronic exchange switches give rise to a set of problems and solutions which are the same as or analogous to those of electro-mechanical exchange switches including the need for latching which can only be electronic.

Marker and guide wire processors do not contain data stores which may hold wrong information. The faults to which they are subject are mostly in the wiring and the effects are not too damaging. An earth fault on a P-wire prevents the trunk from being selected. If the P-wire is disconnected, either circuitry prevents the trunk from being selected or, if selected and connected, the connection will not hold, and this is detected by the control which attempts to find another path. These are security features produced by system evolution and known to be adequate. Earthed and disconnected guide wires similarly result in trunks not being selected for use. If the data in a map in memory are in error, the errors are not corrected by data from the trunks themselves. The memory of connections made and released must be of the temporary or period kind and thus prone to corruption. It has to be validated periodically by the processor, using a low priority program which, for example, applies release data to all those cross-points and relays that its store shows should be free. In time memory faults if any become corrected. Faults in the switches such that they do not operate or release as instructed have to be detected by check tests during connection processing, a requirement common to all systems. There is also a maintenance aspect. To put a trunk or switch out of service because it is faulty or for other reasons, data have to be written into the map in memory store, whereas with guide wire and most other wired logic path search methods, the mere removal of equipment is enough. Fault location also presents problems for both systems, the balance of advantage if any being problematical. Security adds still further problems. It is unusual to duplicate the whole of wired logic path search and connection equipment. Much of it is simple wiring and equipment with very low probabilities of failure, and faults which occur are not disastrous in their effects. The more complicated map in memory processing concentrated in a unique common processor requires at least duplication of the whole.

Clearly there are profound differences between wired logic and map in memory methods of path search and connection with respect to system operation, speed of processing, in the fault and maintenance and security aspects and possibly in respect of the trunking which may be used. The saving effected by the map in memory method, of structure members by the elimination of marking and P-wires through the exchange switches and of processing equipment associated directly with the switches, is to some extent offset by the costs of making the cross-points latch and of processing the release of connections and validating the memory. Magnetic latching is difficult with reed relays containing more than two reed contacts, which is of no significance so long as two wires through the switches are enough but costly where four-wire switching of message transmission circuits is a transmission requirement imposed on an exchange. Although there are differences, both wired logic and map in memory path search and connection can be designed to be satisfactory in operation and choice depends on which fits in the better with other features of exchange design and operation.

9.4 Signal and Data Transmission

In Figure 9.3 the line equipment circuit processors are on signal paths over which the common processor is able to determine the states of the exchange line loops, and in the circuit processors, cut-off relays if any are operated and released by operational data from the common processor and transmitted over paths which include stores md to remember the data for as long as they are needed for the operations to be effective. The concentration and interconnection exchange switches have only operational data transmission connection to the common processor, the transmission including memory. The originating or.j. and terminating t.j. junctors contain equipment which it is not economic to provide individually to each exchange line in its line equipment. All the junctors include message transmission equipment such as line current feeding bridges, and also loop signal transmission to the common processor. At least some junctors will contain transducers including memory md to convert data from the common processor to tone, ringing or other signals to the exchange lines. If some or all the subscriber stations use press-button v.f. number sending, some at least of the originating junctors will need v.f. signal-message to data transducers. Originating junctors in groups also provide for transmission between the exchange lines and the common processor, of signals for coin-box and party line stations and private meters and possibly some other purposes, the transmission including where needed transducers and stores: similarly terminating junctors are on transmission paths between the common processor and the exchange lines, for ringing and other signals The exchange line signals required are at least those listed in Table 9.2. Those inward to the exchange originate at stations, and are transmitted over the exchange lines to the common processor via the line equipments or junctors to which the lines are connected through the exchange switches. Those outward from the exchange originate from the common processor and are remembered in and transmitted from connected junctors, or, in the case of tone signal-messages, they may originate in other exchanges and be received over incoming junctions to which the exchange lines are connected.

Table 9.2   Exchange line and junction signals and data transmissions

Exchange lines — station to exchange	Junctions — o.g. end to i.c. end
Loop open and closed — call, dial, clear, answer, clear back X party calling Y party calling Press-button digits Coin-box coins	Call Clear Directory number digits 0—9 Supply charge Ring and re-ring Call trace c.l.i. calling line identification Coin-box calls
exchange lines — exchange to station	junctions — i.c. end to o.g. end
Tones — ring, busy, n.u. Ring Ring X party Ring Y party Coin-box control Private meter operate p.a.b.x. with d.d.i. — loop call directory number digits	Proceed to send Directory number digits All digits received Tones — busy, n.u. Answer-meter Answer-non-meter Clear back Release Meter pulse Ring back Plant busy Coin-box coin control Call trace c.l.i.

The signals and data to be transmitted between exchanges over or for the junctions between them are also listed in Table 9.2. The tones may be sent as signal-messages over the message paths but otherwise the signals and data are too numerous to be transmitted over the two wires of message paths without interfering with message transmission. Data links have practically unlimited capacity for data transmission and are shown in Figure 9.3 as providing data transmission for groups of junctions which, having only message transmission to satisfy, are not equipped with junctors to which they have physical connection. Nevertheless for every junction in a group with data link transmission there is a virtual junctor comprising data storage accessed by the common processor in the same way as incoming and outgoing junctors and their stores are accessed, namely by the equipment number addresses of the exchange switch trunks to which the junctors are connected. It may seem that data link data transmission in conjunction with stored program logic causes functions and operations detailed in chapter 5 for the control of cross-bar exchanges no longer to apply, nor some series and parallel divisions of the processing deduced in chapter 7. In fact the operations and divisions can be traced in the stored programs, of which the virtual junctors for junctions are one illustration.

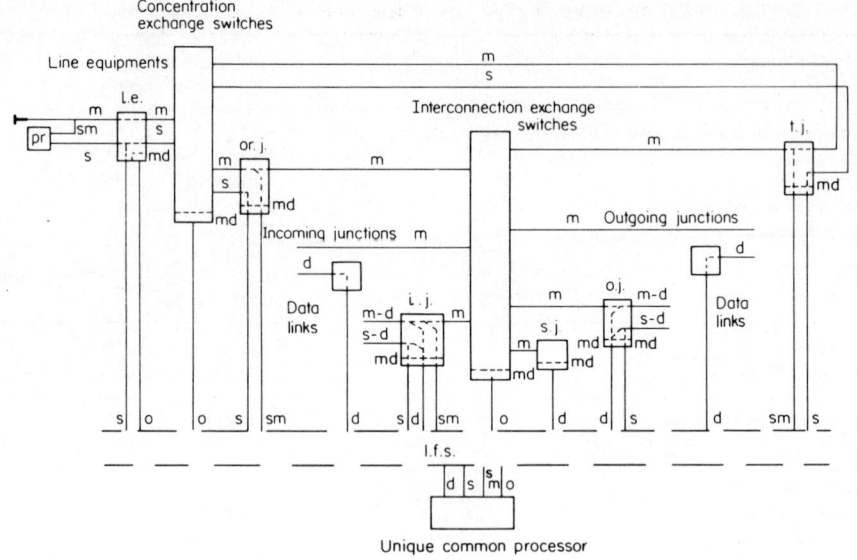

Figure 9.3   Ultimate exchange system with SPC

Data links are the equivalent of manual order wires with the stored program logic processors behaving remarkably like the operators had to. The detailed operation can be different in different systems. That illustrated by Figure 9.4 includes all the essential operations. Each junction of a group served by a data link is allocated a junction number in the group; the number is the same at both ends and is semi-permanently written into the virtual junctor at both ends along with the address of the data link for the group. A data link provides bothway transmission over separate paths in the two directions of transmission, each path having a transmit terminal at one end and a receive terminal at the other. Figure 9.4 shows only one direction of transmission: the two directions operate independently of one another. When data for a call over a junction have to be sent, the common processor is either already processing the virtual junctor for the junction or it has or it can obtain the equipment number of the junction by which it can access the virtual junctor. From the virtual junctor store the processor ascertains the address of the data link for, and the junction number of, the junction. Using this information the common processor accesses the data link transmit terminal and if it is free it writes the number of the junction and the data to be transmitted for the call. These are sent, under the control of the terminal, over the data link at a rate suited to its transmission characteristics, and stored in the distant receive terminal. Even if the transmission characteristics of the link are no better than those of the junctions which it serves, the link can transmit all the data for a large group of junctions and still have time to spare. Therefore the transmit and receive terminals have storage for only one transmission each, independent of the data transmission. When ready to send, the link equipment takes the data from the store to free it to receive another transmission, and thus one transmission can be in process of being received from the common processor while the previous one is still being sent. If when the processor accesses a data link terminal, it finds the transmit store already occupied, it delays the sending until the next cycle of the program which it is at that moment processing.

Associated with each data link receive terminal is a semi-permanent store in which the equipment numbers of the junctions served are written with their junction numbers as addresses. Call data and a junction number being received over a data link, they are stored in the receive terminal store and the junction number applied to the semi-permanent store extracts the junction equipment number. Periodically the common processor scans the data link receive terminals under its control and reads the stores. The period between the readings of each store must be less than the time taken for one transmission of data over the link in order to be sure that the receive terminal store is empty in time to accept the data of the next transmission. When the processor finds a terminal store with received data in store, it uses the junction equipment number output from the semi-permanent store to write the data into the virtual junctor of the junction concerned. As shown in Figure 9.4 the semi-permanent store would be of the wired type. More probably it would be a section in the bulk store of the common processor; the common processor would read the data and the junction number from the receive terminal store and use the information to extract the junction equipment number from the bulk store as a separate operation.

Some junctions may not have data link facilities, or they give only limited service and do not require the facilities or other reason. For these junctions, signal and data transmission is provided over them as m—d and s—d transmissions between o.j. and i.j. junctors. The common processor uses the equipment numbers of the exchange switch trunks to which the junctors are connected as addresses to access the junctors for sm, s and d transmissions and to access their stores for data storage and retrieval. Service junctors and their stores are similarly accessed. The junctors are shown in Figure 9.3 with only d data connections to the common processor

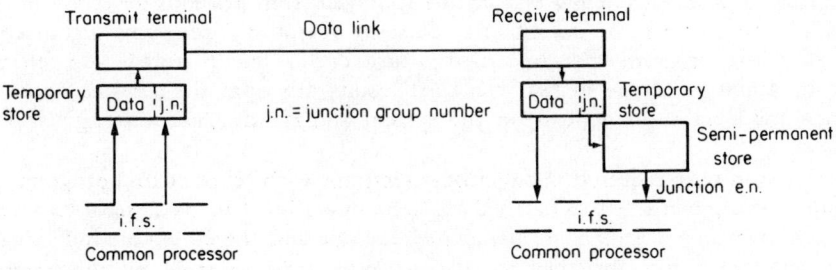

Figure 9.4   Data link data transmission

although some may require s and o connections. The d data from the common processor to the junctors have to be remembered by memory md in the junctors.

The movement of data and their uses within the exchanges are described in detail in the next section.

9.5 Operation of Ultimate SPC Exchange

9.5.1 Structure and Data Stores

Exchanges with control by unique common processor using stored program logic and embodying all the features which that form of control makes possible and desirable, are generalized in Figure 9.3: that kind of exchange represents the ultimate of machine exchanges back in principle to the manual exchanges from which they started, namely a minimal structure with maximal common control.

The structure of the exchange of Figure 9.3 comprises exchange switches for the interconnection of message m transmission lines terminated on the exchange, all the lines transmitting signal-messages sm and exchange lines signals s as well, over paths associated with the message paths: the structure further comprises exchange line equipments and junctors providing, in the most economical ways, message line transmission equipment such as common battery line current feeding bridges and for signals in both directions of transmission, via the interface switches, between the unique common processor and the lines. The structure also includes data link terminals for the transmission of data between exchanges over data links and between the terminals and the unique common processor via interface switches. In particular applications some of the processing memory and logic equipment may be located in the line equipments and junctors but in the general case of Figure 9.3 all the processing is assumed to be performed by a unique common processor with access to a bulk common store or stores providing semi-permanent, temporary and period storage of digital data. The system operation consists of the reception by the processor of data including signals from or for the transmission lines, one line at a time, the processing by stored program logic of those data and others retrieved from the bulk store, and the transmission of data from the unique common processor to processors in other exchanges and of operational and other data to the exchange switches, the line equipments and the junctors. The structure and its operation in principle follow and can be followed from previous descriptions. The detailed operations to be described at those of operational programs organized by the executive program and concerned with data movement, storage and retrieval. The programs are basic in that the final results are what the services need: the details and operations depend on the system and the designer, those given being typical.

Data stores are required in functional sections, each relevant to a program, to a group of equipments or to a particular system operation. The data are of two kinds. Those shown in Figure 9.5 are data which are not time shared but are individual to lines and to trunks for their circuit processes, and accessed by the common processor as it is processing data for the circuits. The columns of the table correspond to the series and parallel division of the processing as in Figures 7.8 and 7.9, and are designated $M0'$, $M1'$, Mj' and $M2'$ in accordance with the notation of those diagrams with the addition of Moj' and Ms' for the outgoing and service junctors. The $M1'$ column includes not only the originating junctors or.j. but also n.u., s.o., s.s.o. and any other originating call junctors. The data for junctions without physical junctors are data for a virtual junctor for each junction. The data of the second kind and shown in Figure 9.6 are common to the equipments of Figure 9.5. The data include $M0''$ and $M2''$ for exchange line and terminating call processing and the $yM1'' + m$ and the similar $zMj'' + m$ for originating and incoming junction register processing, and $M1'''$ data much of it in trunk and exchange line translation tables for call connection processing, together with, for path search and connection, a map of the exchange trunking and a record of the map trunks in use. The data of Figure 9.6 also conclude the programs and data incidental to current processing, and tabular information from which call data and equipment numbers of junctors involved in call processing can be obtained. Equipment numbers are derived via call processing which provides an address in the equipment number e.n. store at which an equipment number is stored and also, if a group of equipment numbers is involved, a second address in the e.n. store at which another equipment number is stored and so on, as previously described.

9.5.2 Data Links

Data link processing proceeds independently of other processing other than association via the executive program in which the data link processing has high priority to prevent loss of data as previously explained. The link processing consists of accessing the link receive terminals in cyclic order, reading from each a junction number and data if received since the last cycle and if data have been received, writing them into the store Mj' or Moj' of the virtual junctor defined by the data link and received junction number. The address of the virtual junctor store is obtained from the d.l.j.n. to e.n. translation table shown in Figure 9.6. The data are processed during the next junctor processing cycle which may be new call detection, supervisory or register processing depending on the junctor and its state of call. In Figure 9.3 the junctions without junctors are shown as in unidirectional traffic groups, which means that for every route between exchanges the junctions are in two separate groups for the two directions of traffic. One bothway data link made suitably secure in operation is, however, sufficient for both groups: for junction economy, very often in practice the junctions are in one group operated for traffic in both directions.

9.5.3 New Calls

The processing to detect and connect new calls originated over exchange lines uses asynchronous cyclic scanning similar in principle to that described in conjunction with Figure 8.6. The common processor uses the equipment number address of an exchange line to connect through the interface switch to the line

Section	MO' exchange line	Ms' s.j.	M2' t.j.	Moj'			Mj'		Ml'
				o.j.	o.j.-virtual	i.j.	i.j.-virtual		or.j., n.u. s.o, s.s.o.
Data type: s.p.	Line		c.o.s.-structure e.n.- conc. sw.	c.o.s. structure	e.n.- data link j.n.	c.o.s.-structure	e.n.- data link j.n.		c.o.s.-structure e.n.- conc. sw.
Period	Meter record								
Temporary	State of line		connected e.l.- e.n. path control c.o.s.						Connected e.l.- e.n. path control c.o.s.
			i.c. last look	i.c. last look	i.c. last look	i.c. last look	i.c. last look		i.c. last look
							Charge rate connected junctor- e.n. path		
				Call State of call					

Figure 9.5 SPC circuit processing data stores

Processing			Tabular Semi-permanent	
Semi-permanent	Temporary	Table	Origin of address	Data
Programs	Current processing	Junctor groups – originating calls	e.l. scanning and data processing	Address in e.n. store – or. j. s.o. s.s.o.
Trunking map $M0''$ $M2''$ $M1'''$-part	Map trunks engaged $yMl'' + m$ $zMj'' + m$ $M1'''$-part	terminating calls	e.l. translation	Address in e.n. store – t.j. s.v.i.
		any call	Data processing	Address in e.n. store – n.u.
		Translation trunk	d.n. from $M1''$ Mj''	Call data, charge rate, address in e.n. store – s.j. o.j. o.j. virtual
		exchange line	d.n. from $M1''$ Mj''	Terminal call data, address in e.n. store – e.l.
		e.n.	Any above in data column	e.n. + address in same store
		Translation – d.l.j.n. to e.n.	d.l. receive terminal	e.n. – virtual junctor

Figure 9.6 SPC common processing data stores

equipment to detect the L state of loop of the line and to read out of the $M0'$ store the line and state of line data for the line including the free or busy state which in the absence of a P-wire becomes a digit in the store. If the line is not looped, or being looped is parked or not permitted to be connected for any reason, the processor changes the address to that of the next line, or embarks on a different program. Otherwise if the line is looped but not connected, the processor initiates the map in memory to connect the line to an originating junctor giving the class of service or service such as s.o. or n.u. tone determined as required by the line and state of line data processing. The junctors being provided in quantities according to the traffic and distributed over the outlets from the concentration switches, the data processing gives a first address in the table of equipment numbers from which the equipment numbers of junctors of the required type are ascertained. To even out the traffic distribution, several addresses in the equipment number table may be given as first address and used in rotation. An equipment number contained in the word read out from the table at the given address is not that of an outlet from the concentration switches but of an inlet to the interconnection switches to which the selected junctor is connected, which simplifies the processing after the junctor is connected and is consistent with the equipment numbers of physical and virtual i.j. junctors for which there are no concentration switches. The processor uses the equipment number as an address to read the $M1'$ store for the junctor to determine if the junctor is free or engaged. If it is not free, the processor uses the address in the word read out of the table of equipment numbers to read another word out of the table and thus to obtain the equipment number of and to test another junctor and possibly others until a free junctor is found or none remains to be tested, when the attempt to connect is abandoned and the scan continued. A free junctor being available, its $M1'$ store word contains the equipment number of the junctor on the concentration switches and this is used for the map in memory search for a free path from the calling line to the junctor. If a free path is found not to exist, the processor may find another free junctor and try again for a free path and possibly yet another before giving up and continuing with the scan. Having found a free path, the processor issues the operational data to the line equipment and to the exchange switches to connect the path and it updates the map in memory data to include the new connection. The processor also writes into the $M0'$ store word for the exchange line that the line is busy, and into the $M1'$ store word for the junctor it writes the new state of call namely busy and requiring register processing to commence; it writes data for the exchange line connected, namely its equipment number and control class of service and call data if any, and finally in anticipation of the release of the connection, it writes the path data comprising the identities of the cross-points used unless they can be deduced from the trunking pattern and the identities of the two ends of the connection. That completes the connection program for the exchange line, and the processor moves on to the next line or program. The subsequent processing of the connection made and the processing of new calls incoming over junctions is dependent on the supervision programs for or.j. originating junctors and i.j. incoming physical and virtual junctors, the supervision programs bringing the register programs into operation.

The addresses of the words in the stores containing M1$'$ and Mj$'$ data for originating and incoming call junctors both physical and virtual, can in practice be and are here assumed to be the same as those of the inputs to the interconnection switches to which the junctors are connected, and further that the same addresses are used by the processor to access the physical junctors through the interface switch. If these assumptions do not apply, translation tables have to be provided and used to link the stores with their equipments and the equipments with their connections. For junctors which are not connected to the interconnection switches, such as those for n.u. tone, convenient fictitious addresses are used. The addresses of the words in the stores containing Moj$'$ data for outgoing junction call junctors both physical and virtual and for Ms$'$ sections for s.j. service junctors, are similarly assumed to be the same as those of the outputs of the interconnection switches to which the junctors are connected.

9.5.4 Register Processes

Register processing is concerned with directory number reception from exchange lines as part of the originating register program, and from processors in other exchanges as part of the incoming junction register program. Both programs are initiated for new calls during junctor supervisory programs; such programs have second priority and their cyclic repetition periods are arbitrary, are not necessarily the same for all groups of junctors and are preferably not greater than 0.1 second for any. The originating register program has first priority because it has to receive from exchange lines directory number digits and thus has to have a cyclic repetition period of not more than 20 msec. The period for the incoming junction register program can be longer and chosen to suit data link operation because the digit sending is under the control of proceed to send data from the registers.

Register processing for an originating junctor or.j. is initiated when a supervisory scan program detects that the incoming loop signal $L = 1$ and the state of call requires register processing to commence. The processing is initiated for an incoming junction junctor i.j. when the state of call of the junctor is free and a call datum is being received: for a virtual junctor, the conditions are a stored call datum from a data link and the free state of call. The processing comprises reading the words of the store yM1$''$ + m or zMj$''$ + m, as the case may be, until an empty one is found: if none is empty, the processor goes on to other processing and tries again during the next cycle of the program. Finding a word empty, it writes into that word any call or other data which the register program will subsequently need and into the +m part of the word it writes the equipment number of the junctor which it is processing at that moment. If the call is from an originating junctor or.j., the processor sends sm data to the junctor to apply dial tone; if from an incoming junctor i.j. it causes an m—d or s—d proceed to send datum to be sent to the distant processor: if it is a virtual junctor which is involved, the processor reads from the data word of the junctor the data link address and the number d.l.j.n. of the junction which the junctor serves, it writes the junction number and the proceed to send datum into the send store of the data link, and a digit into the junctor word to indicate that the datum has been sent. It does not write either if the data link send store is already in use and unable to accept the proceed to send datum but leaves both operations to be executed on a later scan. Thereafter and until the register is made empty again, the junctor at the address stored in the + m part of the word will receive register processing as well as supervision, corresponding to both $P1'$ and $P1''$ processing of Figure 7.9.

The originating register program provides for the words in the $yM1'' + m$ store to be read cyclically. A word being read, the junctor address if there is one in the + m part is used to read the $M1'$ store word for the junctor and to access the junctor to sense the incoming data. A change of data since the last cycle being detected, the data are processed, new data written and operational and other data transmitted as determined by the processing before going on to the next register store word or other program. The register processing is mostly concerned with the receipt and storage of incoming directory number digits. The data in the register stored words are similar to those of the stores of the register of Figure 7.2 and the processing of incoming rotary dial digits is the stored program equivalent of the wired logic of Figures 7.4 and 7.5. The processing of received press-button digits is different but the final number storage is the same. One program capable of receiving and processing either kind of directory number sending is used, the kind to be expected for individual calls being either predetermined and given by the line data of the calling line or determined by the first digit to be received. The incoming junction register processing of junctors using the $zMj'' + m$ store, receives data by data link or other means specified in the junctor data: each directory number received is stored and the proceed to send datum or signal is sent to the distant exchange to cause the next digit to be sent if there is one.

It was described for the wired logic processing of Figure 7.9 that the received digits are translated during every cycle of the register program except when the translators are actually engaged in call connection processing and that in consequence the register program is simplified and translation made possible after the receipt of any quantity of digits for each translation independent of every other translation. A processor with stored program logic produces the same result but does not translate the received digits during every register cycle. The translation program is put into operation only when a digit is received complete and stored: for each call the incoming digits are so translated until a conclusive translation results after which the translation program is not executed again except during call connection. Conclusive translation provides route and rate and other call data if any, and also data which the register needs and stores to complete the processing for that call, such as the total quantity of digits to be received or the minimum to be expected for that call, and after what quantity of digits the call connection processing is to be put into operation. When the second quantity of digits has been received, the processor writes the address of the register concerned into a hopper store of calls awaiting call connection processing and which is part of the current processing store of Figure 9.6.

When call connection processing has been executed, a datum to that effect is written into the junctor store. If register processing is then complete as is usually the case for calls terminating in the exchange, register processing is ceased by erasing all data from the register store. If further processing is required, usually directory number sending to a distant register store, the termination of register processing follows the completion of such processing. With the cessation of register processing for a call, supervision remains.

9.5.5 Call Connection

Call connection is executed by a third priority program which starts with the reading of an address out of the hopper store of addresses of registers awaiting call connection. The address is used to read the data stored for the register selected, and the received directory number digits and other stored data are again translated. Translation produces an output from the trunk translation table which specifies either data for a group of junctors and an address in the table of equipment numbers at which the equipment number of a junctor in the group is written, or that some quantity of digits the last to be received are to be applied to the exchange line translation table to find the address of an exchange line. In the first case the word in the table of equipment numbers at the address given is read and the store of the junctor at the equipment number thus obtained is read to determine if the junctor is free. If not free, the address read out with the first equipment number is used to obtain another equipment number for the reading of a junctor store and so on until either a free junctor is found or none remains to be tested. A free junctor being found, its equipment number together with that of the incoming call which the register is processing and is the start point for path search, are used by the path search in the memory program to seek a free path between the junctors. If none exists, another free junctor is found and so on until a free junctor with a free path is obtained, or none existing or search being abandoned, busy tone is sent to the caller. Having identified a free path to a free junctor, into the store of the junctor is written call data if any for the call being connected and into the store of the incoming call junctor is written call data which includes for an originating junctor the charge rate for the call; also written is the updated state of the call and the path data needed for the eventual release of the connection. The processor issues the operational data needed to make the connection effective and then proceeds to the next hopper address or the next program. Connection being made to an outgoing junction, further processing continues with a periodically repeated register program for the interchange of data with the distant exchange. During each cycle of the program the processor reads the store of the incoming call junctor, using the address in the $yM1'' + m$ or $zMj'' + m$ store, and finding from that store the address of the connected outgoing junction, the processor reads the store of that junction and is able to send and receive data for the junction. If the junction is one with an o.j. junctor, the processor accesses the junctor to send and receive data. If the junction is one for which there is a virtual junctor store, the processor reads data received for the junction when it reads the junctor store, the data having been written into the store during a scan of the data link receive terminals: it sends data for the junction by writing the data into the transmit terminal of the data link together with the junction number of the junction, the processor reading the number and the address of the data link terminal from the junctor store.

If the connection is to be made to an exchange line via a terminating junctor t.j., the directory number of the called station is obtained from the received digits stored in the $M1''$ or Mj'' store of the register controlling the connection. The directory number digits having been applied to the exchange line translation table, the data read from the table give call information common to all the lines where there is more than one with the directory number, and an address in the e.n. store at which will be found the equipment number of an exchange line. The processor reads the line and state of line data of the exchange line thus determined, and possibly of others obtained from the table of equipment numbers, to find a free called line if there is one. A free line being identified, its line and state of line data are processed to determine an address in the e.n. store at which the address of a junctor providing the appropriate terminating service is to be found. If that or another junctor out of the table of equipment numbers is free, the $M2'$ store of the junctor gives the equipment number of the junctor on the concentration switches and that together with the equipment number of the exchange line to be connected and the map in the memory data, enables the processor to find a free path if there is one for the connection of the line to the junctor. If no free path data results, the processor will try again at least once more with another free junctor if there is one. Having found a free path from a free called line to a free junctor of the specified class of service, the processor seeks a path from the junctor to the calling side junctor, and only if a complete path exists does the processor operate any of the exchange switches. If a complete path is not found on the first attempt, at least one more attempt is made before giving up and connecting busy tone. For connections which are successfully completed, the equipment number of the exchange line, its control class of service if any and the data of its connection path to the terminating junctor are written into the terminating junctor store and the equipment number of the terminating junctor is written into the store of the incoming call junctor to which it becomes connected.

Whatever connection is instructed, during a subsequent supervisory scan of the calling circuit, the processor should check that the connection has in fact been established. For that purpose, the junctors and the junctions without junctors may be arranged to present a loop to the message path, the loop being detected in the calling circuit junctor if it has one, and communicated to the processor over a d data path. It is a feature of associated message and data including signal transmission that there is a high probability of the message transmssion being satisfactory if the data transmission is, and that the data transmssion is automatically checked by the processing. With data link data transmission, it is possible to process to completion a call over several junctions in tandem only to find no message transmission because of a fault in one of the junctions. There is no easy solution to this problem. One method and probably the best is to use the whole path between the terminal junctors found on all complete connections for a transmission test comprising the sending in both directions of signal-messages, the connection being accepted only if the signal-messages are correctly received. The problem is less difficult when the transmission involves multiplexing as in p.c.m. transmission where there is always some transmission even by channels which are idle of messages.

9.5.6 Supervision

For supervisory processing the incoming call junctors, that is, the or.j., i.j. physical and virtual junctors, are periodically scanned cyclically independently of any other processing and whether or not the junctors are engaged on calls. The outgoing o.j., t.j. and s.j. junctors are supervised only when connected for calls and as part of the supervision program for the incoming call junctors to which they are connected through the exchange switches. Supervision consists mostly of signal, signal-message and data transmissions over or for lines engaged on connections, call accounting and the releasing of the connections as the calls are terminated.

Supervision processing for a junctor comprises reading the data for the junctor out of the relevant store and, for a physical junctor, sensing the data then being received. The data incoming to or.j. originating and t.j. terminating junctors are signals s: those for physical i.j. and o.j. junction junctors depend on the signal and data transmissions designated m—d and s—d. The stored data for the physical junctors include those last look which were incoming during the previous cycle and with which the present incoming data are compared to detect changes since the last cycle. Data incoming to a virtual junctor are written during data link processing and processed during the next supervision cycle. Supervision processing consists of ascertaining if any of the data for a call have changed since the last cycle: if not, the processing is moved to the next junctor or program but if any have, all the data available are processed, new data are written and if the processing requires data to be transmitted to other processors or equipments or operational data to the structure, such transmissions are initiated by the common processor. Signals and data transmitted to physical junctors for transmission over lines have to be remembered in the junctors for as long as they take to transmit. Data to be transmitted for a virtual junctor are written into the transmit terminal of the data link serving the junctor, the junctor store providing the processor with the address of the data link and the junction number of the junction to be written into the link store together with the data to be transmitted.

The store of an incoming call junctor includes digits allocated to store the address of an outgoing call junctor to which the incoming call junctor is or will eventually be connected for the call in progress. Using the address during the incoming call junctor supervision program, the common processor accesses and supervises the outgoing call junctor at the same time. Thus the unique common processor has signal and data transmission and reception facilities continuously with incoming call junctors and with outgoing call junctors connected to incoming call junctors and can relay signals and data from one side of a connection to the other to provide what amounts to transmission through the exchange comparable with transmission through the exchange switches. Signal and sm transmissions to and from stations are limited only by the transmission capabilities of the exchange lines.

With data link transmission between exchanges the data which can be transmitted are practically unlimited.

The supervision signals and data are included in Table 9.2. Most are self-evident but some need explanation. Charge recording is the most important and the subject of the next sub-section. The plant busy datum gives the processor the opportunity where the facility exists, of attempting to set up a connection by an alternative route when the first route is found to be congested. The busy and n.u. tone data allow a required tone to be applied at the originating exchange to free possibly very expensive junction plant which would otherwise be held until the calling party released the connection.

9.5.7 Call Accounting

Call charges can be recorded by message register type metering, by automatic ticketing and by manual ticketing, any or all of which may be found in any one system. For message type metering, a charge rate is required for each call. The charge rate, which may vary with the time of day and the day of the week, is supplied with the translation which results in a connection being made through the exchange, and is stored in the store of the incoming call junctor. If that junctor is an incoming junction junctor and it has received the 'supply charge' datum from the calling exchange and the called line answer signal, then the common processor will either ticket the call or supply back to the calling exchange meter pulse data at a rate determined by the charge rate, for each datum to add one unit to the meter record for the calling line. If the incoming call junctor which receives the charge rate is an originating junctor, then from the time that it receives the called line answer-meter signal, the common processor uses the equipment number of the calling line and the charge rate stored in the junctor to add units periodically to the meter record for the line, which is a period store in the $M0'$ exchange line data store, at a rate appropriate to the call. If the charge rate is not stored in the junctor but meter pulse data are received from a distant exchange, each datum as it is received is stored in the outgoing junctor and subsequently used by the processor to add one unit to the meter record. Meter pulse data communicated over junctions may also be used to collect the charges for operator rendered services, reversed charge calls and generally for any service for which the charge is not determined by the data supplied by the calling line in conjunction with the called line answer signal. If the calling line is a party line, the parties have separate records and the record to be augmented is decided by the X or Y party datum received and stored at the beginning of the call. Meter pulse recording is a first priority program the period of which is a maximum of 0.5 seconds. Supervision is a second priority program of lesser period, 0.1 second being assumed. Meter pulse recording may have its own first priority program cycles or be included in the supervision program with the safeguard that if interruptions to the supervision program prevent part of it from being executed for approaching 0.5 second, the processor will execute, and with first priority, only the meter recording part of the supervision program.

The record of metered charge units in the $M0'$ store uses equipment number addresses to store the charges incurred by the lines individually, as a desirable and even essential feature. The alternative of directory number addresses encounters difficulty with large groups of lines all with the same directory number: the charges for all the lines aggregated on one number store creates difficulties of overloading of the store. Charge record stores can be and preferably are read by remote control from a distant accounting centre which has the data and the machinery to convert the equipment numbers to directory numbers. On the other hand, directory numbers are more convenient for operators, for ticketing, c.l.i. calling line identification and some other purposes. To have both equipment and directory numbers in the same system means two tables to translate either from the other. Translation from directory to equipment numbers is inescapable in order to make connections defined by received directory number digits. Some systems use only directory numbers for all other purposes; the line data store read with equipment number addresses contains the directory numbers of the lines which numbers are stored in the originating junctor stores instead of the equipment numbers. Directory numbers are then available for charge recording, c.l.i. and any other purpose, and the congestion which is liable to occur with digital stores per directory number is usually avoided by magnetic tape storage. Call accounting data are stored on magnetic tape in one of two ways. In one, all the details of a call are assembled in the incoming call junctor store and transferred to tape when the call is successful and completed, the charge having been determined as for meter type recording. In the other, the number, equipment or directory, of the calling line and directory number of the called line are recorded at the end of register processing, and the number of the calling line is recorded if and when the call is answered and if answered, recorded again when the call is cleared. With the second method the tape also contains standard clock time recorded on another track and against which the commencing and terminating times for the individual calls can be compared to determine their durations: and with the call details also available, the cost can be worked out for billing, and details of the calls supplied with the bills if required, by machinery at an accounting centre independent of the exchange which is thus relieved of call charge processing. Tape recording has some attractive features but it is inconvenient in some ways mostly concerned with services which need to know the cost of calls, even low cost calls, as soon as they are incurred, which applies to pay stations and lines with private meters or coin-boxes.

Magnetic tape recording presents few difficulties at high level exchanges concerned only with long distance calls and operator services for which bulk billing may not be acceptable or even possible for very expensive calls. The call data given to the high level exchange to make a connection has to be supplemented by the identity of the calling line given with the call data or in response to a c.l.i. datum sent back to the originating exchange by the high level exchange. The c.l.i. demand is also made for operator call tracing and allied operations for which only the directory number is satisfactory identification of the exchange line at the end of the connection; and the line may not be the calling line as called lines need to be identified sometimes. On the other hand, call trace as a separate operation is a maintenance and fault tracing aid very similar to c.l.i., but for which equipment numbers are required not only of exchange lines if involved but also of all the equipments in a connection. Thus the operations involved in charge recording, operator and administrative control and maintenance facilities are inter-related and the arrangements adopted are a matter of agreement between the administration and the system designer.

9.5.8 Other Programs

The programs and operations which have been described are those having the highest priorities and taking up most of the time of the unique common processor. Service junctors, each group of which requires its own program, are of the same importance but have not been described in detail. Service junctors resemble in their characteristics, as was previously mentioned, either exchange line or junction junctors according to their functions. Other programs such as those concerned with routine resting and fault diagnosis are used only exceptionally as the need arises or have a low priority and become used only during periods of low traffic intensity.

The problem of satisfying programs of different priorities, different repetition rates and taking different times for execution, was discussed in section 2.13. If the processing is fast enough for it to have time to spare, as it may be for an electronic system with wired logic, the problem may be solved using only clock generated pulses to interleave in time the various processes as in Figure 2.14. If the processing is slower and pressed for time, the programs have to be arranged as is common with stored program processing, in a priority hierarchy imposed by an executive program. The operation of the executive program is governed by clock timed 'interrupts' to control the repetition rates of the programs except as they may be modified by interruptions of programs of lower level in favour of those of higher level in the priority hierarchy. Generally a program to be interrupted is allowed to go on to the finish of processing for the one circuit with which it is at that moment engaged and is interrupted before it can commence the processing for another circuit, in order to avoid partially completed processes: and commonly the point reached in the program cycle is remembered so that when the program comes to be put into operation again, it is restarted at the point where it was interrupted.

9.6 Programs and Security

Figure 9.3 conveys the extreme simplicity to which the structure of an exchange can be reduced, and the description of the operation of the system exposes the complementary complexity of the programs and the programming. The ingenuity and opportunism once associated with circuit designers is now transferred to the programmer, or rather to a team of programmers. One thousand man-years has been quoted as the programming effort expended on one early system, the cost of which though clearly high would even so not be very significant when spread over all the lines of a large enough network. Functional specification techniques and high level languages have drastically reduced the effort required at some loss of storage and time efficiency, but with advantage in respect not only of reducing programming time but also of aiding comprehension which greatly assists programmers who have to make changes and additions to already established programs which they did not write themselves.

The very large quantities of semi-permanent storage needed means that the cost per bit has to be low and this is possible only with some form of bulk store. The words of a bulk store all have the same quantity of digits but the same is far from true of the programs and data to be stored. For reasons of economy the programmer has to arrange the storage to minimize the wastage of storage digits. This requires a combination of programming and addressing so that data requiring many fewer bits than there are in one word are stored more than one to a word, and those requiring more bits than in one word are stored in more than one word.

The quantity of temporary storage required is less than that of the semi-permanent storage but still needs to be in bulk form and has problems of organization for minimum wastage.

The programs written for a system constitute a library from which programs needed for individual exchanges can be drawn. Some stored data are individual to exchanges and must be produced specifically for each exchange; the translation tables for example, and some such as line data are written into the stores during the normal course of the exchange operation.

The executive program links all the individual programs in an hierarchical order such that the frequencies of operation of the programs satisfy the real time requirements of the system. To be able to do so for the largest exchanges presents some problems of timing, in fact the real time requirements could not be satisfied by straight-forward processing of one line and one call at a time. Parallel processing is one solution which is applied particularly to scanning for changes of data requiring further processing. It will be realized from the examples given that each scan, of exchange lines for new calls and of circuits for supervision, discovers very few lines or circuits for which the data have changed since the last scan and therefore most of the scan time is ineffective. By suitable arrangement of the data transmission and storage, and the programs, it can be arranged that a quantity usually of eight or sixteen lines or circuits are processed in parallel simultaneously but only to detect changes of data, the addresses of those in which data changes have occurred being stored in a hopper for processing later. Another solution is parallel processing by more than one processor, with lock-out or other means of preventing interference between them. Parallel processing by processors which otherwise operate as unique common processors is usually associated with the problems of security.

The magnitudes of the common processing equipment and the interface and their corresponding fault liabilities, and of the consequences of, in some cases, even single component failures, are such that at least duplication of all equipment is necessary. Triplication and majority decision which quickly both detects and locates faults, is not common because of the cost of three processors and the complication of micro-synchronism of three processors in parallel. Other security techniques, some making use of the unique characteristics of stored program processing, are practised with as far as possible no more than duplication in the interests of economy. The measures taken are aimed at the detection and location of all faults including those in the structure and for which the urgency of clearance and security are much less than for those in the highly concentrated control equipments. For fault detection, reliance is placed on check tests during normal operation and routine tests during periods of slack traffic. Faults which cause processing to stop are not difficult to detect. Others corrupt data in transmission and in store, produce incorrect operational outputs or prevent operational outputs from being effective. Check tests included in the normal control programs check that data transmitted have been correctly received and operational data have been obeyed, some of the tests being dependent on structure members or data transmission, or both, built-in especially for the purpose. Details of any test which is not satisfied including the equipments which are or at least should be involved are recorded, and analysis of the records over periods provides at least clues to the locations of most of the faults. Routine tests are carried out by stored programs additional to those of normal operation of the exchange and are designed to exercise and verify the operations of every part of the structure and the control. Failures are recorded and analysed as for check tests.

Faults the existences of which are detected but their precise locations are not known present some problems of tracing. Some faults in the structure members are fairly obvious but some require the co-operation of the program to find, and there is always the possibility that stored data are in error. Traditional methods of manual diagnosis being in most cases impossible, there being nothing which can be observed and nothing made to work except through the program, diagnostic programs have to be devised, to do the work of the manual fault finder, including the taking of remedial action if the fault would otherwise have too great an effect on service. The fact that practically no exchanges will be continuously attended by maintenance personal means that the measures necessary to keep the exchanges continually in service must be taken automatically. Diagnostic fault location is dependent on the correct operation of the program even though a fault is known to exist and thus includes the additional problem of ensuring that the program is not affected by the fault. Fault detection, circumvention and clearing thus adds to the complication and cost of the equipment but with the possibly more than compensating advantage of reduction in the manual effort required. Inevitably the last part of fault clearance is manual, namely the exact identification of the faulty equipment and its repair or replacement, but made easier for the majority of faults by the diagnostic testing having located them to particular units and often to particular components.

With security obtained by redundancy, the question arises of how the excess of equipment is to be organized. The interface switches do not need to be more than duplicated but either or both the stores and the common processor may be wholly or partially triplicated. The semi-permanent memory is of such a size that it has to be manufactured in sections. With two complete memories and a spare section, two memories in working order are maintained by substituting the spare section for a working section which goes faulty, the substitution being made by wiring and switches and not manually. If security is not satisfied by such means, three stores may be provided in sections and by comparison of data within the sections, faults may be located as well as detected. Then by switching out of service faulty sections, the risk of less than effectively two complete and correctly working stores is low. What has to be ensured is that there is always available one combination of fault free interface switch, processor and store. Generally this means variable interconnection of the units by switches controlled by test and diagnostic programs. With one such combination on-line and taking all the load, when a fault occurs in a working unit a fault-free equipment is substituted for it. If it is a semi-permanent store which is substituted, it will already have the necessary data written into it or the data can be written from a tape or other reserve semi-permanent store. A substituted temporary store may contain the data necessary to the continuation of existing calls because provision has been made in the system for the spare units to be kept up to date with the processing data. Calls in progress are affected if data are lost due to the effect of the fault which has occurred or because the provision as described has not been made. Corruption of meter stores has particularly to be avoided although failure to record legitimate charges for short periods is tolerable. Alternatively the load may be shared among as many fault free processors as can be assembled by combinations of fault free units, mutual interference between processors being avoided by the executive program. With one or more processors out of service, the load is shared among the remaining units still working. This arrangement benefits from the quicker and more certain location of faults in working equipments than in stand-by equipments depending on routine testing for fault detection; and if faults cause calls to be affected, the quantity so affected is less than with a single processor. The hierarchical system controls the programs carried out to adjust the load to the processing capacity, which means that programs of low importance suffer under fault conditions but also that the normal load, the sharing of which is only occasionally and for short periods affected by faults, can be somewhat greater than it could be if continuously unshared.

The programming has to be arranged so that circuits and equipments involved in calls for which the processing is interrupted by faults, will eventually restore to normal operation, which is analogous to the requirement well known in wired logic systems, that no matter how circuits are operated or misoperated, they must not lock-up and cease to function.

Faults may occur in programs as written. An unanticipated set of circumstances occurring and no provision having been made for it in the programming, processing halts for want of instruction or continues on irrelevant instructions. The probability of this kind of fault decreases with time until it eventually becomes negligible in well-established systems, but this may take several years.

Auxiliary equipment is required to load the program into the processor store, to write data into the data stores, to enable changes to be made to the programs stored and to the stored data, to read the stored data for checking purposes and to make apparent the results of routine and diagnostic tests for immediate action and record. Generally this means a punched paper tape or magnetic tape store for the program and the semi-permanent data, a means of preparing tape by hand, a means of reading tape into the relevant stores under the control of a stored program, and a printer of some kind to print out information under the control of a stored program. Most if not all of such equipment has to be duplicated. Commonly a teleprinter is used for tape preparation and output recording.

9.7 Characteristics and Costs

The problems and possible solutions involved in the design of exchanges have been sufficiently conveyed for the general characteristics and costs of different systems to be apparent.

Referring to Figure 7.8, the division of the processes required of an exchange between series and parallel connected processors is necessary to economy of data storage M and logic processing P and data transmission equipment. The result in a cross-bar exchange is many processors of many different kinds, none performing more than a small proportion of the total processes and because time division time sharing is excluded, none required to perform processing beyond the durability limits of electro-mechanical equipment. Evolution matched the requirements to the capabilities of the equipment including the fact that administrations did not offer or they actively discouraged services requiring many or complicated and expensive processes. Qualitatively the cost of cross-bar exchanges varies with the size of the exchange as indicated in Figure 9.7. Economy of processing no longer being necessary, semi-electronic systems make use of time division time sharing to effect an economy in processing P equipment as was deduced in section 7.4. The economy is effective in reducing the cost of large exchanges but because of replication for security and common equipment like pulse generators, there is an irreducible minimum cost which makes small exchanges more expensive than the equivalent cross-bar exchanges. This is indicated qualitatively in Figure 9.7 by a curve for the cost of semi-electronic wired logic exchanges which crosses the cross-bar curve at some point. Semi-electronic systems based on wired logic have over cross-bar systems the service and operating advantages summarized at the end of chapter 7 but which, although not negligible, do not weigh heavily in a decision for a change of system in an already established network of cross-bar exchanges.

The simple substitution of the distributed wired logic processors of a semi-exchange by a unique common stored program logic processor, as in Figure 9.1, increases the cost without any compensating advantage: use has to be made of the distinguishing feature of stored program logic, that economy of processes need no longer be an objective. Exploitation of this feature leads to the ultimate in exchange development of minimal structure, maximal common control systems symbolized by Figure 9.3 and with processes greatly increased on account of

(a) program handling, including loading, checking and changing;
(b) program organization, to fit uniform length store words;
(c) executive programming, to organize the separate programs for real time operation;
(d) space division time sharing achieved through the trunking;

(e) security;
(f) routine testing for fault detection;
(g) diagnostic programs for fault location.

The program storage required on account of the additional processes is commonly more than half as much as is required for the effective system processes. The minimum cost of exchanges with stored program logic used in this way is high because the common processor and its auxiliary equipments and most of the program storage all replicated for security are independent of the size of the exchange. On the other hand, although the series and parallel division of processes is the same whether wired or stored program logic is used and is, for electronically controlled exchanges, as represented in Figure 7.8(d) and confirmed by Figures 9.5 and 9.6, with stored program logic many of the circuit processors are virtual and the physical ones tend to fewer types because some variations between processors in the same rank in the trunking can be provided by the programs in the common processor. Hence large exchanges with stored program logic control can cost less than if wired logic controlled, which is represented qualitatively in Figure 9.7 by a curve which crosses those for wired logic semi-electronic and cross-bar exchanges. The curves of Figure 9.7 may include not only the installation capital costs but also the capitalized maintenance costs which are less for semi-electronic systems than for cross-bar systems. The curves are qualitative and their cross-over points not theoretically deducible because, although their general characteristics are not likely to change, their relative magnitudes are dependent on the available components and manufacturing techniques and these continuously change and to the advantage of new systems.

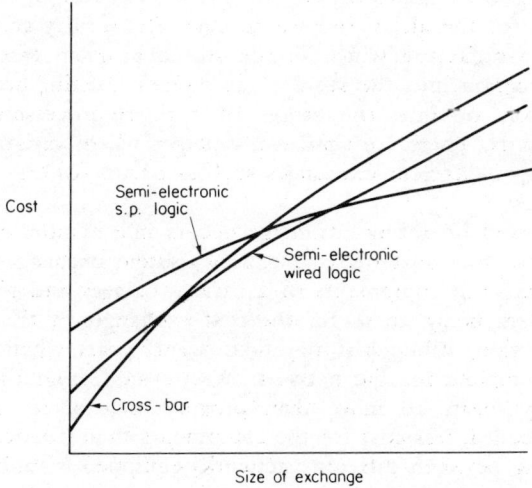

Figure 9.7   Relative costs of exchanges

The high minimum cost of exchanges with stored program logic has caused economic difficulties of application in practice which are explained in the next section, 9.8, one effect of which is to promote ways of reducing the costs of small exchanges. Possibilities in this direction without departing from the principle of the ultimate system of Figure 9.3 are also included in section 9.8. There are also in practice semi-electronic systems based on wired logic and designed to be economical for small exchanges, the facilities being limited to what is possible within the cost limit. These are used in rural and other low traffic density areas, and in some networks for the smaller exchanges in conjunction with a different system for the larger exchanges. Two different systems in the same network are apt to cause planning difficulties and are best avoided if possible. It must also be taken into account that the cost comparisons which can be made and have been made are not of systems which are equivalent in all other respects. The flexibility of stored program logic gives to exchanges with control based on that kind of logic advantages in manufacture, installation and operation, and in the making of changes and additions to the services and facilities provided by already installed and working equipments. Stored program logic also makes it possible to provide, for the customers, service facilities which wired logic systems would have difficulty in realizing, and if provided would bring their costs up to those of a stored program controlled system. These differences in conjunction with the continuously reducing differences of cost are commonly the basis for a decision to use stored program logic exchange systems despite the initial difficulties.

Manufacturing advantage arises from stored program logic when one kind of processor is required in several variations which can be provided by one design of equipment using different programs, and when processors are needed in small quantities which happens for some services and for special circumstances. Both require the writing of programs the cost of which is not very different from that of designing circuits for the alternative wired logic. The facility and economy are in the subsequent manufacture which in the stored program case means only the writing of the program into the store of an already existing processor, usually a common processor, or into the stores of a micro-processors of equipments individual to circuits. There are similar advantages where variations occur in the facilities required of different exchanges such as occurs for a system exported to different territories.

Manufacturing and operating advantage occurs in a number of ways, of which interworking with other systems is one. New system exchanges usually have to interwork with existing equipments of a variety of ages and processing requirements, the problem being acute for the first exchanges of the new system but reducing to extinction, although it may take twenty years, when conversion to the new system is complete for the network. With stored program processing, such interworking may mean no more than programming or if special conversion equipments are needed, less cost for the equipments than if wired logic had to be used. Interworking between different networks equipped with different exchange systems is a continuing requirement but generally a minor problem.

Manufacturing and operating advantage dependent on the use of stored program control is particularly important for p.a.b.x.s many of which are sold individually to customers who are offered a range of facilities from which to choose, who not infrequently require facilities individual to themselves or, in addition, changes and additions to the facilities after the p.a.b.x. is installed and working: also the ability to interwork with public exchanges not all of the same system may be required. For some p.a.b.x.s, the stored program processor is a computer performing other functions besides the control of the exchange, the conomy of which is thus benefited.

The installation of equipment for a new exchange or for the extension of an existing one involves a considerable amount of testing by the construction staff and by the customer to ensure that the equipment operates satisfactorily before acceptance. The labour required can be much reduced in exchanges with stored program control by programs specially written for the purpose and used only during the installation period, the programs being written into the program store as required and in place of permanent programs not needed at the time.

In these and possibly other ways stored program control systems have advantages over wired logic systems, resulting in cost savings in the installation and commissioning of exchanges and their extension as required for growth. Other advantages during the lives of exchanges are concerned with the possibilities of processing changes which may be required and their being made without physical changes to the equipment, but only to the programs or to stored data. The data changes which occur most frequently during the life of an exchange are those in respect of directory numbers, exchange line classes of service and other subscriber line data, due to the connection to the exchange of new lines and to changes and cessations of service to existing lines. As demonstrated by the systems of Figures 8.1 and 8.4, exchange line data changes may be made in basically wired logic systems with a facility equal to that of stored program systems. Changes of other data not accompanied by equipment changes, such as the routings of calls which affect trunk translations, are infrequent and made to a common M''' store without difficulty whatever its construction. Most of the changes are consequent upon new equipment installation of which changes of data by wiring or other physical means are a small part and some merely incidental to installation, as with guide wires. Changes in processes and processing did not occur very often in established public exchanges with wired logic once the systems reached stability, and the cost of making the changes was not significant compared with that of the total capital and operating costs over the lives of the exchanges, a situation which was possibly the result of the difficulty of making changes. On the other hand, it seems likely that as each fundamentally new engineering technique becomes established, the services and facilities become expanded up to a new limit which remains stable until the engineering technique changes again. Flexibility is valuable during the transition but less so when stability has been established. Techniques are always changing in detail: fundamental changes occur rarely, that from electromechanical to electronic being as profound as that from manual to automatic.

For a change to be made to the processes and processing of an exchange control system by a change only of the stored program, the change must affect only the control and not require a change or addition to the structure as is nearly always the case for structure classes of service and not infrequent in other circumstances. Other reasons for structural changes include that the data transmission required by or in the processors involved, do not exist or are insufficient for the new requirements. The usual method of control of programs stored is through paper or magnetic tape. For a new program or program change, the necessary program writing is undertaken and committed to tape at a central point and distributed, possibly by post, one tape to each of the exchanges concerned. At an exchange, the tape is run through a tape reader which in conjunction with auxiliary equipment, writes the program into the program store. The man making the change satisfies himself that the amended program is working satisfactorily, and he treats each of the duplicated or more processors in this way. The equivalent change to wired logic processors would most likely involve no more than three plug-in common processors which are removed one at a time and replaced modified or a new processor substituted and correct operation in service checked. Although not all of the issues involved are covered by the discussion, the conclusion is clear, that substantial differences in the initial costs of exchanges are not easily recovered by economies after installation. The reducing costs and prospects of low ultimate costs of electronic equipment affect wired logic processors substantially even if not to the same extent as those of stored program processors may be, and the low ultimate costs of electronic equipment affect the structures of the exchanges as well as the control and provide as much argument for electronic as against semi-electronic exchanges as for stored program against wired logic. The most powerful argument for stored program operation lies with the service facilities which it makes possible but are difficult and expensive with wired logic: but that, as previously remarked, does not mean that only stored program logic should be used.

9.8 Practical Application

The elegance of the concept of a unique common processor controlling a minimal structure exchange, as in Figure 9.3, combined with the feasibility of meeting all of the requirements of future systems, might have been expected to have rendered all previous exchange systems immediately obsolescent. It did in the sense that practically all development was channelled in the new direction, but outside the country of its origin and some others with many large exchanges, it did not in practice. That it did not was due not only to high cost initially of exchanges based on stored program logic but also, as will now be explained, to the problems and difficulties of introducing a radically new system into already well-developed networks.

Telecommunication systems grow steadily in all countries, doubling in size every five to fifteen years or so. The networks have to be planned for minimum costs, which involves the relative costs of transmission and switching plants among many other factors. The planning at any one time has to take account of the existing state of the network as well as the expected growth. As a consequence, the main features of networks, the exchange locations and cable routes in particular, are decided for many years ahead but the plant is provided in increments at smaller intervals. Generally it is economic for exchanges to start with a quantity of equipment which by successive additions multiplies to several times the initial installation. Thus there is a general tendency for the average size of exchanges to increase, which is offset to some extent by the creation of new exchanges as existing ones become unable to grow larger and new towns and communities are built and so forth. New techniques and developments and changing material and labour costs have the effect of changing the costs and the relative costs of line and exchange plant, and have to be taken into consideration. For example, because p.c.m. line transmission reduces the cost of junction line plant, it is conducive to an increase in the size of exchange areas but not very strongly so long as two wires are needed by every exchange line: a reduction of exchange plant costs which may well be achieved ultimately for exchanges of all sizes would have the opposite effect as would an increase in copper and aluminium prices and in labour thus increasing local line costs. Although individual exchanges may grow quickly, the statistical distribution of exchange sizes changes slowly. Figure 9.8 is representative of what might be expected of an industrialized country up to the end of the present century, which is about the time that it would take for a new system completely to supersede a system already established in a network. The average size of exchange might increase two or threefold in that time, with a fairly uniform distribution over a wide range of sizes all the time.

The economics of a new system have a short term aspect, which is what the administration has to pay during the first few years of transition, and a long term aspect, which is what the administration will have spent to establish the new system and will continue to spend, compared with having persisted with the old system. If the new system costs less than the old one for exchanges of all sizes, the long term justification for change to the new system is not difficult. If the new system costs more than the old for some exchanges, the long term economics cannot be more favourable than the integrated value of the product of the number of lines connected to exchanges of different sizes and the cost per line. With the distributions of exchange sizes of Figure 9.8, the integrated costs are, to a first approximation, proportional to the areas under the cost per line curves given in Figure 9.9 to correspond with those of Figure 9.7, and they do not indicate a notable reduction in costs by the use of electronics. The transitional costs have a short period administrative component for the retraining of staff and the replanning of the network and a more important extra cost for interworking between the equipments of the old and the new system. The interworking costs reach a maximum after a few years and then slowly decrease to extinction with the replacement of the last of the old system. To pay for the transition and to justify the upheaval incidental to change on the basis of costs alone, requires the long term economics to show a substantial difference in favour of the new system.

The transitional costs also have to take account of the fact that statistical stability in the installation of the new system will not be reached for some years. Generally an installation is an increment of growth, not a complete exchange, and although the increments tend to increase as the ultimate sizes of the exchanges

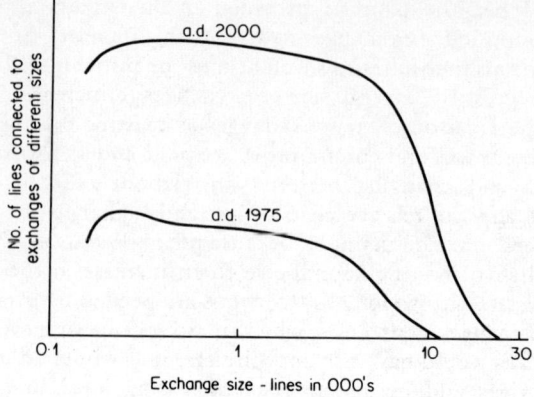

Figure 9.8   Distribution of exchange sizes

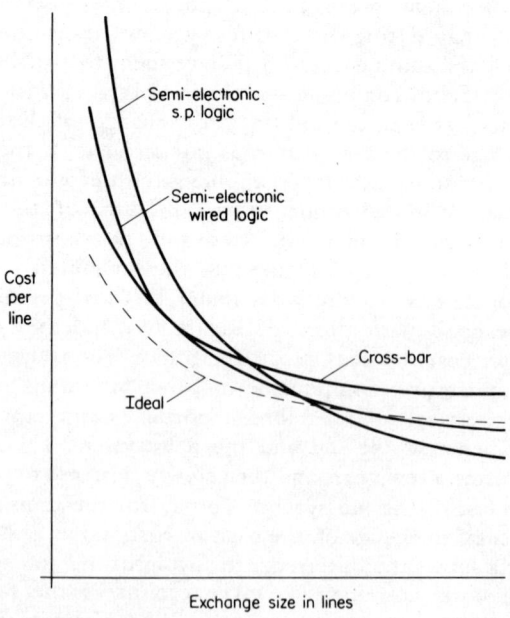

Figure 9.9   Exchange system costs per line

increase, the rate of increase is less. As a result, the initial installations of a new system tend to incur the higher costs per line associated with exchanges at the lower end of the range of exchange sizes, and which puts systems with a high minimum cost per exchange at a still further disadvantage.

The curves and data quoted to illustrate the economics of the various kinds of system are qualitative and not to be taken literally, one reason being that single curves are insufficient to represent much of the data: but they are valid enough to demonstrate that one cogent reason for resistance to the installation of semi-electronic stored program systems at their early cost levels was the high rate of investment needed during the first years of transition from existing systems. Other possible reasons include the consideration that if a high rate of investment in the early years is accepted to gain advantage later, the situation might have changed by then because of further developments. Although a system with minimal structure and maximal control may set a limit to the theoretical concept of an exchange, it does not imply a unique design of exchange on which no further improvement can be made nor that a system of that type is necessarily cheaper and to be preferred to some other. The possibility of wholly electronic exchanges in the future with which is linked the problem of switching p.c.m. channels at multiplex level could be enough reason for waiting to see the course of system development.

System means of improving the economics of stored program controlled exchanges have as their main feature exchanges sharing processors operating on data for more than one exchange instead of being individual to exchanges. Referring to Figure 9.3, the connections between the interface switch and the unique common processor are data transmission circuits which may be data links several kilometres long if their transmission characteristics of data rate and transmission time are satisfactory. Separating the processors from the exchange structures as in Figure 9.10, the data links between processors need have no physical relation to the junctions but to the processors with which the data link terminals may be co-located. The exchange data storage, figures 9.5 and 9.6, may also be co-located with the processors, which if at an administrative centre which controls some of the stored data, the trouble and expense of controlling the data at remote exchanges are avoided. Thus the concept emerges of firstly, a network of message transmission circuits provided for a given public service and interconnected by minimal structure exchanges, the network having been designed for minimum costs without restraints imposed by control costs: secondly, the control of the network being vested in processors which do not have to be individual to exchanges but located at strategic points and interconnected for the exchange of data by a network of data links; and thirdly, the two networks thus created being interconnected by data links, every exchange having at least one data link in operation to a processor in the control network to achieve which in the presence of faults, exchanges generally have to have at least two data links to at least two processors. The quantity of processors is reduced not only by being shared by more than one exchange but also if several are concentrated into one control centre, the replication necessary for security is reduced. This, sometimes called exploded network, concept extends the ultimate in exchange systems to the ultimate in network planning: but ultimates are theoretical

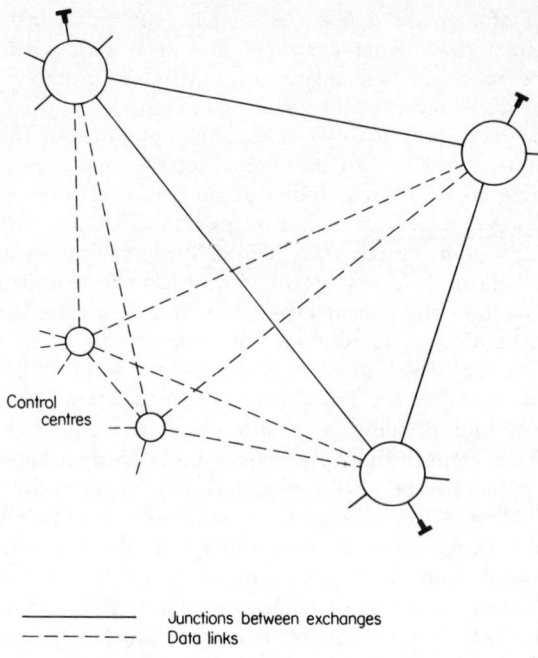

```
―――――― Junctions between exchanges
― ― ― ― ― Data links
```

Figure 9.10   Exploded network

concepts not necessarily the cheapest and best practical solutions to problems, and the present one is not without practical and economic difficulties.

The need to progress smoothly from an existing network and set of design principles to a network designed on different principles imposes considerable restraints during the early stages of transition and which the necessity of interworking between the old and the new switching equipments intensifies. It is not economic to design and provide separate transmission networks for message transmission between exchanges, for data transmission between the exchanges and the processors and between the processors. All three and the line plant for the existing exchanges have to share cables and other line plant suited primarily to message transmission which engages most of the plant and is the basic reason for its provision. As a result, transmission circuits directly suited to the kind of data transmission may not readily exist in all places nor be economically provided. Consider for example exchange line scanning for new calls as illustrated by Figure 8.5. With the processor at a control centre, the exchange would contain the interface switch i.f.s. addressed by a duplicate of the address register LAR in the control centre with, between the interface switch and the processor, data link facilities for the transmission of addresses and operational data o to the exchange and of loop data L to the processor. Given that the scanning rate may not be less than 10^5 lines per second to satisfy the operational requirements, the data digit transmission rate has to be of the order of 10^6 digits per second with a transmission time of the order of one microsecond at most, both of which are well beyond possibility for the transmission lines normally available. By parallel processing of lines, parallel transmission of data over the data link and modified addressing, the operational requirements can be satisfied but at increased equipment cost. Digital p.c.m. line transmission, each channel of which operates at 64 k.b.s. digit rate with a transmission time of about 5 μsec per kilometre, makes the problem much easier. The increase in the use of that form of transmission for junctions and which is taking place for line transmission economy reasons, leads to the belief that digital p.c.m. transmission will eventually be available between all exchanges independent of switching needs and exchange design can be based on that assumption. The consequences of that eventuality on the structure and control of exchanges is discussed in sections 10.4 and 10.5. Still further reduction in the data transmission requirements can be made by transferring processes from the control centres to the exchanges. Scanning for new calls, for example, may be performed by a local processor with the transmission to the control centre of the equipment numbers of lines requiring to be connected. Wired logic processors or micro-processors may be used for the transferred processes which generally are routine operations never likely to be changed during the lifetime of the system, leaving the more complicated and variable processes to the concentrated control. In these ways the control and the data transmission can be adapted to one another.

A practical application of the principle of Figure 9.10 and the modifications described is the special case of main exchanges with structure and common control associated in the conventional way of Figure 9.3, together with satellite, generally small exchanges with structures mainly controlled, over data link connections, by neighbouring main exchanges. In this way coupled with the continuously decreasing manufacturing costs of stored program processors, the economical as well as the technical problems of stored program control of semi-electronic exchanges are most likely to be solved.

Chapter Ten
ELECTRONIC EXCHANGES

10.1 Introduction

Electronic apparatus used in the control parts of exchanges has, on the operations and characteristics of the exchanges, the profound effects which have been described in previous chapters. Used in electronic exchanges for the structures as well as for control, electronic apparatus has equally profound effects on the message, signal and data transmissions and some additional effects on the control operations. The full import has not yet been explored and established, which is one reason why electronic exchanges are only briefly mentioned. Another and more practical reason is that the space available in one volume is not sufficient for more detailed treatment. That future problems and possible systems to supersede those currently in production should be known at least in principle is a reason for the following sections in which are examined some likely forms in the future of exchange switches, peripheral equipments and exchange systems and their control.

10.2 Exchange Switches

Switches, the switched outputs of which are linearly proportional to the inputs (true of all metal contact switches), are designated analogue. Switches, the switched outputs of which are stable only in the n-states of input digital transmissions (true of some electronic switches), are designated digital. Analogue switches may be used to switch analogue or digital transmitted messages and data but digital switches are limited to the switching of digital transmissions. All of the exchange switches described so far are designated space switches because they interconnect circuits identified by their positions in space, connections being made by the operation of cross-points on a path between the circuits. Electronic switching makes a different kind of switch possible and called a time switch. Space switch analogue cross-points continuously operated may be used to interconnect analogue or digital transmissions: space switch analogue cross-points time division multiplexed, a possibility limited by switching speed and durability to electronic switches, may be used to interconnect analogue unmultiplexed or t.d.m. digital transmissions if certain conditions are satisfied: also with certain conditions satisfied, digital switch cross-points will switch unmultiplexed or t.d.m. digital transmissions: and finally, t.d.m. digital transmissions occupying different slots in different multiplexes may be interconnected by an electronic digital time switch. These various switches and their operations will become clearer from the following descriptions of an electronic analogue, an electronic digital and an electronic time switch.

Gas discharge tubes and other electronic devices have been used for electronic switches but only semi-conductor devices are now of interest. As cross-points in co-ordinate arrays similar to those of cross-bar and reed relay exchange switches, they form electronic exchange switches in ranks connected in series by trunking arrangements also similar to those of cross-bar and semi-electronic systems and for which Figures 5.3 and 6.4 are examples. Figure 10.1 shows bi-polar transistors used as cross-points: each being capable of transmission in only one direction, two transistors are needed per cross-point in paths separate for the two directions of bothway transmission, which is thus four-wire with single wire and earth return unbalanced transmission in the two-wire paths. Cross-points using other kinds of transistors and semi-conductor devices are capable of switching and transmitting two-wire balanced or unbalanced transmissions, but are generally less useful than the four-wire unbalanced kind which also has the merit for the present purpose of being easier to describe and understand. The two transistors of a cross-point of Figure 10.1 are switched to the off, which is the non-transmitting, condition by earth potential = logic 0 applied to the point X of the cross-point, and they are switched on by a potential of +v = logic 1 at point X. The point X would be one of the points X of Figure 6.5 if guide wire is the system of path search and connection used. A cross-point is off when the base potentials of its transistors are more negative than slightly below earth, the potentials being derived from resistor potential dividers between the point X at earth and a power supply at $-v$. A cross-point is on, which is equivalent to the closed condition of metal contacts, when the base potentials are sufficiently above those of the emitters: this means that for the first transistor in the path the base potential is just above earth but progressively higher potentials are needed for the succeeding transistors in order that each shall provide sufficient collector to base potential difference for the transistor immediately preceding it to function satisfactorily. The switching potential requirements cause some difficulties and circuit complications for more than two or three cross-points in series. With a cross-point switched to the on state in every rank of switches of a complete path, current input at one end appears at the output end only slightly diminished by losses in the transistors, provided that the collector to base potential of none of the transistors falls too near to zero. The transmission through the switches is analogue.

The switched analogue four-wire transmission of the cross-points of Figure 10.1 may be used in many ways. The cross-points forming space switches and continuously operated, connections may be made directly between four-wire analogue and digital transmissions the currents for which flow only in one direction. If the currents are bi-directional, a constant bias or pedestal current has to be added so that the currents through the transistors are unidirectional. The four-wire transmission through the switches may be continued directly from and to the channels of four-wire transmission lines which are usually junctions but may be

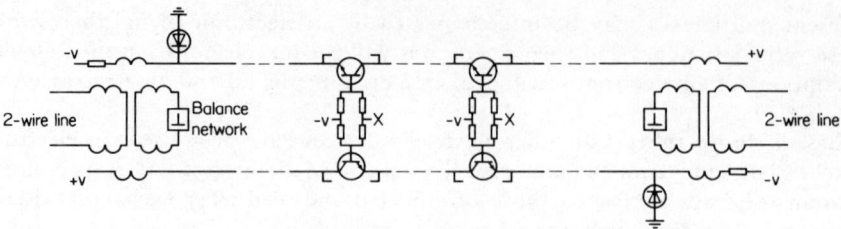

Figure 10.1   Electronic analogue exchange switches

for example lines to teleprinter stations; or the four-wire transmission through the switches if analogue may be transformed to two-wire transmission, as shown in the figure, by four- to two-wire terminations and line balancing networks. In Figure 10.1 the constant current bias for bidirectional current transmission is supplied from the −v power source via a resistor, to the emitter of an operated cross-point or to a diode connected to earth and which catches the bias current when no cross-point to which the resistor is connected is switched on. The resistor is also the terminating impedance for the four-wire transmit channel. The impedance transformation between the collector circuit of the last transistor in the four-wire path and the external line termination makes available transmission gain which may be used to off-set the losses of the cross-points, transformers and terminations, and depending on the four-wire to four-wire return losses with external lines connected, compensation may also be made for some of the losses of the external lines. Unlike metal contact cross-points which may be interposed between external lines and junctors in which transmission bridges for feeding D.C. current to the lines may be located and thus space division time shared among a greater quantity of lines, circuit reasons prevent electronic cross-points from being so connected. Every exchange line and analogue junction has to be terminated in a transformer, which increases the cost of the line equipments individual to each line and results in at least two transformers in every analogue connection through the exchanges. Because losses in the line equipments can be made up by gain through the switches, transformers and other equipment of higher loss than is otherwise usual may be used to reduce the size and cost of the line equipments. Interference to which telecommunication external lines are liable from power and other circuits, can produce current and voltage surges which would damage unprotected semi-conductor signal and message transmission devices. Protection to line transistors such as L in Figure 7.1, may be given by diodes able to absorb the interference and limit the currents and voltages reaching the vulnerable devices. Line surges are much attenuated by terminal transformers but some protection is usually needed for the semi-conductors having direct connection to the exchange sides of such transformers.

The interconnection of non-multiplexed analogue transmission circuits by space switch t.d.m. operated cross-points is illustrated in Figure 10.2 which is Figure 10.1 with the addition of cross-points individual to the circuits to be switched and interposed between the circuits and the exchange switch cross-points. Suppose the cross-points between the two four-wire analogue transmissions to be operated not continuously but cyclically, for 1 μsec every 125 μsec, that is 8,000 times per second, being usual. Then what is received as output at the end of the connection is not the continuously varying input current but samples of the input current, which is pulse amplitude modulated p.a.m. analogue transmission, superimposed on constant amplitude pedestal pulses due to the applied bias current. The input transmission is recovered almost exactly at the output by demodulation of the p.a.m. pulses and amplification by equipment not shown in detail in Figure 10.2, provided that the repetition frequency of the pulses is somewhat greater than twice that of the highest frequency component of the input transmission, the frequency being limited by a low-pass filter. In the present example, transmissions containing frequences up to the accepted limit for telephone speech currents, namely 3.4 kHz, may be switched. Many, usually thirty-two, such transmissions, their pulses being uniformly interleaved in time, time division time share the trunks and cross-points between the cross-points individual to the switched circuits, with considerable economic advantage because n-channel t.d.m. switches are able to carry n times as much traffic as their unmultiplexed equivalents and t.d.m. exchanges are to that extent smaller and less costly in respect of exchange switches and trunks than unmultiplexed exchanges. The transmissions switched by time division operated analogue switches do not have to be analogue: they may be digital. If the switching time division is not synchronized with that of the digital transmission, as to be described with reference to Figure 10.3, the minimum frequency of the p.a.m. pulses is related to the maximum time allowed for a change of digital state: the exchange switch transmission being in fact analogue, the switched digital transmissions are treated as if they were analogue. This kind of exchange switching is useful for digital transmissions like start-stop teleprinter for which the instants at which digital changes may occur are not predictable at the receiving points.

Unmultiplexed trunks included in connections through exchanges may contain transformers or other equipment for analogue transmission, for example of tones as in Figure 7.9, which is clearly not possible if the trunks are multiplexed. Equivalent arrangements when the trunks are multiplexed are described in section 10.5 to follow.

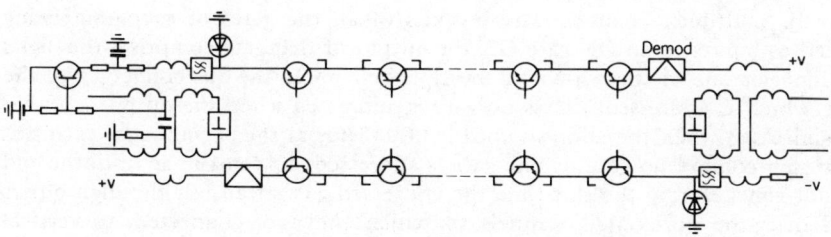

Figure 10.2   Analogue t.d.m. exchange switches

Digital switches suitable only for digital transmissions use, as cross-points, multi-input semi-conductor AND or NAND gate circuits commonly of the Figure 6.1 kind but shown symbolically in Figure 10.3 as AND gates G1. The cross-points are assembled into space switches which may by continuously operated to interconnect transmission circuits carrying digital transmissions of any digit rate within the switching capacity of the cross-points. The cross-points may also be time division time shared: most often the digital transmissions switched by time division time shared cross-points are themselves time division multiplexed and the switch time sharing is synchronized with the transmission multiplexing. Figure 10.3 illustrates p.c.m.–t.d.m. transmissions switched by t.d.m. switches, the transmission and switching multiplexes being synchronous with one another.

In Figure 10.3, analogue channels of unidirectional transmission and numbered 0 to 31 are time division multiplexed by cyclic connection to a common channel, using for each channel a low-pass filter and a both-way conduction gate the output of which is p.a.m.–t.d.m. pulses communicated to the common channel. Usually the repetition frequency of the multiplex cycle is 8 kHz. The amplitude of each pulse is pulse coded by a coder to a binary number of eight digits, 256 discrete amplitudes thus being available from which the one nearest to the amplitude of a pulse to be transmitted is chosen and its corresponding binary digits transmitted. In the figure the binary number transmitted for channel x during one cycle of the multiplex is shown, and that the digits of the channel $x - 1$ immediately precede those of channel x and the digits of channel $x + 1$ follow in successive slots of the switch multiplex. The p.c.m.–t.d.m. output transmission is applied as one input to every gate in a row of gates G1 which in Figure 10.3 have three other inputs. The gates are the cross-points of a t.d.m. digital space switch having a maximum, in this particular example, of seven horizontal p.c.m.–t.d.m. channels and rows of gates G1. For each column of gates there is a cross-point address store comprising words 0 to 31 each of three binary digits. The store words are read cyclically and in synchronism with the p.c.m. multiplexes, all of which are in synchronism with one another. At the beginning of each slot time, one word is non-destructively read out of each cross-point address store to a store CA with three complementary binary outputs to which the three address inputs of the column cross-point gates are connected each to a different combination of the digits. A gate is addressed when its three address inputs are all at logic 1: when addressed what appears at the gate output is the p.c.m. channel digits applied to the fourth input, the input connected to a p.c.m. multiplex channel. The outputs of all the gates of a column being connected as inputs to an OR gate G2, the output of that gate comprises the digits of a channel of one of the horizontal multiplex channels, the one connected to the gate G1 which is addressed, unless no gate is addressed when the output is 0. No gate is addressed when the address word is 000. Thus at the beginning of each slot time, in each column no gate or one gate is addressed and remains so until the end of the slot time. During the slot time the addressed gates transmit the digit pulses received over the horizontal channels to which they are connected, to vertical channels to which they are connected. The words of the cross-point stores of addresses are written into the stores by the exchange common control, to control the connections to be made by the switch.

The output pulses of the vertical channels may be decoded to p.a.m.–t.d.m. pulses by a decoder, the channels demultiplexed and demodulated back to the original analogue transmissions, as shown in the figure for one vertical channel. It will be seen that if only space switches are used in any number of ranks in series to switch p.c.m.–t.d.m. channels, the only connections which can be made are between the same channels in different multiplexes. In other words, there are thirty-two separately switched networks or exchanges with no means of interconnection between them. Connections between them require at least one rank of time switches, also as shown in Figure 10.3. A time switch comprises a digit store, in the figure of thirty-two words 0 to 31, each word having eight digits for the eight digits of one of the channels of a thirty-two-channel multiplex transmission. The word order 0 to 31 corresponds to the channel order 0–31 of the output multiplex channel. At the beginning of each slot time, the word corresponding to that time is read out of the store to a 'digits out' store DO from which the digits of the word are transmitted serially in time over the output channel to the next switch or other equipment. During the slot time the digits incoming from the gate G2 are stored in a 'digits in' store DI from which they are transferred at the end of the slot time to a selected word in the digit store. The word is selected by an address in a channel address store CHA, the store having thirty-two words each of five digits. The words in the channel address store are written in by the exchange common control and read out non-destructively and continuously in multiplex cyclic order. Thus any incoming channel or slot can be selectively switched and transmitted to any channel or slot of the output channel, which is termed slot changing by time switching. By its means, switched connections between different channels in two multiplexed transmissions can be made as well as between the same channels to which t.d.m. space switches are limited.

The equipment shown in Figure 10.3 provides for transmission in one direction only. Duplication of the transmission channels, cross-points and time switch digit stores is necessary for transmission in both directions although with some common addressing equipments. The trunks in parallel are the four-wire bothway trunks of switches which can be connected in ranks in series to form any size of exchange in the same way as cross-bar and reed relay switches, and electronic as in Figure 10.1, are connected except that the trunks, cross-points and ranks of switches required to switch a given quantity of traffic are, because of the multiplexing, much reduced in quantity compared with unmultiplexed transmission.

The input and output trunks of the switches do not have to be terminated in multiplexing and demultiplexing equipments as in Figure 10.3. They may be terminated in p.c.m.–t.d.m. junctions to other exchanges, which means that an analogue transmission once converted to p.c.m.–t.d.m. transmission can be switched and transmitted over any number of junctions of any lengths without suffering impairment other than that inherent in the coding to p.c.m. transmission at one end and demodulating back to analogue from multiplex pulses at the other

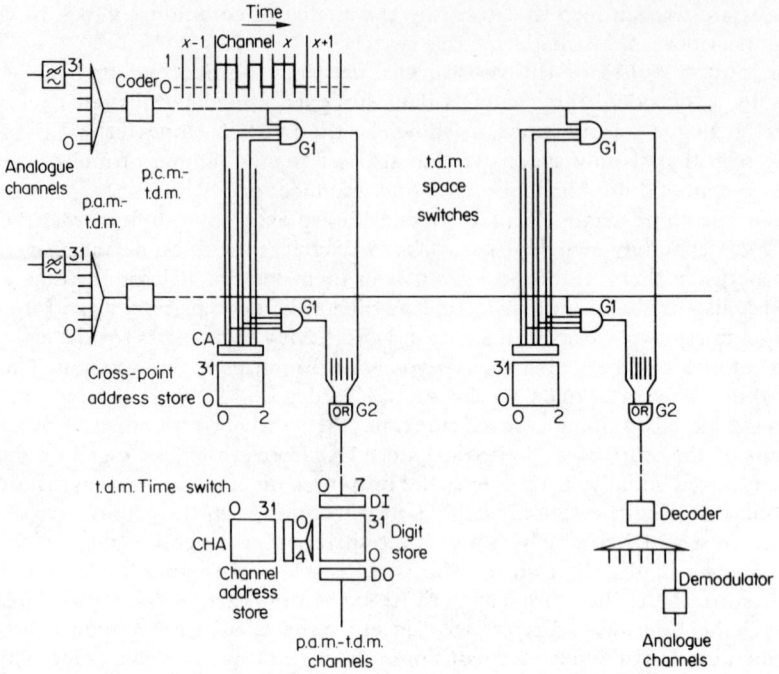

Figure 10.3   Digital t.d.m. space and time switches

end. Under these conditions distance has no effect other than echo time on transmission and this feature, in combination with the economy of the switching, makes p.c.m.–t.d.m. transmission and switching attractive as the basis for future network equipment and planning.

Clearly for p.c.m. multiplex channels connected by t.d.m. exchange switches, the digits transmitted through the switches during the channel time are not limited to those of the message transmission and digits for data may be included. In Figure 10.3, through the switches and still within the p.c.m. transmission slot time may be transmitted eight digits from the coder and one or more data digits from an associated data transmission path, the message and data transmissions being separated again after switching. Using synchronizing pulses occurring every n frames, that is cycles, of the transmission multiplex to divide the frames into n-frame super frames, a data path comprising a digit per frame becomes an n-digit transmission path within a super-frame of a super-multiplex, the significance of a transmitted digit being dependent on the frame in which it occurs within the superframe. With $n = 16$, all of the signals and data of Table 9.2 can be transmitted via a message path associated channel through the exchange switches, at a transmission rate of 500 digits per second which is more than adequate for the purpose. The data transmission of message path disassociated data links with stored program common control processing, as in Figure 9.3, is thus extended to message path associated data transmission not dependent on stored program or common control processing. Transmission of this kind through the exchange switches is not easily continued on message path associated paths of junctions between exchanges and disassociated data links have to be used. The data links have terminals as in Figure 9.4 but operated differently because the exchange control transmits data to and receives data from the terminals not via interface switches but via the exchange switches, the data links being designated order wires because of the difference. Usually at least one channel of every thirty-two channel p.c.m. transmission is given up for common use as a data transmission path for the other thirty-one channels, which use may include that of an order wire for those other channels. As such it is a form of message path associated data transmission very convenient and economical to use for junctions with p.c.m. multiplex transmission.

Equally clearly, analogue p.a.m. multiplex channels connected by t.d.m. exchange switches cannot also carry data transmissions as just described for p.c.m. transmissions. To achieve equivalent data transmission capability requires data paths associated with but separate from the message paths which is too expensive for unmultiplexed space switches but not for multiplexed switches. The reduction in the cost of exchange switching resulting from the time division operation of the switches reduces to a like extent the cost of message path associated transmission of control data through the switches and thus removes the major limitation of unmultiplexed switching. The data transmitted over the separate paths may be digital and equivalent to that described for p.c.m. multiplexes and with the same framing and super-frame multi-framing to make it possible to operate the two kinds of transmission in series through the two kinds of switches in series.

It will be recalled that an advantage of stored program unique common processor control of semi-electronic exchanges is that it does not depend on message path associated data transmission through the exchange switches — transmission which is unable to meet all the requirements of the future. This advantage is no longer unique to stored program logic when the exchange switches are time division multiplexed, the data transmission through which and over order wires between exchanges is equivalent to that of stored program controlled data links between the exchanges.

Plainly when multiplexing is introduced into exchange switching it has effects on exchange system design as profound as those which multiplexing have had on the design of transmission systems.

10.3 Peripheral Equipments

Metal contact switches impose little restriction on the currents and voltages which can be switched. Those which have come to be used and to which reference was made in section 2.2, to power the station microphones and ring bells, operate coin-box mechanisms and so forth, have been chosen to suit electro-mechanical equipments not specially designed for low power operation. Thus voltages of tens of volts and currents of tens of milliamperes are in use and these are difficult if not impractical for electronic switches and equipments to handle. In consequence, if electronic exchanges are not to be handicapped by practices which were developed for exchanges which they supersede, a new range of peripheral equipments suited to the characteristics of electronic equipment is required. The cost and administrative effort to make such changes in practice has been given as a major reason for the adoption of semi-electronic rather than electronic exchanges.

Figure 7.1 shows how exchange line current can be derived via a transistor L which performs the functions of a line relay by responding to the state of the line loop. Figure 10.2 shows how with a line transformer per exchange line the transistor may be connected. State of line loop data may be communicated as in Figure 7.1 to a common processor via an interface switch, which is always the case during new call scanning cycles. Rotary dial, arythmic code dial, clearing and any other loop signals may be communicated in the same way to a common processor or, the exchange switches having a message path associated data transmission path through them, loop data may be transmitted to a circuit processor in the connection by using, as in Figure 10.2, the transistor L collector current to provide the constant bias current needed by the switches. Current of 10 to 15 milliamperes from a 30 volt exchange battery, transmitted over the line to the station equipment, is sufficient for specially designed microphones and, if needed, transistor amplification of received messages and signal-messages. It is also sufficient for press-button arythmic code or v.f. number sending equipment, and for tone sounders as substitutes for magneto bells. A ringing signal-message sent from the exchange is amplified at the station and applied to a loud-speaker type transducer of a tone sounder to produce a sound of sufficient power. The tone sounders of party lines may be responsive to signal-messages of different frequencies in order that the parties may be called selectively. Exchange lines equipped with hands-free loud-speaking telephones which electronic exchanges may be expected to make more commonly used, may be called through the loud speakers, by calling tone or simulated bell ringing or even by speech by the calling party.

For coin-boxes, private meters and other equipments which the exchange line d.c. power is insufficient to operate, transducers usually termed conversion equipments individual to each line have to be used, in the exchange or at the stations. Signal-messages on the message path or data over a message path associated path are sent via the exchange switches and over the line to activate an equipment at the station to apply power from a local supply to peripheral equipment: or data are transmitted over interfaces switches from a common processor to activate conversion equipment in the exchange, the conversion equipment applying to the line d.c. or a.c. power from a common supply in the exchange.

Peripheral equipments suited to electronic exchanges have been produced for some electronic p.a.b.x.s and special exchanges and some experimental public exchanges but much development still remains to be done on both peripheral and conversion equipments.

10.4 Structures

That a system to switch both analogue and p.c.m.–t.d.m. lines in the same exchange may comprise a semi-electronic section for the first, an electronic section for the second and a control system common to both, is readily apparent from the expositions already given. There are exchange system features available to electronic systems which being dependent on the high operating speed and amplifying properties of electronic devices, are not possible in semi-electronic systems or in mixed semi-electronic and electronic systems. Message transmission and exchange control are concerned and in conjunction with the development of devices and manufacturing techniques which are continuously in progress, it is to be expected that ultimately electronic exchanges will be required to interconnect analogue and digital transmission lines in whatever proportion network planning decides to be the most economical in practice, and it is to that end that the present analysis and description is applied.

The switching of analogue transmissions by electronic switches involves, in addition to that of cost, problems of attenuation, distortion and cross-talk, and of line impedance balacing at four- to two-wire terminations. Transmission through an electronic cross-point causes a loss which although small cannot be neglected because of the very large numbers liable to be in series in any one connection. The loss has to be made up by gain as was described with reference to Figures 10.1 and 10.2. Gain adjustments for circuits individually not being practical, exchange lines and junctions have to be treated in classes with nominal gains individual to the classes. Both the losses and the gains being subject to random variations, the overall losses through exchanges have to be treated statistically, in which respect standard deviations of much less than 1 dB are achievable and the mean overall losses of some connections through the exchanges can be negative, that is they can provide gain to offset some of the losses of the external lines. Because the nominal losses can include those of transmission bridge transformers and the like in the switched circuits, the overall transmission is generally better than that of networks with semi-electronic exchanges having no amplifying means of compensation for losses. Non-linear distortion due to the switches and line transmission equipments such as transformers is generally not a difficulty. Attenuation distortion is produced by items of transmission equipment included in the transmission path and as these tend to be more numerous in electronic exchanges, in the form of transfomers and frequency filters, design to satisfy specified limits tends to be more difficult and expensive than for semi-electronic systems. Cross-talk is a major problem for analogue transmission particularly when time division multiplexing is used. Balanced transmission requires two electrically identical paths to carry equal currents in opposite directions. Switches in the paths must be identically the same, which is more or less but not exactly true of the electrical properties of electronic switches although practically true of integrated circuit pairs of transistors close together on the same chip. Because of the cost and practical problems with balanced transmission, unbalanced transmission is commonly used through electronic exchange switches and has some problems. Because the path lengths within the switches are so short, cross-talk arising within unbalanced switches generally does not exceed an acceptable limit, but that due to the trunks between switches and from the switches to the line equipment terminals can be serious. For that reason twisted wire pair or co-axial conductor is used for the trunks, so that currents of different trunks do not have to share a common path as they would have to do if a common earth return was used by all of them. Co-axial conductor is used when the outer conductor is an effective screen for the currents in the centre conductor, the screening depending on the frequency content of the currents. Speech frequencies are not high enough to derive any benefit from co-axial conductor, those of p.a.m.−t.d.m. are, and the screening is practically complete for p.c.m.−t.d.m. transmission. Consequently the path length is not a difficulty for multiplexed transmissions, but the unbalanced path length for speech current switching is limited to metres of wire pair, which is sufficient for some p.a.b.x.s but not for large main exchanges The path length can in some circumstances be increased by series transformers at the ends of the paths to produce more nearly balanced transmission over the length. In addition to cross-talk between different transmission paths, cross-talk also occurs between the channels of the same p.a.m.−t.d.m. transmission due to attenuation distortion along the path length and which increases with path length, the limit for which for both reasons is tens of metres and sufficient for the likely use of p.a.m.−t.d.m. transmission within any exchange. The switching of p.c.m.−t.d.m. and other digital transmission by analogue or digital switches presents little difficulty although attenuation, distortion and cross-talk within the exchanges can lead to digital errors in reception.

Not all electronic switches are one-way transmission devices as are the bi-polar transistors of previous figures but the necessity for amplification to make up the losses, linearize the transmission or other reason, make analogue bothway transmission over two-wire paths within public exchanges difficult if not impossible. Hence four-wire switching and transmission through electronic exchanges is to be expected for analogue as well as for digital messages for which it is essential, with four- to two-wire terminations between the switches and two-wire external lines. The overall transmission of complete connections between two-wire lines depends on the balance return losses obtainable at the terminations: the balance networks being individual to lines, the return losses depend on the effort spent on constructing the networks. With moderate effort, overall transmissions better than those of two-wire switched networks are possible, notwithstanding that two-wire switched networks commonly use four-wire switching of long distance circuits to improve the transmission of long distance connections. Advantage also accrues from four-wire transmission in other respects. For example, multiway connections can be made, for three-way conversations and multi-party conferences, without transmission loss or side tone difficulties and very simply as illustrated by Figure 10.4. In that figure, the four-wire transmit channel of each of four lines transmits via amplifiers to the four-wire receive channels of the other three lines.

The economy of p.c.m.−t.d.m. transmission in respect of line plant provision presages its eventual use for all lines exceeding in length some very short distance. This possibility together with the switching and transmission advantages of transit

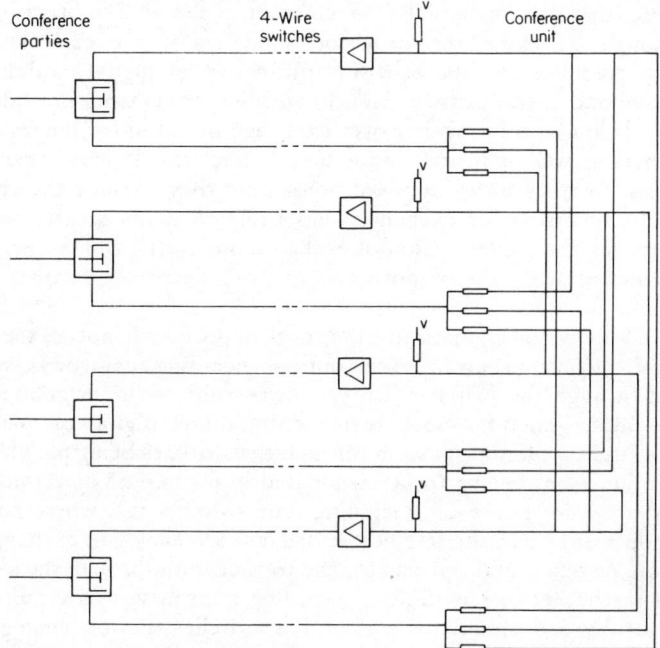

Figure 10.4 Multiway connection

connections between digital multiplex channels being made digitally, by space and time switches as in Figure 10.3, prompts the concept of all transit connections between junctions being made digitally, analogue junctions being terminated on p.c.m. terminals just for that purpose: and because multiplexed transmissions are cheaper to switch than unmultiplexed, the concept is extended to exchanges in which all the exchange switches are digital and time division multiplexed, all analogue circuits terminated on the exchanges being equipped with p.c.m.–t.d.m. terminals. Regarding the first concept, if there were no influences other than line plant economy, there would always be some analogue junctions, because the economics of junction plant provision are affected not only by the lengths but also by the quantities of junctions on the routes, and by the more serious effects on service of faults on multiplexed than on unmultiplexed junctions. Most of the analogue junctions would, however, terminate on local exchanges and carry short distance traffic which could be excluded from the digital switching. Thus all digital switching of junction transit connections at exchanges in the higher hierarchic levels of networks is to be expected and has already begun in practice. Regarding the second concept, not only must all analogue junctions be terminated on time division transmission terminals, but so must all service lines and all exchange lines.

It is possible that the manufacturing costs of p.c.m.–t.d.m. terminals will be reduced enough to make the provision individually for each analogue line economically possible: and the provision ultimately of digital modulation equipment in telephone instruments is not impossible, to extend the advantages of digital transmission to transmission over exchange lines and for the transmission of high speed data as well as for telephone speech. Such developments cannot be seen clearly enough for present assumption, nor should they exclude the consideration of others. The quantities of exchange lines being so much greater than those of other lines makes the concentration of exchange line traffic the key problem. Once traffic is concentrated, the importance of cost decreases relative to that of performance.

Figure 10.5 shows diagramatically the most important if not all the options for exchange line traffic concentration, the options including analogue as well as digital transmission through the switches. On the lower right of the diagram are analogue and p.c.m.–t.d.m. junctions both switched by t.d.m. digital or analogue interconnection switches, the transmission through the switches being p.c.m.–t.d.m. and the analogue junctions having to be terminated in p.c.m.–t.d.m. terminals. Transit connections between trunks and including time switches t.s., where necessary, are made via loop trunks. On the left of the diagram are analogue exchange lines and concentration switches, and options for the transmission through the switches. The first option is whether or not the exchange line transmissions are pulse amplitude modulated into and demodulated out of the switches. If not, analogue unmultiplexed space switches are used for concentration, with trunks to the digital transmission interconnection switches, the trunks including analogue to p.c.m. transducers. Connections are made between exchange lines by analogue loop trunks. If, however, the exchange lines are terminated on p.a.m. terminals, the next option concerns coding and decoding. If there is none, the transmission into and out of the switches is analogue p.a.m.: if coded, the transmission is digital. The coding may take several forms which in the diagram are generalized into two types, single digit and multi-digit. Single digit coding, assumed to be some form of delta-modulation, comprises one binary digit per coding cycle and is transmitted by one of two electrical n-states, usually either current and no current or positive and negative current. Coding comprising more than one digit per cycle is assumed to be conventional p.c.m. Finally, the modulated, coded or not, transmissions may be or may not be time division multiplexed. It is assumed, however, that unmultiplexed p.a.m. or p.c.m. transmissions are not useful and would not be used in practice, and it is noted in passing that the order in which the options are presented is not necessarily that in which the corresponding operations are carried out in practical systems.

With the exchange lines equipped for p.c.m. transmission, the concentration switches for exchange line traffic may be uniform with those for interconnection switching and the transmission uniform with the junction transmission. The trunks between the concentration and interconnection switches do not need transducers as do trunks between dissimilar transmissions. The exchange switch costs are low but the cost of equipping every exchange line and analogue junction with a p.c.m.

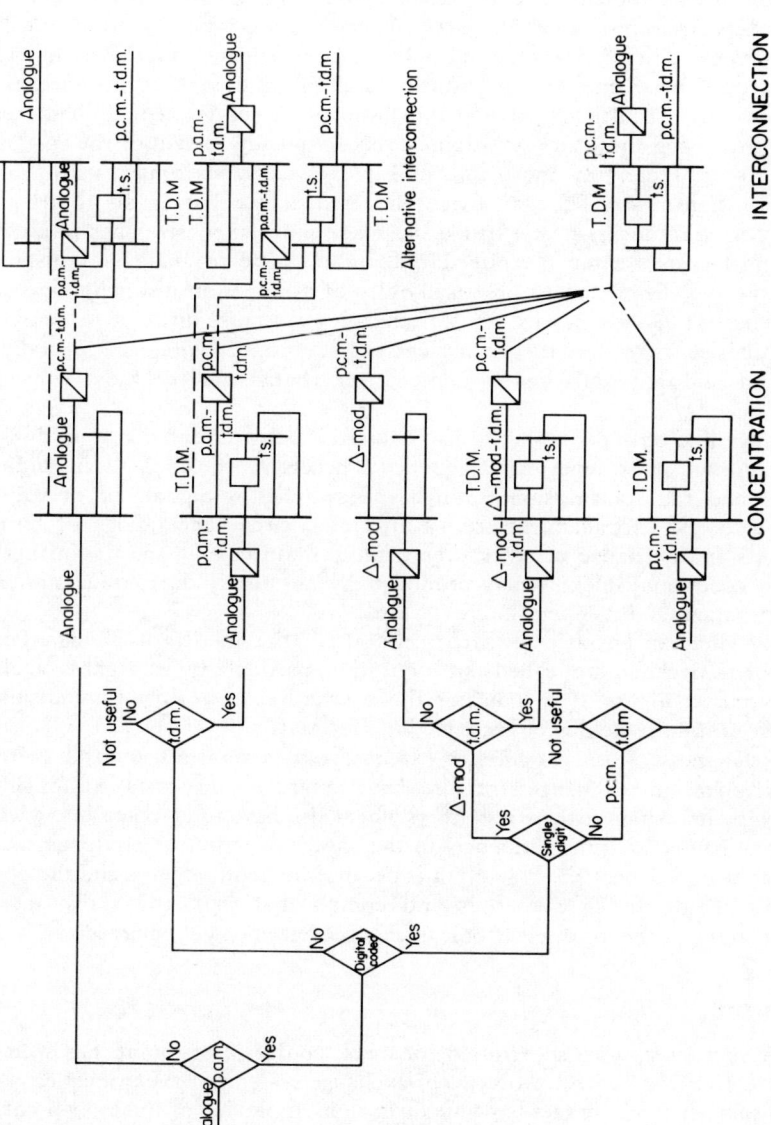

Figure 10.5 Electronic exchange system structures

terminal in order that it might be switched is high, the cost being divided between the common coding and the individual channel equipment, that of the channel equipment being the greater mostly because of the filter, transmit or receive required by each channel. The cost may be reduced by omitting the coding to leave p.a.m.–t.d.m. transmission through the concentration switches which must be analogue type, coding being required only for the analogue junction terminations and the trunks between the concentration and the interconnection switches, the p.a.m. and t.d.m. multiplexes being synchronous. A greater reduction in cost is made by increasing the p.a.m. multiplex cycle frequency to reduce the cost of the terminal filters. Preferably the p.a.m. multiplexes are synchronized with those of the p.c.m. transmission but at twice the frequency, that is at 16 kHz. The p.a.m./p.c.m. transducers in the trunks are made more expensive but the exchange line terminal costs are much reduced and the analogue to analogue transmission through the switches rendered practically free of attenuation distortion due to the multiplexing. Time switches t.s. for slot changing are more difficult to design and more expensive than for the equivalent p.c.m.–t.d.m. switches. Generally demodulation and remodulation equipment per channel is the most convenient solution.

Exchange line terminal costs are also reduced if delta modulation is used instead of p.c.m. A high pulse repetition frequency is necessary, the coding is individual to terminals and the transmission through the switches is digital, all of which is conducive to low cost manufacture by integrated circuit technology of terminal equipments and exchange switches, even without multiplexing, and the further cost reduction which multiplexing may produce may not justify the extra complication of the equipment and its operation.

If analogue concentration switches are used, the merits of all-digital interconnection switching are called into question particularly as regards analogue junction traffic. Figure 10.5 indicates the alternatives of analogue unmultiplexed and p.a.m.–t.d.m. switching of analogue junction traffic.

Only the message and to a lesser extent the data transmission and switching aspects of electronic exchange structures have so far been discussed. Although they are of major importance, there are others which also have to be taken into account, the control having a strong influence on the choice of structure equipment as will be indicated in section 10.5. Possibly it is because the requirements and the options to satisfy them are so numerous and complicated that universally accepted solutions to the problems of electronic exchanges have not yet emerged.

10.5 Control

Electronic cross-points substituted for metal contact cross-points has itself little or no effect on the control problems of exchange systems or the principles of their possible solutions, nor in fact has time division multiplexing of the cross-points and trunks. The trunks used in trunking systems typified by Figures 5.3 and 5.4 may be unmultiplexed or multiplexed and the generalizations of systems based on wired and stored program logic processing, namely those of Figures 7.10, 9.1 and 9.3, still apply. Exchange switches and trunks n-channel time division multiplexed provide n switching networks in parallel, one for each channel, with n times the traffic carrying capacity of the identical network using unmultiplexed trunks and switches. Time switches have to be included additionally to interconnect the separate networks. The important differences between multiplexed and unmultiplexed transmission through exchange switches are practical ones. The time sharing by multiplexing of the trunks and cross-points reduces by a large factor the quantities of both required to carry a given traffic, which leads to a reduction in the number of ranks of switches required, and to a reduction in the cost and size of the structure. Departures from the straight forward trunking of Figure 5.3 to others such as Figures 5.4 and 6.4 to save trunks and cross-points at the cost of additional control become of little and possibly negative value; in fact the quantities of trunks and cross-points are sometimes increased to gain simplification of the control, by ensuring that path search and connection between known starting and terminating points cannot fail for want of a free path between them. Of great economic and system significance is that time sharing and the resultant reductions in the quantities of equipment required also apply to circuit processors individual to multiplexed trunks. The system of Figure 9.3, for example, may use t.d.m. switches and trunks according to one of the systems of Figure 10.5, the or.j. originating and t.j. terminating junctors then serving not one call at a time but all the calls at one time over the channels of a t.d.m. trunk, and without the cost and complication of an interface switch and data transmission equipment needed for the normal methods of time sharing of a common processor among unmultiplexed junctors. The economy of time sharing also applies to path search and connection circuit processors, the guide wire being particularly suitable: but data storage and processing for map in memory path search are not reduced.

It is mentioned in section 7.6 that time division time shared processors for semi-electronic systems are not able to send or to receive, via the interface switch, data in the form of signal-messages and that this disability has two main consequences. Not being able to receive signal-messages produces the necessity, in particular for press-button v.f. signal-messages, for some form of space division time sharing of transducers and to which reference has several times been made. This disability does not necessarily apply to signal-messages on digital t.d.m. transmission channels because detection by digital processing by a common processor may be possible: but if for electronic exchanges new types of peripheral equipments have to be designed, arythmic code sending of control signals and data, as described in section 3.3.2, is the obvious choice and no problem of signal detection arises. Directory number digits, coinbox coins and all other data are readily and cheaply sent by such means and received and identified in time division time shared common processors. The second consequence is that a time division time shared processor unable itself to send signal-messages, sends instead data which have to be transduced in the circuit processors to the required signal-messages, the transducers including bi-stable circuit stores and tone modulators, as seen in Figure 7.9 for dial and n.u. tones. Signal and datum transmissions from such processors to unmultiplexed circuit processors also have to be instructed by data to control circuit stores and switches, as in Figure 7.9 for the DB, A, Po and MK data. Continuous output devices like bi-stable circuits are not appropriate to a multiplexed trunk circuit processor which is serving all the channels of the multiplex. The equivalent device is a binary digit in a channel store which, being read every multiplex cycle, determines for each cycle whether or not the processor is to send for the channel an instructed signal or datum over the message path associated data transmission path or a signal-message over the message path. To illustrate multiplex channel data sending, Figure 10.6 shows a small part of a multiplex circuit processor connected to a trunk between t.d.m. space switches of the kind in Figure 10.2, the processor including a data sending store DS of n words for the n channels of the multiplex. The words are written into the store by the circuit processor or a common processor, for the circuit processor to control the sending of data for each word until it is erased. The circuit processor writes into the store data derived from its own processing or received from another processor over a message path associated data path. At the beginning of each slot time, the processor reads the appropriate word out of the store: if that indicates that digital data are to be sent over a data path, the processor applies the data to the base circuit of a transistor Tr1 which transmits the data over the data path, and if a signal-message is to be sent on the message path, the processor selects an appropriate one of possibly several transistors Tr2 and operates it during the whole of the message transmission time of the slot. With p.a.m.–t.d.m. transmission the data and message paths are separate in space as they may be with p.c.m.–t.d.m. transmission although separate in time on a common channel is more likely. The pulse code or pulse amplitude modulation required for digital or analogue transmission respectively is provided by tone generators common to transistors Tr2 which transmit the tones. For analogue transmission the tone generators are applied directly in the emitter circuits of the transistors, and for p.c.m. transmission they are applied via coders as in Figure 10.3: a coder with a continuous analogue input produces p.c.m.–t.d.m. digital output continuously for every channel of the multiplex. The sending of signal-messages in this way means that no equipment is needed to be individual to a circuit other than a word in a data store, the economy of which produces still further economy in that it leads to a reduction in the varieties of processor required to send dial, pay, busy, n.u. and other signal-message tones and also to send the signals of present systems, ringing and so forth, if they become signal-messages in electronic systems. One kind of processor for all classes of calls, originating and terminating, thus becomes a possibility and the ultimate in simplicity and development of exchange structures is reached.

Considering that all the exchange switches and the trunks between them are time division multiplexed, the generalized ultimate SPC system of Figure 9.3 is still valid, the only difficulty being in the unique common processor receiving data from a multiplexed channel, such data being available only during the slot time of the channel in each multiplex cycle and not at times under the control of the processor. Exchange line signals s through the concentration switches to or.j. and t.j. junctors are the data mainly involved. The sending of data including operational data presents no difficulties as they have to be stored and the storage acts as a buffer between the

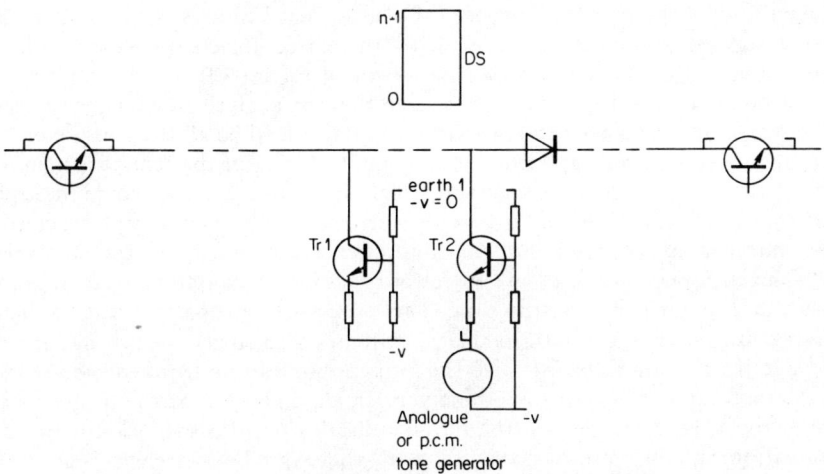

Figure 10.6   T.D.M. circuit processor data and signal-message transmission

control and the required operation. For example, the common control has to write into the cross-point and channel address stores of Figure 10.3 and the channel data stores of Figure 10.6, to make connections through the switches and to send data respectively, which it may do at any time within the exchange switch multiplex cycles except briefly as one slot ends and the next begins. The difficulty with received data is similarly resolved by the circuit processors writing during each multiplex cycle and into channel stores the present states of the received signals s, for a common processor to be able to read at any time.

Again considering that all the exchange switches and trunks between them are time division multiplexed, electronic wired logic processors as circuit processors individual to trunks are able to complete their operations within the slot times of the multiplexes, and it becomes economically possible to go back to the generalized system of Figure 7.10 modified only in that it represents an electronic system and some parts of the common time division time shared processors for the circuit processors have become integral with the multiplex circuit processors. It was remarked in section 2.13 that a machine processor working at infinite speed could operate any set of operational programs linked by any executive program and satisfy the control of an exchange of any size. An electronic t.d.m. exchange with wired logic can approach this situation, using time interleaved common clock controlled synchronous cyclic programs typified by Figure 8.1 and not needing program hierarchic levels and interrupts to keep up with real time processing. The processors being simpler and the processing more distributed and less constrained, security problems are less severe than those of stored program logic, and common processors triplicated and in micro-synchronism provide much of the fault detection and location without recourse to lengthy routine testing and diagnostic programs. All the basic facilities are economically provided in this way, and all the other facilities demanded of exchanges of the future may be by a stored program logic processor supplementing the processing of the wired logic processors, with the prospect of achieving the ideal cost characteristic of Figure 9.9.

It will be seen from the descriptions given that the analysis of exchange systems and the design of exchanges have to be looked at afresh when all the components of an exchange are electronic and time division multiplexing of the structure members as well as the of the control becomes possible. Circuit processors become multiplex time shared and required in quantities only a fraction (approximately the reciprocal of the number of channels in the multiplexes) of those of the equivalent unmultiplexed processors: they are of fewer types in correspondingly fewer groups and they are simpler in construction. The processors are part of the exchange structure along with the t.d.m. exchange switches also so reduced in quantity of equipment by the multiplexing that the cross-points can be trunked individually. The structure costs shrink to something very small, and most of the problems and the cost are in the common control, in the multiplex terminals for modulation and demodulation of the transmissions so that they can be switched, and in the peripheral equipments. Exchanges become much smaller in physical size and capable of almost any service or facility which may be required of them. The outstanding problem is their economical introduction into already well-established networks, by achieving a cost characteristic resembling but showing lower costs than those of cross-bar exchanges, to approach the ideal indicated in Figure 9.9, and this may well be possible only by electronic exchanges. Semi-electronic systems with wired logic having insufficient advantage over cross-bar systems and those with stored program logic having economic difficulties of practical application, t.d.m. electronic systems exploiting both forms of logic may be more successful and could, it is arguable, have been achieved without going through the semi-electronic phase. Technical progress is sometimes said to be like a man going up a spiral staircase: he keeps coming back to where he started but higher up. Exchange systems having made one convolution of the spiral, from minimal structure through maximal and back to minimal, in progressing from manual through electro-mechanical automatic to stored program logic controlled semi-electronic systems, there is no reason for them to stop just there: from which point of view the next step to electronic changes which the chronology of Figure 2.15 suggests to be imminent, may be expected to be up and round the spiral.

Chapter Eleven
CONCLUSION

There are or will be switched telecommunication services for telephone, teleprinter, data, video and other forms of communication. One feature common to them all is the transmission of electric phenomena over distance and for that reason alone, the services of the future need to be integrated into one comprehensive system to achieve maximum efficiency and economy of plant provision. Although the analyses and descriptions given in the foregoing chapters have been related specifically to the telephone service, much is common to all services, and in many ways the telephone is the most complicated of all and the individual problems of other services less difficult to work out.

There was a clear objective and incentive behind the development of the first telephone exchanges, namely the exploitation of the newly invented telephone. It is not difficult to appreciate how and why in the then state of electro-technology, manual switching came to be used in the first exchanges. The clear objective and incentive to the invention of step-by-step automatic exchanges was the elimination of manual operators for the most commonly used services, in the interests of speed of service and continuous day and night service, and costs below those of manual exchanges were eventually achieved. Cross-bar systems promised more reliable and quicker service to the users with some improvement to the services and operational advantages to the administrations, and in time became competitive with step-by-step systems. It is doubtful that all of the advantages which cross-bar systems are now known to have were clearly foreseen when the cross-bar switch was first conceived. The first attempts to apply electronic devices and techniques to the construction of exchanges were aimed at electronic exchanges. Semi-electronic exchange developments came later but at this distance of time it is difficult to be sure what the objectives and incentives really were, other than the usual exploratory research and development on how to use new devices and techniques, in respect of which for data storage the cathode ray storage tube and magnetic drum were major steps and the transistor for logic processing, and all appeared in the 1940s. Before that time greater reliability generally of exchange operation was being sought, the reed relay being invented as part of this movement to improve the reliability of metal contact switching. Electronic switching was seen as a possible means of still greater reliability. Also at that time, message transmission was much in the mind of telephone engineers, improvement particularly of long distance connections was known to be highly desirable and possible with four-wire switching and amplification in the exchanges, and electronic exchanges were seen as a possible means of satisfying these conditions. To do so with metal contact exchange switches was known to be prohibitively expensive. Frequency division multiplexing having made a revolutionary difference to line transmission, it was thought by analogy that a similar effect should exist for switching and seemed to be found in time division switching. It is a matter of history, however, that research and development became concentrated not on the transmission and structure aspects of exchanges but on the control. Possibly the expansion of the services to the users and the advantages to the administrations and the maintenance forces consequent upon the greater data storage and processing capability of electronic equipment, were seen to be possible if not in detail. Hopes which are always entertained of reduced costs were not conspicuously fulfilled by the semi-electronic systems which first appeared in practice. The major achievement has been stored program logic processing and the consequential system developments and characteristics which have been described, with an economy, clear objective and incentive which now seem to be established but may appear in a different light with the passage of time. Message, signal and data transmission improvements are an objective and incentive for the electronic exchanges now being pursued.

Telecommunication exchange systems having by no means reached finality in their development, many new and exciting changes are to be expected in the coming years. Whatever they may be, it is hoped that the analyses and description which have been given in the foregoing pages are sufficiently basic and objective as still to be valid and useful.

INDEX

Abbreviated dialling 179
Accounting centres 180
A-Digit switch 122
Administration centre 179
Allotter 88
Alternative route 154
Answer 10
Answer back 10
Arythmic code 69
Asynchronous t.d.m., wired logic 242
    stored program logic 244
Attenuation 15
Auxiliary common processor 255, 258
Availability 102
    full 102
    limited 102

Baudot 7
Bell, Alexander Graham 3
Bell, magneto 15
Billing, bulk 59
Binary, complementary 176
Binary decisions 36
Bi-stable circuit 176
Bits 8
Bridge, diversion 120
    transmission 111
Bus-wires 134, 238
Busy 26
Busy, all lines 43
    plant 284

Call 10
    data 163
    effective 59
    local 10
    originating 16
    state of 22
    terminating 16
    tracing 285
    waiting 179

Calling line identification c.l.i. 180
Calling message 7
Calls, barred international i.b. 79
    barred long distance l.d.b. 79
    barred outgoing 100
    delayed 29
    lost 29
Camp on busy 179
Centre, accounting 179
    administration 179
    maintenance 181
    operator assistance 179
Channel 5
Charge, rates 167
    reversed 60
Charges 59
Charging, bulk 59
Check tests 288
Circuit processor 26, 48
Circuit switching 1, 10
Class of service 60, 76, 218
    control 77
    group 96, 159
    structure 77
Clear 17
Click test 26
Coin-box 8, 72, 78
    coins 72
Common processor 24, 46
Common processors, mutual interference 49
Communication 1
Computers, major 252
    mini- 252
Conference connections 179, 310
Connection 10
    check 282
    reserved 183
    third wire 26
Connections, holding and releasing 169
    incoming junction 104

junction tandem 104
local 104
outgoing junction 104
Connector 85
Contact bank 80
Contact trees 156
Continuous service 11
Conversion equipments 308
Cord 20
circuits 30, 49
Cost of service 59
Cradle 17
Cross-bar switch 130
auxiliary contacts 131
Cross-bar systems 63
Cross-points 131
electronic analogue 301, 309
electronic digital 304
Cross-talk 15
Customer 8
Cut off relay K 24

Data 179
call 165
class of service 165
exchange line 91
high speed digital 11
junction 271
links 272
memory 35
operational 23
period 51
semi-permanent 49
state of line 92
storage 51
temporary 51
transmission between registers 126, 134
transmission on bus-wires 243
Design, apparatus 1
exchange system 1
system 1
traffic 2
Destination common wires 238
Diagnostic programs 288
Dialling, press-button 68
rotary 70
through p.a.b.x. 66
Dials 66
Direct dialling-in d.d.i. 66
Directory 20
number 20
Disabled subscriber 170
Distribution frame, intermediate 32
main 32
Dwell of processor 198

Early morning calls 179
Earphone receiver 8
Eccles—Jordan 174
Electrical states 2
Electric telegraph 2
Enquiries 8
Enquiry, p.a.b.x. 76
Equipment engineering 105
Ericsson 500-line switch 82, 117
Exchange 8
codes 43, 106, 119
generalizations 46
lines 8, 9
line signals 271
local 9
numbers 20
private branch p.b.x. 9
switches 10, 20, 22
transit 9
Exchanges, electro-mechanical 12
electronic 12, 63, 66
manual 12
quasi-electronic 66
semi-electronic 64, 66
Executive program 54
Extensions p.b.x. 9

Facilities 35
authorized 254
Fault detection 288
Fault tracing 285
Ferreed 182
File 144
indexed 35
Flashing 45
recall 46
Flexibility 48, 159, 254, 292
Frames, t.d.m. 306
t.d.m. super 307
switching 136
Freefone 60
Frequency division multiplex f.d.m. 5

Gate 174
circuit 174
Generalized exchange, cross-bar 139
manual 48
semi-electronic, wired logic 230
stored program logic 255
step-by-step 107
register-translator controlled 123
Grade of service 29
Grading 102
Group class of service 96, 159
Group marking wires 238

Group selector   103, 104
Guide wire systems   186

Head-set   20
Hierarchy of exchanges   12
High speed digital data   11
Hook   17
Hopper   199

Identification, calling line c.l.i.   180
I.D.F.   32
Indicator   16
Instruments   8
Insulation testing of lines   200
Interface   22
    switch   22
Interrogator-marker   186
Inverter   175

Jack   20
Jumper   32
Junctions   8
Junctor   132
    incoming i.j.   132
    local l.j.   134
    multiplexed   315
    originating or.j.   132
    service s.j.   133
    terminating t.j.   133
    virtual   275
Junctor classes of service   137

Keith   85
Keith line switch   88
Key, ring   20
    speak   20

Last look digit   201
Latching switches   268
Levels of exchanges   12
Line   3
    finders   88
    four-wire   6
    party   19
    plant, local   9
    surges   302
    two-wire   6
Line data storage   195
Line drawings   179
Line relay, transistor   308
Line relay L   20
Lines, exchange   8, 9
Linked numbering   119
Local exchanges   9
Lock-out   142

Logic, human   52
    stored program   52
    wired   52
Loop   17
    signalling   17
Loudspeakers   8
Loudspeaking telephones   178

Magnet, bridge   130
    select   130
Magnetic tape   236
    call accounting   285
Magneto generator   15, 16
Maintenance   181
Majority decision   235, 287
Manual hold   180
Map in memory   266
Marker, cross-bar   139
    delayed dial control   114
Marks   139
M.D.F.   32
Message   2, 18
    address   7
    analogue   4
        transmitted   4
    calling   7, 10
    control   7
    proceed-to-send   7
    proceed-to-send-the-address   10
    proceed-to-send-the-message-proper   10
    register, see Meter
    switching   1, 6
    transmitted   3
Message proper   3
Meter   42
    call charges on   42
    private   62, 73, 79, 99
Metering over junctions   110
Microphone   8
Micro-processors   252
Midnight lines   254
Molina   63
Multiple   23
    appearance   23
Multipling   23

N-symbols   2
Network, exploded   297
    planning   294, 297
New systems economics   295
Night service   94
No answer   21
Number, block   27
    directory   20
    engaged   21

324

equipment 27
exchange 20
group 155
night service 26
regular 26
unobtainable 21

Office 8
destination 7
telegraph 6
One call at a time 143, 182
Operator 9
A and B 43
Operator phrase, all lines busy 43
have you finished 40
go ahead 37
no answer 21, 41
number engaged 21
Order wire, electronic 307
manual 43

p.a.m. analogue transmission 303
p.a.m.—t.d.m. switching 303
Park 239
Parking 89
Party lines 19, 26, 72, 167
Path one-only selection 184
Path search and connection 138, 186
faults in 269
guide wire 268
map in memory 266
Pause, inter-digital 106
inter-train 106
Pay stations 8
p.c.m. message path associated data path 306
p.c.m. super-frames 306
p.c.m.—t.d.m. transmission 5, 304
Peg 28
Permanent glow PG 28
Personal, person-to-person call 60
Pictures 179
Picture telegraphy 11
Plant growth 11
Position, switchboard 23
Preference 254
Pre-selectors 85
Press-button dialling 68
Private branch exchange p.b.x. 9
Processing, memory stored program 51
speed 250
two-phase 212
Processor, circuit 26, 48
common 24, 46
costs 251
data 23

peripheral pr 47
t.d.m., register 224
supervision 223, 225, 228
translator-marker 225
Program 36
executive 54
interrupts 57
operational 36
Public services 1
Pulse code modulation p.c.m. 5
Pulse coding 4
P wire 74

Rank of processors 214
Ranks of switches 135
Rate, flat 59
measured 59
Receiver 6, 14
Reed relay 182
Register 117
junction i.c.r. 132
originating o.r. 132
Register-translator 119
in series 126
Release, called party 90
calling party 90
datum 127
first party 90
last party 90
forced 91
Re-ring 166
Reversed charge 60
Reverted pulses 117
Ring back 166
Ringing 16
circuit 78
re-call 179
signal 16
delayed 166
party line 96, 165
time shared 166
Ring trip 78
without metering 92
Route relays 160

Scanning, asynchronous 198
synchronous 198
Scratch pad memory 176
Section, switching 139
Selector, final 85
group 103
one only o.o.s. 88, 139
Service, cost of 59
disabled persons 75

interception s.v.i. 94
observation s.o. 100
special s.s.o. 100
temporarily out of 94
Services 35
Shift registers 204
Signal 16, 18, 23
answer 17
call 17
clear 17
exchange line 270
off hook 17
on hook 17
ringing 16
voice frequency v.f. 111
Signal-message 18, 23
Slot 199
Sounders 7
Space division 52
State, of the call 22
of the structure 22
Station, attendant 8
called 10
calling 10
Stations 8
Step-by-step 101
systems 63
Store, bulk 176
period 51
semi-permanent 51
temporary 51
Stored program 236
Strowger 48, 58, 63, 65, 84
Structure 22
maximal 34
minimal 34
minimum 34
state of 22
Subscribers 8
telephonic abilities 178
Supervision 21, 223
processes 219
Switch, analogue 300
cross-bar 130
digital 300, 304
electronic analogue 301
exchange 10, 20, 22
interface 22
space 300
multiplexed 300
time 300, 305
two-motion, Strowger 80
clutched motor 80
500-line 82
Switchboard 23

Switching 6
circuit 10
four-wire 178
logic 35
message 6
Synchronous cyclic t.d.m. 241
Synchronous t.d.m. wired logic 237
System 1
national 1

Table of equipment numbers 261
Table in memory 259
Tape, magnetic 236
paper 7
Telecommunication 1
bothway 6
point-to-point 6
unidirectional 6
Telegram 3, 6
Telegraph, offices 6
services 1
Telephone set 17
Teleprinter answer back 10
Teleprinters 10
Telewriting 179
Telex 10
Temporarily out of service t.o.s. 94
Test and plug-up 28
Tests, check 288
routine 288
Ticketed calls 59
Time, division 52
measurement 203
sharing 58
acyclic 199
parallel 214
space division 52, 215
time division 52, 216
series 216
Tone generators, t.d.m. 316
Tones 66, 67
Tone sounders 308
Traffic, engineering 30
concentration, exchange line 312
Transfer, follow me 75
incoming call 75
pre-arranged 75
Transit exchange 9
Translation, arbitrary number 119
d.n. to e.n. 28
e.n.c. to d.n. 99
program 280
Translator, exchange line 143
trunk 143
Translator-markers 142

Transmission, analogue 4
  bandwidth 5
  bothway 6
  class of 5
  digital 3
    asynchronous 5
    n-state 3, 5
    synchronous 5
  electric 4
  four-wire 310
  lines 8
  message path, associated 18
    disassociated 18
  multiplex channel data 316
  over junctions 108
  p.a.m.—t.d.m. 303
  p.c.m.—t.d.m. 5, 306
  point-to-point 6
  t.d.m 7
  unidirectional 6

Transmitter 6
Trigger circuit 174, 176
Trunking, folded 183
Trunking diagram 105
Trunk offering 180
Trunks, loop 49
  operating 49
Two-wire line 6

Ultimate exchange control 274
Ultimate exchange structure 316
Ultimate network planning 297
Uniselector 80
User 8

Videophone 11
Von Neumann 236

Watson 15
Wiper 80